半导体金属氧化物纳米材料：
合成、气敏特性及气体传感应用

邓勇辉 袁凯平 邹义冬 罗维 ◎ 著

Semiconducting Metal Oxide Nanomaterials:
Synthesis, Gas Sensitive Property and Gas Sensing Application

上海交通大学出版社
SHANGHAI JIAO TONG UNIVERSITY PRESS

内容简介

随着互联网、物联网和人工智能技术的发展,传感器在新技术、新装备、新业态等各方面都发挥着越来越重要的作用。作为传感器中的一类重要成员,气体传感器在感知各种环境的气体种类和浓度,探测有毒有害、易燃易爆等气体的诸多应用中展示了积极且关键的作用。发展高端、高性能气体传感器具有重要实际意义和广泛应用前景。在各种气体传感材料中,半导体金属氧化物材料应用最为广泛,且日益受到各学科领域的关注。本书围绕半导体金属氧化物(SMO)型气体传感器,对敏感材料以及传感器件进行了全面介绍,并重点综述了 SMO 敏感材料的传感机制、传感材料改性方法、工作特点以及应用情况。

本书专业性强、深入浅出,适于不同领域、不同应用层次的读者。本书可作为相关专业的本科生和研究生的教学参考书,也可作为致力于半导体气体传感器研究和开发的工程师的参考资料和工具书。

图书在版编目(CIP)数据

半导体金属氧化物纳米材料:合成、气敏特性及气体传感应用/ 邓勇辉等著. —上海:上海交通大学出版社,2024.5
 ISBN 978-7-313-30449-0

Ⅰ.①半… Ⅱ.①邓… Ⅲ.①氧化物-金属材料-半导体材料-纳米材料-研究 Ⅳ.①TB383②TN304.2

中国国家版本馆 CIP 数据核字(2024)第 059259 号

半导体金属氧化物纳米材料:合成、气敏特性及气体传感应用
BANDAOTI JINSHU YANGHUAWU NAMI CAILIAO:
HECHENG QIMIN TEXING JI QITI CHUANGAN YINGYONG

著　者:	邓勇辉　袁凯平　邹义冬　罗　维			
出版发行:	上海交通大学出版社	地　址:	上海市番禺路 951 号
邮政编码:	200030	电　话:	021 - 64071208
印　制:	上海颛辉印刷厂有限公司	经　销:	全国新华书店
开　本:	710 mm×1000 mm　1/16	印　张:	24
字　数:	467 千字		
版　次:	2024 年 5 月第 1 版	印　次:	2024 年 5 月第 1 次印刷
书　号:	978-7-313-30449-0		
定　价:	198.00 元		

版权所有　侵权必究
告读者:如发现本书有印装质量问题请与印刷厂质量科联系
联系电话: 021 - 56152633

PREFACE 序言

随着人工智能、通信技术及大数据的快速发展,万物互联已经成为大势所趋。小型化、智能和网络化的传感器已构成物联网、工业 4.0 和智能制造的重要基础。在众多传感器中,化学传感器是一类对特定分子、离子等具有敏感响应特性并能够将浓度转换为电信号进行检测的电子元器件。化学传感器已逐渐发展成为研究领域最多、应用潜力最大的传感器件之一,在工业生产、环境监测、食品安全、生物医学、航空航天等诸多领域发挥了不可替代的作用;气体传感器被誉为超越人类嗅觉极限的"电子鼻",能够精准识别环境中的痕量气体分子。其中化学电阻型气体传感器因其依赖半导体敏感材料与待测气体之间的相互作用(吸附、扩散、催化氧化、脱附等)而成为化学、材料科学、物理、微电子和数学等多学科交叉的研究热点。20世纪 60 年代半导体金属氧化物(SMO)被成功用作为气体传感器的核心敏感材料,并由此诞生了费加罗气体传感器,开启了 SMO 气体传感器在有毒有害气体检测、易燃易爆气体、疾病呼出气体分析、食品安全、农药残留监测等领域的应用。半个多世纪以来,基础研究领域和产业界的交互发展,加速了 SMO 气体传感器的研发和应用。当前,如何进一步提升 SMO 化学气体传感器的灵敏度、选择性和长期稳定性等关键指标,是实现智能化、便携化及低功耗高可靠性 SMO 气体传感器的关键。

众所周知,SMO 化学气体传感器性能主要取决于半导体金属氧化物材料的微纳结构、组成及表界面性质。通过精准调控材料组分、微纳结构、表界面化学微环境、微观形貌能够实现传感器传感性能的优化,进而实现传感器灵敏度、选择性、响应速度及长期稳定性的提升和调控。因此,从化学材料合成及器件设计角度出发,实现对材料组分、微纳结构及表界面效应的调控是实现传感器综合性能提升的关键,同时还需要厘清SMO气体传感器性能与敏感材料性质之间的构效关系,并以此来指导智能气体传感器的开发。此外,在可穿戴以及人机交互电子设备领域,从传感器结构设计出发来解决器件乃至整机的功耗问题,进而实现气体传感器在智能时代的快速发展和应用。

本书从SMO气体传感器的发展历程出发,围绕传统SMO气体传感器在室内外空气监测、疾病诊断和食品安全领域的应用,总结了SMO气体传感器的特点,系统介绍了SMO的理化性质;从传感器机理出发,对比讨论了p型和n型半导体的气敏响应特性,详细概述了异质结结构、杂质掺杂、贵金属敏化及晶粒尺寸对传感器性能的影响;从材料形貌设计到组分调控,从合成方法到微结构设计,从材料合成到传感性能,分别从不同的角度和层次解析了传感器性能与SMO材料性质之间的构效关系。本书还介绍了新兴多孔结构尤其是介孔结构的SMO材料在气体传感器领域的应用,深入结合了作者研究团队十多年在该领域的研究成果和对于SMO气体传感器的理解,结合行业未来发展趋势由浅入深地阐明了多孔SMO在发展新一代气体传感器方面的优势。

特别值得一提的是,本书重点介绍了不同类型半导体金属氧化物的气体传感器的最新进展,尤其是深入讨论了近年来出现的新型气体传感器及优化方法,如表面拉曼增强的气体传感器、有机晶体管场效应气体传感器、光学气体传感器及脉冲驱动的气体传感器,并指出不同类型传感器的优缺点及面临的挑战。从传感性能和低能耗气敏器件发展的应用需求出发,本书还对传感器器件设计、封装集成、电路设计、算法及机器嗅觉(电子鼻)应用展开了深入的讨论。在此基础上,本书对气体传感器未来的发展方向进行了展望,包括敏感材料性能调控与气体传感机制、感知器件设计与传感器应用开发。本书还为读者提供了广泛的研究实例,从基础研究层面为传感器发展提供了相关思路,同时从行业发展等角度为传感器的商业化提供了重要技术支撑。因此,该书的出版将有望推动新一代智能气体传感器的研制及商业化,并带动和促进相关产业的快速发展和可持续升级。

赵东元

中国科学院院士、复旦大学教授

2024年4月

FOREWORD 前言

气体传感器作为物联网（IoT）和互联网的重要组成部分，在工业生产、环境监测、医疗诊断、国防安全、食品健康等领域具有重要应用价值。根据敏感材料及传感机制的差异，气体传感器可划分为多种类型，其中半导体金属氧化物（SMO）是应用较为广泛的敏感材料之一，具有高灵敏度、低成本、合成便捷的特点，满足了多种技术和市场对气体检测的要求。但在实际应用中，SMO传感器还存在一些提升空间，因此半导体金属氧化物气体传感器的研究开发越来越受到各个学科领域的关注。高性能SMO气体传感器的快速发展要求能够精确控制半导体金属氧化物纳米材料的组成、形态、比表面积和电子结构特性。从化学合成和材料应用的角度来看，如何实现对SMO的组成、微纳米结构和界面性能的良好控制，以及如何理解这些因素与气敏性能之间的关系，是发展现代智能气体传感器的关键。

本书总结了SMO气体传感器的特点（主要是基本特性）、气体传感器的原理、气固界面气敏催化机制以及各种类型的SMO气体传感器。本书围绕半导体金属氧化物气体传感材料展开，尤其关注了SMO纳米尺度敏感材料的现状、提高性能的策略以及其在各种领域中的应用，展望了SMO气体传感器的未来，包括材料设计与气敏机理、纳米器件与结构设计、应用发展等。本书在SMO气体传感器的最新发展及应用物理学、材料科学、纳米电子学等各种应用方面提供了广泛的例子，便于读者理解。

本书从气体传感器原理出发，阐明了气体传感器性能与材料微纳结构之间的构效关系，由浅入深地揭示了材料结构、组成及形貌对材料电子结构等理化性质的调控，从微观至宏观，从分子水平至原子水平，涵盖化学、材料、高分子、微电子等学科。本书为深入理解气体传感器的工作形式和作用机制奠定了基础，有利于广大读者尤其是科研人员把握当前气体传感器敏感材料的不足及未来发展方向，推进新一代智能化、便捷式、高灵敏半导体类气体传感器的发展。

著　者

2024 年 1 月

CONTENTS 目录

第1章 半导体金属氧化物气体传感器概述 1
  1.1 半导体金属氧化物气体传感器的起源 1
    1.1.1 金属氧化物气体传感器的种类 2
    1.1.2 提高金属氧化物半导体气体传感器性能的方法 3
  1.2 半导体金属氧化物气体传感器的应用 3
    1.2.1 半导体金属氧化物传感器在室外空气质量评价中的应用 4
    1.2.2 半导体金属氧化物传感器在室内空气质量评价中的应用 5
    1.2.3 半导体金属氧化物传感器在疾病诊断中的应用 7
    1.2.4 半导体金属氧化物传感器在食品安全中的应用 9
    1.2.5 半导体金属氧化物传感器在农业生产中的应用 11
  1.3 半导体金属氧化物的物理特性 13
    1.3.1 半导体金属氧化物的定义 14
    1.3.2 半导体金属氧化物的潜在性能 14
    1.3.3 半导体金属氧化物的物理基础 15
  参考文献 22

第2章 半导体金属氧化物气体传感器的传感机理 30
  2.1 纯金属氧化物半导体 30
    2.1.1 n型金属氧化物 32
    2.1.2 p型金属氧化物 33
  2.2 金属氧化物异质结 34
    2.2.1 n-n异质结 34
    2.2.2 p-p异质结 38
    2.2.3 p-n异质结 39
  2.3 掺杂金属氧化物半导体 40

2.4 贵金属敏化金属氧化物 ·· 45
2.5 晶粒尺寸的影响 ·· 51
2.6 气体传感器评价标准 ·· 53
 2.6.1 灵敏度 ·· 54
 2.6.2 工作温度 ·· 54
 2.6.3 选择性 ·· 56
 2.6.4 稳定性 ·· 59
 2.6.5 响应-恢复时间 ·· 60
 2.6.6 检测限 ·· 60
参考文献 ·· 61

第3章 半导体金属氧化物：形态和传感性能研究 ························ 66
3.1 形貌和结构对气体传感的影响 ·· 66
 3.1.1 颗粒大小的影响 ·· 66
 3.1.2 颗粒晶相的影响 ·· 68
 3.1.3 表面几何构型的影响 ·· 69
 3.1.4 晶体孔隙度和晶粒间接触面积 ···································· 70
 3.1.5 颗粒的聚集 ·· 71
3.2 金属氧化物传感材料的合成方法 ······································ 71
 3.2.1 溶胶-凝胶法 ·· 71
 3.2.2 水热和溶剂热法 ·· 73
 3.2.3 自组装方法 ·· 75
 3.2.4 微乳液介导的合成 ·· 80
 3.2.5 化学气相沉积法 ·· 82
参考文献 ·· 82

第4章 半导体金属氧化物：组分及其气敏性能 ·························· 90
4.1 双金属氧化物异质结 ·· 91
 4.1.1 p-n 异质结 ·· 91
 4.1.2 n-n 异质结 ·· 96
 4.1.3 p-p 异质结 ·· 101
4.2 贵金属修饰 ·· 102
4.3 元素掺杂 ·· 111
 4.3.1 非金属元素掺杂 ·· 111
 4.3.2 金属元素掺杂 ·· 113

         4.3.3 稀土元素掺杂 ·· 115
    4.4 与碳材料(石墨烯、碳纳米管)复合 ································· 117
    参考文献 ··· 121

第5章 半导体金属氧化物：微结构与传感性能 128
    5.1 半导体金属氧化物的基本特征 ······································ 128
    5.2 结构类型和典型架构 ··· 128
    5.3 晶粒尺寸与多孔结构 ··· 140
    5.4 比表面与异质结界面 ··· 147
    5.5 晶体结构和内部缺陷 ··· 152
    参考文献 ··· 157

第6章 气体分子和金属氧化物半导体界面相互作用模型 166
    6.1 吸附和脱附 ··· 168
         6.1.1 经典的吸附与脱附理论 ····································· 168
         6.1.2 MOS材料气体传感中吸附与脱附的影响 ················· 171
    6.2 气体扩散 ·· 187
         6.2.1 经典扩散理论和模型 ·· 187
         6.2.2 MOS气体传感器的扩散模型 ······························ 197
    6.3 结论与展望 ··· 209
    参考文献 ··· 211

第7章 提高气敏性能的新方法 225
    7.1 光学气体传感 ··· 225
    7.2 表面等离子体共振增强气体传感 ·································· 229
    7.3 脉冲驱动气体传感 ··· 234
    7.4 场效应晶体管气体传感器 ··· 236
    参考文献 ··· 239

第8章 不同类型的常用半导体金属氧化物气体传感器 241
    8.1 电阻型气体传感器 ··· 242
         8.1.1 电阻型气体传感器的器件结构与制作 ···················· 243
         8.1.2 敏感材料 ·· 243
    8.2 MEMS平台气体传感器 ··· 244
         8.2.1 MEMS气体传感器的器件结构与制作 ··················· 245

  8.2.2 传感材料 …………………………………………………………… 246
 8.3 场效应晶体管型气体传感器 …………………………………………… 251
 8.4 传感装置和材料 ………………………………………………………… 252
  8.4.1 纳米线场效应晶体管型气体传感器的结构与制备 ……………… 252
  8.4.2 传感材料 …………………………………………………………… 253
 参考文献 ……………………………………………………………………… 256

第9章 气体传感器集成技术：电子鼻 …………………………………… 263
 9.1 电子鼻：传感器阵列 …………………………………………………… 263
 9.2 信号-数据分析技术 ……………………………………………………… 266
 9.3 温度梯度法 ……………………………………………………………… 269
 9.4 电子鼻的集成制造 ……………………………………………………… 272
 9.5 电子鼻的应用 …………………………………………………………… 272
  9.5.1 电子鼻在食品工业中的应用 ……………………………………… 273
  9.5.2 电子鼻在环境监测中的应用 ……………………………………… 276
  9.5.3 电子鼻在呼吸系统疾病诊断中的应用 …………………………… 278
 9.6 结论与展望 ……………………………………………………………… 281
 参考文献 ……………………………………………………………………… 281

第10章 半导体金属氧化物传感器的应用 ………………………………… 287
 10.1 挥发性有机物传感器 …………………………………………………… 287
  10.1.1 乙醇气体传感器 …………………………………………………… 287
  10.1.2 丙酮气体传感器 …………………………………………………… 292
  10.1.3 甲醛气体传感器 …………………………………………………… 295
  10.1.4 苯系物（苯、甲苯、二甲苯）传感器 …………………………… 299
 10.2 环境气体传感器 ………………………………………………………… 303
  10.2.1 二氧化碳传感器 …………………………………………………… 303
  10.2.2 氧气传感器 ………………………………………………………… 304
  10.2.3 二氧化硫传感器 …………………………………………………… 306
  10.2.4 臭氧传感器 ………………………………………………………… 308
  10.2.5 氨气传感器 ………………………………………………………… 309
 10.3 有毒气体传感器 ………………………………………………………… 311
  10.3.1 一氧化碳传感器 …………………………………………………… 311
  10.3.2 H_2S 传感器 ……………………………………………………… 313
  10.3.3 二氧化氮传感器 …………………………………………………… 316

10.4 可燃气体传感器 ………………………………………………………… 319
　　10.4.1 甲烷传感器 …………………………………………………… 319
　　10.4.2 氢气传感器 …………………………………………………… 320
　　10.4.3 LPG 传感器 …………………………………………………… 322
10.5 其他气体传感器 ………………………………………………………… 324
参考文献 ………………………………………………………………………… 331

索引 …………………………………………………………………………… 343

第1章
半导体金属氧化物气体传感器概述

1.1 半导体金属氧化物气体传感器的起源

自20世纪50年代被首次提出以来，半导体金属氧化物(SMO)就用作电阻性气体传感器的敏感材料被广泛研究。学者普遍认识到半导体材料近表面的电学性质会因附近气氛或吸附分子组成的变化而产生显著差异[1-7]。此后，这种半导体传感器由于具有低成本、结构简单、能在线监测以及可靠性高等特点而被广泛应用于实时控制系统和各种实际应用场合，该类电阻式气体传感器的研究受到了极大的关注[8]。1968年，半导体型气体传感器作为家用气体泄漏探测器(首例该类传感器是Figaro TGS)首次投入实际应用。人们很快认识到，在300～450℃的工作温度下，具有高比表面积的金属氧化物气敏电阻器对单一气体的选择性仍然较差[7-10]。但是，它们可以作为家庭中天然气泄漏的早期警报以及用于许多工业生产过程的报警装置。自那以后，人们一直致力于改进半导体气体传感器的性能并降低其基本功耗，为提高这种器件的选择性和灵敏度以及降低其操作功率做出了巨大努力[11-13]。这些研究主要涉及探索响应机制[14]、选择最合适的氧化物组成、制造两相"异质结构"[15]、添加金属催化剂颗粒和设计合成具有纳米结构的敏感材料来提高气体传感器的各种性能[16]。对半导体传感器研究的热点都聚焦在理解气体敏感机制、确定敏感元件和材料的最佳组成及组合方式等方面[17]。近年来，研究人员在理解可用于气体传感的金属氧化物性质方面取得了显著进展，并发现了一系列新的纳米材料合成方式，这为进一步提高传感器的各种性能铺平了道路[18]。

表面吸附/解吸理论和催化理论属于半导体表面物理学(空间电荷层、表面状态)的范畴，为半导体气体传感器敏感机制的研究提供了重要的科学基础[19-20]。吸附等温线是将给定气体组分表面覆盖描述为其组分分压的函数，被吸附的离子产生功函数偏移、表面电荷或偶极层，从而导致半导体域中的空间电荷层发生显著改变。因此，在低于500℃的工作温度下，控制检测气体在半导体金属氧化物表面发生的反应通常涉及表面氧物种浓度的变化，如O_2^-、O^-或O^{2-}，它们在较大的工作温度范围内都是稳定存在的[21]。这些表面氧物种离子可以被看作电子陷阱,能够从金属氧化物中提取电子[22]。其中,在 p 型金属氧化物半导体中,吸附的表面氧物种能够再次充当表面受体,从材料价带中捕获电子,从而导致近表面区域中载流子浓度显著增加[23]。因此,p 型氧化物对暴露于空气中的还原气体和氧化气体的响应与 n 型半导体材料所表现出的规律相反。根据上述机理,半导体材料根据其对还原或氧化气体的电阻响应不同可分为 n 型和 p 型两大类,如表 1.1 所示,但此分类模式仅适用于半导体材料的近表面区域[24]。

表 1.1 当还原或氧化气体引入空气中时,近表面层
(显示 n 型或 p 型特征)的电阻响应

近表面层特征	还原气体	氧化气体
n 型	本征电阻减小	本征电阻增大
p 型	本征电阻增大	本征电阻减小

对于 n 型半导体,由于电子从半导体材料导带的离子化供体中被引出,所以晶粒间表面的载流子浓度降低,并且形成对电荷传输的势垒。即使引入低浓度的还原气体(如一氧化碳),也会与表面吸附氧发生反应,释放不带电分子(如二氧化碳)的同时使电子返回导带。在 300~450℃的正常工作温度范围内,主要的表面氧离子是 O^-,这导致测量的电导率增加到与还原气体局部浓度相关的程度。如果空气中的这种 n 型结构暴露于低浓度的氧化气体而不是还原气体中,则会发生诸如 NO 分子的竞争性吸附行为,总体结果是吸附在氧化物表面的载流子浓度增加,测量的电导率降低。

1.1.1　金属氧化物气体传感器的种类

在过去的 40 年中,各种各样的金属氧化物半导体电阻型气体传感器被用来检测环境中痕量浓度的待测气体[25]。禁带宽度为 2.0~4.0 eV 的二元氧化物(如 TiO_2、Nb_2O_5、Ta_2O_5、ZnO、SnO_2 和 WO_3)在空气环境中引入少量气体时表现出 n 型氧化物半导体的行为[26]。特别地,SnO_2 敏感材料由于具有高电子迁移率、高稳定性和可检验气体种类较多等优点而成为众多研究人员的研究重点。在所有的金属氧化物气敏材料中,SnO_2 敏感材料起着相当重要的作用,如果没有 SnO_2,半导体气体传感器的进程将会受到很大的阻碍[27]。相比之下,迄今为止,使用 p 型氧化物半导体(如 NiO、CuO、Co_3O_4、Cr_2O_3 和 Mn_3O_4)制造的气体传感器受到的关注相对较少,并且制造这种半导体电阻型气体传感器的相关研究仍处于开发的早期阶段[28-29]。

基于金属氧化物的纳米复合材料在气敏领域中发挥着重要的作用,因此该类材料在气敏材料发展中具有广阔的应用前景[30]。通过引入另一种纳米结构的金属氧化物($Me^{II}O$)来提高金属氧化物($Me^{I}O$)的敏感性能已经得到大量研究的证实,这是开发用于非均相气敏复合材料($Me^{I}O - Me^{II}O$)的一种很有前景的方法。金属氧化物($Me^{II}O$)可以是催化改性剂或结构改性剂。目前,许多研究者正致力于研究具有不同金属比的上述材料,以开发对目标气体具有优异响应性和选择性的传感器[31],他们也正在研究基于稳定的混合型导电金属氧化物和复合金属氧化物作为具有优异气敏性能的半导体传感器,主要包括 $SnO_2 - In_2O_3$、$SnO_2 - ZnO$、$SnO_2 - WO_3$ 和 $In_2O_3 - ZnO$[32]。

1.1.2 提高金属氧化物半导体气体传感器性能的方法

随着科学技术和合成工艺在各个领域的进步与发展,需要综合考虑传感器的可靠性、选择性、灵敏度、检测下限和成本,这使得开发传感器的工作变得更加具有挑战性。越来越多的科学研究致力于改进金属氧化物气体传感器的综合性能,比如减小感测装置的尺寸、最小化生产成本,以及从快速响应和高灵敏度、选择性、稳定性和可行性等方面提高传感器性能,上述方法在气体传感器技术领域中受到了极大的关注。此外,开发和探索应用于高效气体传感器的新结构、新材料也非常重要。提高性能的办法主要如下:① 开发具有纳米结构的新材料以增强金属氧化物半导体气体传感器的性能。该方法使用具有不同形貌的各种金属氧化物半导体,如纳米板、薄膜、纳米颗粒、纳米棒、纳米管、纳米纤维、纳米线和空心球等[33]。② 使用纵向氧化物 p-n 结、夹在 p 型和 n 型氧化物半导体纳米颗粒之间的纳米复合材料和用 p 型氧化物半导体电阻器修饰的一维 n 型氧化物半导体作为高性能气体传感器[34]。③ 通过结合次级元件和利用交流电而不是直流电的其他类型传感器来提高材料的传感性能。多年来,人们已经证明了将诸如金、铂、钯、银等贵金属纳米颗粒负载到金属氧化物的表面,可以有效地降低传感器的工作温度,提高检测下限、灵敏度和选择性[35],而这种发展模式主要以经验或试错法为基础。为了进一步发展和创新半导体气体传感器,基础研究包括气体传感机制和传感器设计原理是必不可少的,但是到目前为止,上述气体传感研究领域的进展仍相对缓慢。

1.2 半导体金属氧化物气体传感器的应用

在许多工业领域和日常生活中,监测气体浓度是一项长期的要求。本节内容期望使半导体气体传感器技术适合于更广泛的应用领域。金属氧化物半导体气体传感器由于体积小、成本低,可作为检测可燃气体和有毒气体泄漏的重要装置,已经得到广泛的研究和开发,并且在空气质量检测、农业生产、工业加工、食品安全、公共安全和通过分析呼出气体进行医疗诊断等各种技术领域中得到了广泛的应用[36]。表1.2总结了金属氧化物半导体气体传感器的应用领域和代表性目标气体。

表 1.2 金属氧化物半导体气体传感器的应用领域和代表性目标气体

应用领域	目标气体
室外空气质量评价	SO_2、NO_x、CO、碳氢化合物、VOC
室内空气质量评价	甲醛、CH_4、天然气、液化石油气
疾病诊断	H_2S、丙酮、O_2、CO_2、3-羟基-2-丁酮
食品安全	H_2S、SO_2、甲醇、三甲醇、乙醇
农业生产	NH_3、Cl_2、H_2S、HF、SO_2、CO、VOC

1.2.1 半导体金属氧化物传感器在室外空气质量评价中的应用

由于诸如柴油、取暖油和各种燃料等石油产品的不完全燃烧而向大气中释放大量有毒气体(如 CO 和 SO_2),环境正受到严重污染。这些气体除了对环境有影响外,还对人类有各种有害影响,尤其是呼吸器官发育不成熟的儿童将面临巨大的危险。如一氧化碳会导致胸痛,并导致人的精神警觉性降低。酸雨会使土壤的养分发生化学变化,导致矿物质的流失。除了上述气体,NO 和 CO 在阳光照射下还会产生对植物和人类都有害的 O_3。科学家正在设计更安全的替代品,并且探索在它们产生有害影响之前,用一种安全的方法对大气中的这些气体进行早期检测。目前,已经开发了用于检测这些气体的各种传感器,其中金属氧化物传感器在用于监测这些气体的短期暴露极限方面具有优势[37]。

Song 等[38]设计了一种独特的 NO_2 传感器,以 3D 氧化锡(SnO_2)纳米管阵列作为传感层,铂(Pt)纳米簇装饰作为催化层(见图1.1)。Pt/SnO_2 传感器通过增强吸附能和降低对 NO_2 的活化能,显著提高了 NO_2 检测的灵敏度和选择性。它不仅对 NO_2 具有极高的灵敏度,检测极限为 107 ppb①,而且能够抑制干扰气体的响应,实现 NO_2 选择性检测。此外,他们还开发了一个由传感器、微控制器和蓝牙单元集成的无线传感器系统,用于实时室内和道路 NO_2 检测应用。该传感器的合理设计及成功演示为未来智能家居和智能城市应用中的实时气体监测奠定了基础。

① 1 ppb 表示 10^{-9}。

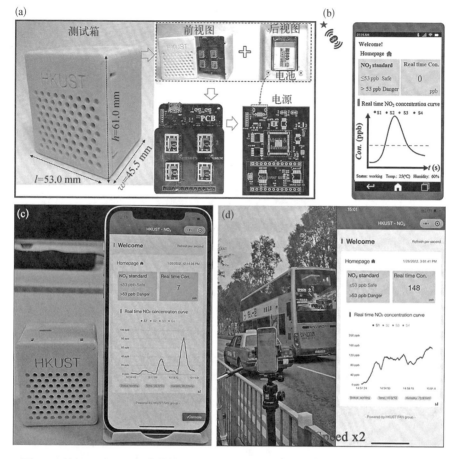

图 1.1 基于 Pt/SnO$_2$ 纳米管阵列的 NO$_2$ 传感器用于实时室内和道路 NO$_2$ 检测[38]
(a)、(b) 可与智能手机通信的无线 NO$_2$ 传感器系统；(c)、(d) Pt/SnO$_2$ 传感器在室内和道路 NO$_2$ 监测的潜在应用

1.2.2 半导体金属氧化物传感器在室内空气质量评价中的应用

室内空气质量对居住者的健康、舒适度和幸福感有着重要影响，如病态建筑综合征、办公室生产力下降和学校学生学习障碍均与恶劣的空气有关。城市中的大多数人大部分时间都是在室内度过的，因此室内空气质量与空气污染物的总暴露量有显著的相关性。常见的室内空气污染物包括挥发性有机化合物（volatile organic compound，VOC）、二氧化氮、臭氧和一氧化碳。这些污染物来源于交通和工业的室外污染物，它们通过通风系统进入建筑物。另外，燃烧的燃料、蜡烛和烟草，以及建筑材料、家具、清洁产品、电子设备、洗漱用品、人和宠物的排放也会造成严重的室内污染。其中，新的建筑产品是特别重要的污染源。如果能够避免干扰问题，在大多数情

况下,这些气体对健康造成威胁的浓度完全能够在金属氧化物传感器达到的检测下限内。在具有挑战性的低浓度下,检测 VOC(如苯或甲醛)十分困难,然而,在实践中,测量 VOC 的总浓度足以满足总挥发性有机化合物(TVOC)指标的检测要求[39]。

Wang 等[40]以中空二氧化锡纳米纤维为骨架,通过水热法成功合成了具有多级纳米结构的二氧化锡纳米纤维/纳米片。通过 X 射线衍射(XRD)、场发射扫描电子显微镜(FE-SEM)、透射电子显微镜(TEM)、X 射线光电子能谱(XPS)和氮气吸附比较了 SnO_2 纳米纤维、SnO_2 纳米纤维/纳米片和 SnO_2 纳米片的微观结构、形貌、化学组成、氧化状态和比表面积。表征结果表明,通过在纳米纤维表面均匀生长纳米片阵列,形成了分级 SnO_2 纳米纤维/纳米片结构。同时,以甲醛(HCHO)为目标气体,他们研究了 SnO_2 基纳米材料的传感性能。与 SnO_2 纳米纤维、SnO_2 纳米片以及 SnO_2 纳米纤维和纳米片的物理混合物相比,基于 SnO_2 纳米纤维/纳米片的气体传感器对 HCHO 气体具有更好的响应、更好的选择性、瞬态响应和痕量检测能力。气体传感器在 120℃对 100 ppm①HCHO 的响应(R_a/R_g)为 57,分别比纯 SnO_2 纳米纤维传感器和 SnO_2 纳米片传感器高约 300% 和 200%。此外,该传感器具有快速响应/恢复性能,对 100 ppm HCHO 的响应/恢复时间仅为 4.7 s/11.6 s。此外,他们还彻底地研究了 SnO_2 分级纳米结构的生长过程和气敏机理。SnO_2 纳米纤维/纳米片的成功制备可归因于纳米纤维上种子的均匀修饰和纳米片的合适生长条件。其增强的传感性能主要来自纳米纤维和纳米片的协同效应、分级结构和较大的比表面积。

探索用于选择性监测甲烷(CH_4)的新型传感材料对于实际应用至关重要,尤其是在干扰气体乙醇(C_2H_5OH)和二氧化氮(NO_2)存在的情况下。在最近的科研工作中,一种新型的核壳型 ZnO/Pd@ZIF-8/Pt 纳米材料被报道具有优异的 CH_4 选择性传感性能[41],如图 1.2 所示。在 230℃ 下,相对于 C_2H_5OH 和 NO_2,ZnO/Pd@ZIF-8/Pt 传感器对 CH_4 的选择性($R_{CH_4}/R_{C_2H_5OH}$、R_{CH_4}/R_{NO_2})分别提高了 13.2 倍和 76.5 倍。催化实验表明,虽然 Pt 对 C_2H_5OH 的催化活性高于 CH_4,但由于 ZIF-8 的催化过滤效应,提高了其对 CH_4 的选择性催化性能。此外,ZIF-8 的筛分效应可以进一步阻碍干扰气体,因为其孔径为 4.0~4.2 Å,其中 NO_2(4.5 Å)被隔绝在 ZIF-8 层外,而 CH_4(3.8 Å)则可以通过。因此,这两种效应协同提高了 CH_4 的选择性。在 230℃ 下,ZnO/Pd@ZIF-8/Pt 对 5 000 ppm CH_4 的响应为 304.6%,响应/恢复时间为 4.0 s/1.5 s,并且具有优异的循环稳定性和长期稳定性,表明其对实际环境中的 CH_4 选择性检测极具潜力。此外,贵金属的气体选择性催化与 MOF 过滤层的组合可以很容易地扩展到其他类型的 SMOs@MOF 传感器中,为高选择性传感器器件的制备开辟了新的途径。

---

① 1 ppm=10^{-6}。

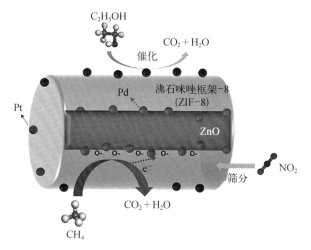

图 1.2　ZIF-8 和 Pt 提高 CH_4 选择性的示意图[41]

1.2.3　半导体金属氧化物传感器在疾病诊断中的应用

呼出气分析在疾病诊断中的应用可以追溯到古希腊,但直到近几年,现代分析技术才使得这种方法的潜在便利性(无创、实时、低成本分析)得以实现。近年来,基于半导体金属氧化物的化学电阻型传感器也被用于检测呼出气中极低浓度的 VOC,Kim 等[42]对这一技术进行了全面整理和综述。在特定条件下,金属氧化物传感器已经被认为是可用于诊断糖尿病、口臭和肺癌的工具。例如,已知硫化氢(H_2S)、丙酮、甲苯、氨(NH_3)、一氧化氮(NO)和戊烷分别与糖尿病、口臭、肺癌、肾衰竭、哮喘和心脏病有密切关系,尤其是使用一氧化氮传感器来监测哮喘已经成功实现商业化[42]。

利用气体传感器检测 H_2S 生物标记物从而间接监测口臭是一项新兴技术。然而,此类 H_2S 传感器要求具有极高的选择性和灵敏度,以及 ppb 级别的低检测限,极具有挑战性。为了解决这些问题,Feng 等[43]开发了一种基于 NiO/WO_3 纳米粒子(NPs)的高灵敏度、高选择性 H_2S 传感器(见图 1.3)。NiO/WO_3 NPs 的合成涉及 WO_3 NPs 的水解,以及随后的水热修饰 NiO NPs 过程。理论上,NiO/WO_3 NPs 有助于形成更厚的电子耗尽层,并吸附更多的氧物种(O_2^-)来氧化 H_2S,从而释放更多电子。2.1% 的 NiO/WO_3 NPs 在 100℃下对 10 ppm H_2S 具有较高的灵敏度($R_a/R_g = 15\ 031$),是原始 WO_3 NPs 的 42.6 倍($R_a/R_g = 353$)。此外,该 H_2S 传感器表现出 ppb 级别的检测限($R_a/R_g = 4.95@0.05$ ppm,100℃)和高选择性。NiO/WO_3 NPs 传感器原型已用于检测模拟呼出口臭,其检测浓度与气相色谱法测出的浓度值极为接近。因此,该研究可为未来智能医疗应用提供实验基础。

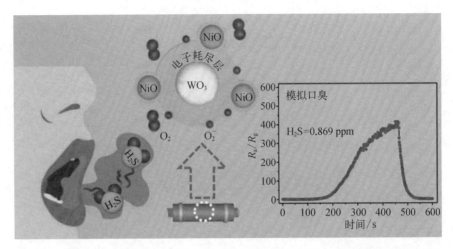

图1.3 高灵敏度、高选择性的NiO/WO$_3$复合纳米粒子检测口臭H$_2$S生物标志物[43]

氨气传感器可用于呼气诊断、环境污染监测和工业泄漏报警,但由于其具有高灵敏度、选择性、稳定性、湿度耐受性和宽浓度范围检测的要求,目前技术挑战仍然存在。Du等[44]通过在商用滤纸上修饰特殊设计的三元纳米复合材料,开发出可集成在手术口罩上的双稳态和耐水纸基传感器,并用于室温下检测NH$_3$(见图1.4)。这种纳米复合材料由聚吡咯纳米层涂覆的多壁碳纳米管框架组成,并进一步修饰铂纳米点。得益于三元组分的协同效应,该传感器对较宽浓度范围(5 ppb-60%①)的NH$_3$表现出超灵敏响应,同时,理论计算证实了其具有较高的选择性。值得注意的是,这种纸基传感器具有出色的抗扭稳定性(0~1 080°)、极好的可切割性和可折叠性。此外,该NH$_3$传感器还可以集成在外科口罩上,用于模拟呼出物诊断,并显示出优异的耐水性。离子色谱法进一步验证了该纸基传感器对模拟呼出NH$_3$、环境污染物NH$_3$和工业NH$_3$泄漏的高检测精度。

受杂原子掺杂工程的启发,Liu等[45]通过一步协同组装策略合成了独特的稀土铒掺杂的介孔氧化钨(Er/mWO$_3$),其表现出独特的六方介观结构($P6_3/mmc$)、可调比表面积(58.1~78.3 m^2/g)、均匀的介孔尺寸和高度结晶的骨架,以及杂原子掺杂孔壁(见图1.5)。Er原子的引入可以显著调整微/纳米结构和频带特性,1.5% Er/mWO$_3$显示出优异的丙酮传感性能,对50 ppm丙酮具有较高的响应值(R_a/R_g=107)、快速响应/恢复时间(9 s/56 s)、显著的丙酮选择性、低浓度检测(125 ppb)、抗湿性和优异的长期稳定性,这些优异的传感性能主要归功于杂原子间质掺杂诱导的独特界面催化位点。这一发现为设计用于环境监测和工业安全的智能气体传感器奠定了基础。

---

① 60%为体积分数。

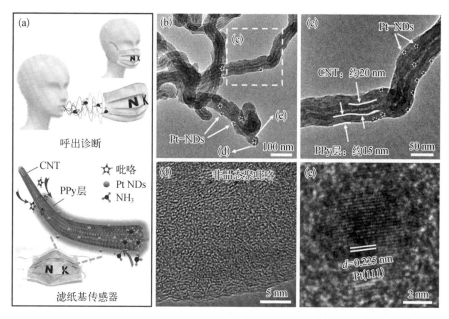

图 1.4 纳米复合修饰滤纸作为选择性检测氨气的可扭曲耐水传感器[44]

图 1.5 智能便携式丙酮检测传感装置及手机浓度信号读出[45]

1.2.4 半导体金属氧化物传感器在食品安全中的应用

食品安全和农业生产对于发展中国家的经济增长至关重要,而过程控制和质量检测是其中的关键步骤。半导体金属氧化物(SMO)传感器具有高通量实时检测的能力,在食品和农业领域掀起了一场传感革命,有望应用于检测食品污染物(如防腐剂、抗生素、重金属离子、毒素、微生物和病原体),以及快速监测食品的温度、湿度和香气。其中,细菌性食源性病原体会导致各种疾病,威胁着全世界公众的健康。迄今为止,研究人员已经提出了用于检测病原体的各种方法。众所周知,微生物[如单核细胞增生李斯特菌(*Listeria monocytogenes*,LM)]可产生物种特异性微生物挥发性有机化合物(MVOC),它可以被当作检测相关微生物的微生物标志物靶标[46]。近年来,检测 MVOC 已经成为一种用于揭示微生物污染的新颖且有效的方法,它可以以无创且快速的方式操作,不需要复杂且昂贵的仪器和专业人员。因此,半导体金属氧化物基化学电阻传感器引起了人们的特别关注,由于它们具有低成本、方便操作、快速响应和恢复以及对目标气体分子可调谐响应的优点,可以通过测量相关 MVOC 的浓度来间接监测微生物污染情况[47]。

LM 作为一种食源性细菌可导致严重疾病,甚至导致虚弱人群死亡。3-羟基-2-丁酮(3H-2B)已被证明是 LM 呼气的生物标记物。检测 3H-2B 是确定食品是否受到污染的一种快速且有效的方法。Wang 等[48]提出了一种基于双金属 PtCu 纳米晶修饰的 WO_3 空心球(PtCu/WO_3)的 3H-2B 气体传感器(见图 1.6),对 PtCu/WO_3 的结构和形貌进行了表征,并用静态测试方法测量了它们的气敏性能。结果表明,经双金属 PtCu 纳米晶修饰后,WO_3 空心球的传感器响应提高了约 15 倍。在 110℃时,PtCu/WO_3 传感器对 10 ppm 3H-2B 的响应值高达 221.2。同时,PtCu/WO_3 传感器表现出对 3H-2B 的良好选择性、快速响应/恢复时间(9 s/28 s)和低检测限(LOD<0.5 ppm)。此外,他们还通过气相色谱-质谱监测反应产物,对其气敏机制进行了研究,其优异的气敏性能可归因于 PtCu 和 WO_3 之间的协同作用,包括 O_2 在 PtCu 纳米颗粒上的独特溢出效应、p 型 Cu_xO 对 n 型 WO_3 的电子耗尽层的调控,以及对 3H-2B 的选择性催化。因此,这项工作为高性能气敏材料的设计、合成和食品安全中 LM 的检测提供了新思路。

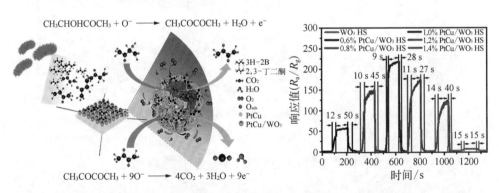

图1.6 双金属 PtCu 纳米晶敏化 WO_3 空心球高效检测 3-羟基 2-丁酮生物标志物的示意图(a)和不同含量 PtCu 纳米晶敏化的 WO_3 空心球对 3-羟基 2-丁酮的响应-恢复曲线(b)(后附彩图)[48]

三甲胺(TMA)是一种有机胺气体,是评价海鲜新鲜度的重要指标。Zhao 等[49]通过超声处理在 WO_3 纳米片上成功负载 Au 纳米粒子(约 4 nm),从而构建了一种超高效的三甲胺气体传感器(Au/WO_3)(见图 1.7)。在 300℃时,该传感器对 25 ppm TMA 的响应值(R_a/R_g)高达 217.72。同时,Au/WO$_3$ 传感器显示出快速的响应/恢复时间(8 s/6 s)、低检测限(0.5 ppm)和高选择性的 TMA 检测能力。此外,他们还检测了大花落叶松在 0~15 天的腐烂过程中产生的挥发物成分,证明了 Au/WO$_3$ 传感器检测 TMA 可以评估大花落叶松的新鲜度。如此出色的气敏性能表明,Au/WO$_3$ 传感器在现场快速无损检测海鲜新鲜度方面具有显著的应用潜力。

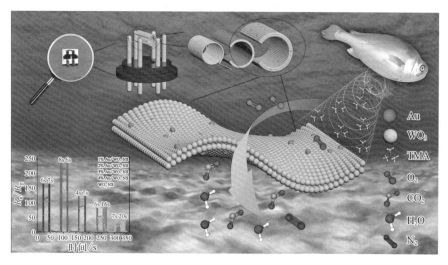

图 1.7　基于 Au 纳米粒子敏化 WO$_3$ 纳米片的高效三甲胺气体
传感器用于快速评估海鲜新鲜度[49]

1.2.5　半导体金属氧化物传感器在农业生产中的应用

现代农业是取代传统农业的社会化和商业化的新型农业模式,其在传统农业基础上依靠先进的技术和设备,以现代管理方法开发和运营[50]。因此,发展现代农业是提高农业生产效率和增强抗灾能力的关键措施。在农业现代化的必经过程中,先进的半导体金属氧化物基气体传感器发挥着不可替代的作用,可以调节和控制作物生长的环境条件,从而摆脱自然条件的约束。例如,基于 SMO 的 CO$_2$ 传感器可用于实时监测温室中的 CO$_2$ 含量,以确定其是否低于作物光合作用的最佳浓度[51]。密封鸡舍的二氧化碳含量也需要检测,以确保低于影响家畜生长发育的最大浓度,否则需要及时通风。此外,NH$_3$ 传感技术对于养殖业也尤为重要[52]。例如,在养鸡业中,鸡的消化系统不能完全消化饲料,导致大量蛋白质通过粪便排泄,通过复杂的化学反应转化为 NH$_3$。然而,NH$_3$ 是影响家禽健康和产蛋的关键因素[53]。一旦 NH$_3$ 浓度超过极限值,产蛋率将显著下降。因此,基于 SMO 的气体传感器的发展在农业生产现代化中起着关键作用。

Chen 等[54]首次合成了直径为 40 nm 的钴(Co)掺杂的 TiO_2 纳米颗粒,并对基于纯锐钛矿相 TiO_2 和 Co 掺杂 TiO_2 纳米颗粒的改进型气体传感器进行了研究(见图 1.8)。他们通过开发具有不同 Co 掺杂浓度的气体传感器,研究了其对 NH_3 的气体敏感性。与在 180 ℃下工作的纯锐钛矿相 TiO_2 相比,Co 掺杂 TiO_2 在室温下对 NH_3 表现出优异的气敏性能,其中 20% 钴掺杂的样品具有最佳性能,对 50 ppm NH_3 的响应值为 14,是纯 TiO_2 的 7 倍。传感器的响应时间和恢复时间分别为 25 s 和 48 s,具有良好的稳定性和选择性。此外,他们基于密度泛函理论(DFT)研究了其气敏性能提升的原因,结果显示与纯锐钛矿相 TiO_2 相比,Co 掺杂 TiO_2 的计算带隙降低了 72%。因此,结构模拟和机理分析证明了 Co 掺杂 TiO_2 气体传感器具有增强的室温 NH_3 传感性能。

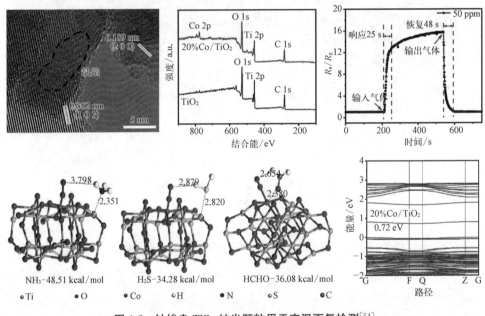

图 1.8 钴掺杂 TiO_2 纳米颗粒用于室温下氨检测[54]

金属氧化物半导体气体传感器在室温下检测挥发性有机化合物是一个巨大的挑战。Xu 和 Zhang[55]通过调节氧化铈(CeO_2)纳米线的表面化学状态,实现了对大米陈化过程中的气体标志物(芳樟醇)的室温检测。在这项研究中,作者通过控制在各种受控气氛下的退火条件,对水热法合成的 CeO_2 纳米线上的氧空位进行了调控。在 5% H_2+95% Ar 条件下退火的样品表现出优异的室温传感性能,对 20 ppm 芳樟醇具有较高的响应值(16.7),快速的响应/恢复时间(16 s/121 s),以及较低的检测限(0.54 ppm)。其增强的传感性能可归因于独特的纳米线形貌、大的比表面积(83.95 m^2/g)以及由 Ce^{3+} 含量增加而产生的大量氧空位。此外,通过监测在不同时期(1天、3天、5天、7天、15天和30天)储存的两个水稻品种(籼稻和粳稻)释放的气体,证明了该传感器能够成功地区分籼稻和粳稻,从而验证了该传感器的实用性。综上,这项研究可以为开发用于监测稻米质量的高性能室温电子鼻设备提供气敏材料设计的新思路。

近几十年来,材料科学和半导体技术的飞速发展极大地推动了气体传感器技术的发展。研究人员在开发金属氧化物气体传感器方面已经取得了很大进展,在改进金属氧化物气体传感器敏感性能方面也取得了显著进展。

(1) 在复合金属氧化物内构建 p-n 异质结是提升气敏性能的重要策略,但其 n 型-p 型平衡受到材料组成、主要氧分压、局部大气中干扰气体的浓度和温度的影响,因此必须注意传感器材料接近临界转换点时的电阻信号变化。

(2) 在传感器材料中掺入第二相组分可以通过限制主相的颗粒尺寸并因此维持高比表面积、添加催化功能和对近表面带结构施加有利的影响来提高灵敏度。

(3) 精确控制单一氧化物和材料组合的微观结构也能大大提高灵敏度。然而,湿气对气体响应的干扰和破坏一直是一个亟待解决的问题。

迄今为止,在用于更苛刻检测应用的功能化金属氧化物传感器上的大部分工作集中于 n 型金属氧化物半导体材料,但 p 型响应材料的等效测试也是值得研究的。通过使用金属氧化物传感器阵列,可以使用模式识别技术来分析信号,从而进一步提高在空气质量监测和医疗应用上所必须的选择性和灵敏度。

1.3 半导体金属氧化物的物理特性

随着工业生产的快速发展,一系列因工业制造或汽车发动机排放的 VOC 和其他有毒气体对环境生态平衡形成了严重的威胁,成为当前环境污染物处理及毒性气体安全监测的重大课题。近年来,以活性纳米材料为依托的传感技术相关研究已经呈现指数型增长,在气体分子尤其是有毒挥发性气体检测中具有重要的应用价值。传感技术有望将低浓度(ppm 和 ppb 级别)的气氛转化为可视化的输出信号,如电信号、光信号和磁信号等,为快速地原位检测粮食安全和环境气体带来了巨大希望。

在纳米材料和纳米技术蓬勃发展的背景下,新型的高效气体传感活性材料层出不穷,如金属氧化物半导体、导电聚合物、金属氧化物-聚合物复合材料等,这些材料均表现出较优异的气体检测性能,但目前相关检测机理和传感行为仍有待深入探究[56]。在上述新型材料中,半导体金属氧化物(SMO)作为一种电阻信号变化显著的特殊材料,能够为目标气体分子提供高灵敏度和选择性,并对低浓度气体展现出较高的信号响应[57]。此外,SMO 材料还具有许多传感应用的天然优势,包括低成本、快速响应/恢复时间、长期循环稳定性、简单易制成器件及热稳定性高等,一致被认为是目前最有发展前景的 VOC 和其他有毒气体传感器材料之一[58-60]。然而，SMO 材料作为气体传感器的相关特征和作用机制，尤其是材料的本征物理属性及理化性质在传感过程中发挥的作用尚不清楚。本节将尽可能全面地讨论和分析半导体金属氧化物的理化性质，希望能够帮助读者较全面地理解半导体金属氧化物气体传感器的基础知识。

1.3.1 半导体金属氧化物的定义

一般而言，作为一种传统的非化学计量比的氧化物，通常半导体金属氧化物的导电性对外界环境的影响十分敏感。主要体现在导电率的变化上，其中导电率因氧化性气氛而增加的半导体属于氧化型半导体，导电率随还原性气氛而增加的半导体属于还原型半导体，导电类型随氧分压的大小而变化的半导体则属于两性半导体。从化学组成而言，SMO 材料属于离子类固体材料，由正金属离子和负氧离子组成，并通过强离子键进行结合。此外，在 SMO 材料内部，"s"电子壳层的电子通常处于充满状态，与非金属氧化物相比，其具有非常优异的热力学稳定性和化学稳定性。相比之下，"d"电子壳层的电子通常处于不完全填充状态，从而可赋予 SMO 材料特殊的光学和电子特性，主要包括可调节的能带结构、高介电常数和新颖的光电响应行为[59,61-63]。在微观结构方面，SMO 材料的纳米结构具有柔性、可控与易加工的特征，维度上可涉及零维、一维、二维和三维结构，形貌层面还包括多孔结构、核-壳结构、块体结构、层-状结构、网络结构和其他结构[64-66]。此外，SMO 材料具有一系列典型的半导体特征，如 SMO 材料的导电能力可随着环境温度的升高而增加，并引起内部电阻率的下降。此外，微量的杂质、特定波长的光照、电场和磁场也能够显著改变半导体材料的导电性。

1.3.2 半导体金属氧化物的潜在性能

相比传统的非金属氧化物材料，SMO 材料具有更多的性能优势，主要包括相对高的化学活性、广泛的适用性、优异的稳定性和潜在的环境友好性。上述优异的理化特性赋予了 SMO 材料巨大的应用价值，目前 SMO 材料在众多研究领域已经得到了重要的应用，包括催化、传感器、能源储与转化、环境修复和太阳能电池等[67-70]。在上述应用中，与 SMO 传感器相关的应用最受关注。SMO 传感器已经开发了几十年并且被视为最有前景的化学阻抗型半导体或催化/热传递型导体。在不同的氧分压[$p(O_2)$]下，SMO 材料能够通过电子结构的变化或响应调节自身的导电行为和导电能力，而这种因外界环境变化引发的导电特性差异赋予了它们用于集成至各种电子设备的能力，有利于未来 SMO 传感材料的集成化和模块化发展[59,71]。从材料设计而言，SMO 材料的应用领域及性能优化将主要依赖于材料的组分、缺陷、晶型、尺寸和形貌等微观结构，上述参数的调控对于气体传感过程尤为重要,直接决定了传感的最大响应值、灵敏度和选择性。

近年来,化学阻抗型 SMO 传感器已经引起了人们的广泛关注,并被认为是新一代最有潜力的环境污染物检测器[57,72-73]。在实际应用中,该类传感器表现出较高的灵敏度、快速的响应/恢复时间、便携式结构、维护成本低和高检测容量的能力。SMO 传感器主要通过调控材料的内部电阻以实现信号的转变,一旦 SMO 传感器接触到还原性气氛或氧化性气氛,其将借助材料与带负电的表面吸附氧组分之间的氧化作用产生明显的电阻转换[74]。当待测分子与 SMO 材料接触时,分子吸附在材料表面导致材料本身的电子或者空穴等发生变化从而产生相应的响应信号,并把这种信号直接转化成能检测的电信号直观地表达出来,通过电信号来识别待检测气体的浓度、种类,对待检测进行定性与定量分析。在这个过程中,影响气体传感特性的主要参数可通过优化 SMO 材料的微观性能进行控制或调节,这些微观性能涉及材料的比表面积、电子/空穴受体密度、团聚程度、孔隙率、表面酸碱特性、辅助催化剂的存在及结晶性等。因此,可控地设计和调节 SMO 材料的理化性质是优化气体传感的关键。

1.3.3 半导体金属氧化物的物理基础

总体而言,SMO 材料的物理性质对于调节气体传感性能具有举足轻重的作用。就物理性质而言,其主要取决于材料的晶体结构、物理缺陷、能带结构、杂质级别、电荷传输和异质结构[75-77]。其中晶体结构主要是由于不同的晶面具有不同的反应活性和选择性,不同的晶型直接决定了材料是对氧化性气氛敏感还是对还原性气氛敏感。一定量的物理缺陷则提供了丰富的氧空位和不同形态的氧组分,这些氧组分将参与分子与 SMO 材料的表面氧化还原反应。能带结构在很大程度上影响了材料内部电子-空穴对的活跃程度,而杂质的引入不仅能够改变能带结构,而且能够提供局部异质界面。相对而言,异质结构对材料的性能影响最为显著,关于异质结的作用将在后续章节着重进行阐述。SMO 材料中主要的电荷载体浓度可通过适当的供体或受体掺杂进行调节,但是,金属氧化物中主要的电荷载体则由掺杂的异价阳离子或非化学计量的氧提供。由此可见,无论是元素掺杂还是多组分复合形成复合材料,物理层面的性能调节都依赖于上述物理参数[78-80]。对于不同的半导体而言,其基本物理特性差异较大,部分半导体因电子构型的差异而形成了全新的理化性质。根据导电类型和电子传输的原理,半导体可以分为 n 型和 p 型两大类,其中 n 型半导体因具有较高的电子-空穴对分离能力而广泛应用于各个领域。

1. 半导体的晶体结构和缺陷

晶体(crystal)通常是由大量微观物质单位(如原子、离子或分子等)按照特定规则形成长程有序排列的结构,具有整齐规则的几何外形、固定熔点和各向异性的固态物质[71,81]。根据微观排列的差异性,半导体的晶体结构可以划分为两种经典类型——单晶和多晶。单晶是指结晶体内部的微粒在三维空间呈有规律的、周期性的排列,或者说晶体的整体在三维方向上由同一空间格子构成,整个晶体中质点在空间的排列为长程有序。多晶是众多取向晶粒的单晶的集合。多晶与单晶内部均以点阵式的周期性结构为基础,对同一种晶体来说,两者本质相同,不同的是单晶属于各向异性,而多晶属于各向同性。相比之下,非晶体具有不规则的形状和不固定的熔点,且内部不存在长程有序结构,但在几个原子尺度范围内存在小范围内的结构短程有序[82-84]。例如,作为一种重要的半导体,SnO_2 晶体表面通常缺少一个或多个相邻的原子,导致较弱或不完全的配位,从而产生丰富的悬空键和不饱和键。此外,Sn^* 的表面还表现出不同的化合价,有利于该类材料参与一系列氧化还原反应,同时暴露出高的化学活性[85-86]。SnO_2 常见的晶体结构是金红石相,其中 (110) 晶面是最稳定的晶面,属于四方或正交晶系,其空间结构为 $P4_2/mnm$ (D_{4h}^{14}),晶胞参数分别为 $a=b=0.4738$ nm、$c=0.3188$ nm[87-88]。每个晶胞由两个 Sn 原子和四个 O 原子组成,并形成 6∶3 的配位结构。特别地,在正常温度下 SnO_2 的能带为 3.6~4.0 eV,具备 n 型半导体的特性,即包含一定数量的内在氧空位和间隙 Sn 原子[89-91]。相反,ZnO 作为一种常见的多功能半导体金属氧化物,具有 $P6_3/mc$ 空间群的六方晶系纤维状结构。其对应的晶胞参数分别为 $a=b=0.3253$ nm 和 $c=0.5213$ nm[64,92]。该六方晶系中,O^{2-} 以六面堆积的方式排列,而 Zn^{2+} 则填补四面体间隙的一半区域,常见的晶面主要包括带正电的 (0001)Zn^* 晶面和带负电的 O^* 晶面[89,93]。综上,不同的晶体结构对半导体金属氧化物的物理化学性能会产生较大影响。

半导体金属氧化物含有不同类型的缺陷,如点缺陷、线缺陷、面缺陷和体积缺陷,缺陷的引入往往有利于调节材料的理化性质,然而某些缺陷却往往被视为杂质而限制材料的应用范畴[77,94]。与光电性质相关的缺陷主要是点缺陷,点缺陷是最简单的晶体缺陷,它是在结点上或邻近的微观区域内偏离晶体结构正常排列的一种缺陷。点缺陷发生在晶体中一个或几个晶格常数范围内,其特征是在三维方向上的尺寸都很小(如空位、间隙原子、杂质原子等),一般处于原子大小的数量级。从维度而言,点缺陷亦称为零维缺陷,它的存在往往与材料的电学性质、光学性质,以及高温动力学过程等有关[95-96]。点缺陷的类型主要包括空位、间隙质点、错位原子或离子、外来原子或离子(杂质质点)和双空位等复合体。例如,典型的半导体金属氧化物形式为 MO,其中 M 代表金属原子,而 O 代表氧原子。当存在杂原子时,可能会出现六种不同的点缺陷,包括由 O 替换 M 原子、M 替换 O 原子、M 原子处的空位、O 原子处的空位、M 或 O 位于原子间隙。然而,随着杂原子的引入,这些杂质原子可以占据 M 原子或 O 原子的轨道甚至是原子间隙,它们由三个不同的点缺陷组成。上述各种点缺陷都可以调节半导体金属氧化物的电子性能,但发生的可能性和调节的位置不同。事实上,M 原子和 O 原子不可能依靠离子化合物之间的强相互作用来交换位置,但这种位置差异可以存在于共价化合物中[97-99]。当 O 原子的电负性大于 M 原子的电负性时,由 O 原子取代 M 原子可以作为电离电子的供体。相反,当 O 原子被 M 原子取代时,电离产生的空穴将被视为供体。如果在初始 M(或 O)位点没有额外的位置,中性 M(或 O)原子将被替换并留下两个空穴(或电子),这将被激发到价带(或导带)并形成非空穴类缺陷效应。

线缺陷是指在一维方向上偏离理想晶体中的周期性、规则性排列产生的晶格缺陷,即缺陷尺寸在一维方向上较长,在二维方向上较短,常见的表现形式是位错(刃型位错和螺型位错),由晶体中原子平面的错动引起。线缺陷的产生及运动主要与材料的韧性、脆性相关。此外,位错缺陷具备一些典型特征,它是实际晶体中广泛发育的一种从微观到亚微观的线状晶体缺陷,且当位错区域具有一定宽度时它不再是几何线[99-100]。位错区域可以和附近区域形成应力场,位错区域的原子的平均能量远大于其他区域的,因此位错不属于平衡缺陷。位错可以在晶体中或晶体表面上或者晶间边界上形成闭合环,但不能在晶体内部形成。相比之下,材料的面缺陷是指在二维方向上偏离理想晶体中的周期性、规则性排列而产生的缺陷,即缺陷尺寸在二维方向上延伸,在第三维方向上很小,主要包括晶界、表面、堆拓层错、镶嵌结构等,材料的面缺陷体现了材料的断裂韧性。在面缺陷中,整个晶体常常被一些界面分隔成许多较小的畴区,畴区内具有较高的原子排列完整性,畴区之间的界面附近存在着较严重的原子错排。通常包含三种不同类型,即小角度晶界、堆叠层缺陷和孪晶。小角度晶界是晶体中的一个小区域,当某些晶体存在取向差异时,等距边缘位错阵列可形成小角度晶界。相反,堆叠层缺陷是通过原子层的正常累积顺序的位错形成的,其包含固有缺陷和外在堆叠缺陷。孪晶则是由两种晶体(或晶体的两个部分)形成的,沿着共同的晶体表面形成镜像对称的方位关系(如特定的取向关系),并且与堆叠层缺陷关系密切。体积缺陷是指材料中某些特定的区域,这些区域在基质晶体宏观尺度上具有不同的结构、密度或化学组成[101-102]。然而,在材料设计中很少通过引入体积缺陷进行电子结构的调节,体积缺陷被视为在材料中引入了额外的组分、混合晶体或活性成分。

2. 半导体的能带和杂质能级

n 型和 p 型半导体的传感性能也依赖于 SMO 材料的能带结构和能级分布。当结晶性材料的维度处于电子耗尽层厚度的范畴时,能带弯曲不再局限于表面区域,而是延伸到大部分晶粒中,这将对 SMO 材料的电子结构和电子或空穴载流子的迁移产生不可忽视的影响[76,103-104]。一般而言,半导体的带隙能量(E_g)是指将电子从基态价能带激发到空导带能带所需的最小能量。一旦吸收能量高于基本带隙能量的光子,激发的电子将离开价带而形成轨道空穴。负电子和正电的空穴可以在外加电场作用下较灵活地移动,它们的最低能态是电子-空穴对静电束缚态^[105-106]。特别地，不同组分杂质的引入能够诱导带内电子能级跃迁，允许从缺陷态到基态的低能电子或光发射。此外，半导体的电子带隙可以通过调整半导体的尺寸、形状和组分来调节。有趣的是，在气体传感器应用中，以大带隙能材料制备的传感器能够在高工作温度下检测，这也表明 SMO 材料具有更高的热稳定性。研究表明，当工作温度高于 300℃时，气体传感器最佳的能带必须高于 2.5 eV。与此同时，在较高的工作温度下，半导体金属氧化物的化学活性对环境温度和湿度的依赖性较小[106-108]。例如，可以设计新型核-壳半导体材料，其中核和壳的价带与导带是交错分布的，从而可以引起电子-空穴对的分离，其中壳层具有最小的导带能，而核层具有相反的特征，有利于气体传感检测。壳层电子与核层空穴的能带分离以及载流子复合常发生在能量低于带隙能区域的界面，而很少出现在半导体高能量区域[105,108]。宽带隙、通用的异质结、高电子饱和速度和高场击穿效应造就了高速/高灵敏度的气体检测装置。

带隙结构直接决定着半导体的诸多性质，包括光吸收能力、载流子分离与复合能力、磁性能和光催化活性等，这些性质将直接影响材料在光催化或光电转化中的应用[104,109]。特别地，光学转换的类型亦取决于半导体的能带结构，且光学转换不需要直接间隙半导体对挥发性气体响应，其可以吸收几微米的所有入射光。同时，直接间隙半导体中的载流子只能通过移动相对短的距离传递给电解质，而间接间隙半导体需要在光学转换过程中改变晶格波动，并且入射光（光子）具有小的动量，其中间接转换需要额外添加光子（晶格振动），因此，它需要更大的厚度（通常约 100 μm）来吸收足够的入射光。在一个具有较短电荷扩散长度的间接间隙半导体中，载流子可在到达电解质之前与空穴重新复合[109-110]。简言之，能带结构对载流子运动的影响取决于传导和带隙，其中载流子的迁移率与其质量成反比。较宽的波段大多弯曲并降低了载流子的有效质量，提高了电荷的运动效率[111]。有趣的是，能带结构也会影响光电响应的光刻和热力学势能，它决定了半导体光吸收的光谱范围和太阳能产氨效率的最大理论值。

能带是半导体的一个基本性质，而另一个重要参数是半导体的杂质能级，它也能对电子或空穴载流子和微结构产生较大的影响，尤其是对电导率的影响[103,112]。在纯净的半导体中掺入一定量不同类型的杂质，并通过对其数量和空间分布的精确控制，可实现对电阻率和少数载流子寿命的有效控制，从而改变半导体的电学性质。通常而言，杂质半导体主要由元素的掺杂或额外组分的引入产生，指在本征半导体中掺入某些微量元素作为杂质，从而使半导体的导电性发生显著变化。掺入的杂质主要是三价或五价元素，掺入杂质的本征半导体称为杂质半导体。在导体中掺入微量杂质时，杂质原子附近的周期势场会受到干扰并形成附加的束缚状态，在禁带中产生附加的杂质能级。能提供电子载流子的杂质称为施主（donor）杂质，即束缚在杂质能级上的电子被激发到导带 E_c 成为导带电子，该杂质电离后成为正电中心,相应能级称为施主能级,通常位于禁带上方靠近导带底附近。例如,四价元素锗或硅晶体中掺入五价元素磷、砷、锑等杂质原子时,杂质原子作为晶格的一分子,其五个价电子中有四个与周围的锗(或硅)原子形成共价键,多余的一个电子被束缚于杂质原子附近,产生类氢浅能级-施主能级[109-110,113]。施主能级上的电子跃迁到导带所需能量比从价带激发到导带所需能量小得多,很易被激发到导带成为电子载流子,因此对于掺入施主杂质的半导体,导电载流子主要是被激发到导带中的电子,属于电子导电型,称为 n 型半导体。由于半导体中总是存在本征激发的电子-空穴对,所以在 n 型半导体中电子是多数载流子,空穴是少数载流子。

相应地,能提供空穴载流子的杂质称为受主(acceptor)杂质,相应能级称为受主能级,通常位于禁带下方靠近价带顶附近。例如,在锗或硅晶体中掺入微量三价元素硼、铝、镓等杂质原子时,杂质原子与周围四个锗(或硅)原子形成共价键时缺少一个电子,因而存在一个空位,与此空位相应的能量状态就是受主能级。由于受主能级靠近价带顶,价带中的电子很容易被激发到受主能级上填补这个空位,使受主杂质原子成为负电中心。同时价带中由于电离出一个电子而留下一个空位,形成自由的空穴,这一过程所需电离能比本征半导体情形下产生电子-空穴对所需能量要小得多[105-106,111]。因此这时空穴是多数载流子,杂质半导体主要靠空穴导电,即空穴导电型,称为 p 型半导体。在 p 型半导体中空穴是多数载流子,电子是少数载流子。在半导体器件的各种效应中,少数载流子常扮演重要角色。综上,在大多数情况下,引入特定杂质可以有效地调节材料的导电性和传质,这有利于气体传感器的应用。

3. 半导体的载流子运输和电子结构

在半导体物理学中,载体也被称为电流载流子,其主要通过电子损失的共价键空位而产生,包括电子和空穴。实际上,不论是 n 型半导体的自由电子还是 p 型半导体的空穴都可以在导电中发挥重要作用。根据该定义,在半导体中携带电流的带电粒子,如电子和空穴,均可视为自由载流子[90,114-115]。相反,当半导体在一定温度下处于热平衡状态时,半导体中导电电子和空穴的浓度保持稳定值,可以认为其是热平衡载体[82,102-103]。半导体载体是一种经典的物理现象,它由三种典型的传输组成,即漂移、扩散和复合,它们会受到各种因素的影响,如电子结构、温度和外加电场,甚至杂质、缺陷和材料紊乱[116-117]。载流子漂移可以定义为带电粒子在外部电场下的运动,并且该运动将基于电子和空穴的相反运动方向产生漂移电流[102-104]。研究发现,载流子迁移率(载流子在单位电场强度下单位时间内迁移的距离)同时影响载流子的产生和传输,这主要取决于晶格散射、元素掺杂和环境温度。

相比之下,载流子扩散是简单且普遍存在的运动,属于不规则的热运动。它指载流子从高浓度区域移动到低浓度区域,诱导载体的内部重排,可能产生扩散电流[118-119]。然而,载流子复合是最受关注的现象,它是一个复杂的过程,这个过程涉及电子/空穴湮灭或消失[68,105]。此外,载流子复合可以分为三类,即直接重组、间接重组和螺旋重组。在处于平衡状态的晶体中,在载流子产生和重组之间存在动态平衡。另外,载流子复合率对载流子的寿命有明显的影响,因此有效控制载流子复合对提高光电性能和气体传感器具有重要意义。

电子结构是半导体的重要特征,半导体金属氧化物主要由过渡金属元素组成,表现出非常好的电学性能[66,89]。半导体金属氧化物的电子结构非常复杂,除了s和p价键轨道外,还有d价键轨道。众所周知,d价键轨道具有丰富的物理和化学性质,可为各种应用,特别是气体传感器提供高的化学活性。

4. p-n异质结

近年来,p-n异质结半导体引起了不同学科或领域的广泛关注。p-n异质结主要由两种或两种以上不同电子结构的半导体组装或重组而成,其中一部分由n型半导体构成,另一部分由p型半导体构成[120-122]。通常,p型半导体具有更高的空穴浓度和更低的自由电子浓度,其主要依赖空穴产生导电特性。而n型半导体则具有更高的自由电子浓度和更低的空穴浓度,其主要由自由电子导电。当n型半导体和p型半导体相互接触时,上述两种不同的载流子能够从高浓度区域向低浓度区域移动,分别涉及自由电子和空穴的运动,其驱动力源于载流子浓度和费米能级的差异[123-125]。这种自发的扩散行为能够导致p型半导体缺失初始空穴并留下带负电的杂质离子,而n型半导体则失去初始自由电子并留下带正电的杂质离子。然而,这些杂质离子通常不能自发移动或扩散,因此,这些带相反电荷的杂质离子将在材料界面产生空间电荷区,定义其为p-n异质结,该异质结结构具有特殊的单向导电性。

随着纳米科学技术的发展,p-n异质结的研究已经取得了重大进展,一系列先进技术被广泛应用于异质结的合成,主要包括机械合金法、扩散、离子注入、溶胶-凝胶法、磁控溅射、化学气相沉积和外延生长,以下是几种方法基本原理的简要介绍[106,126]。机械合金法常在n型半导体表面引入一些特殊的杂质,使用高温熔融或局部熔融的方式引入杂质,然后冷却并晶化形成p-n异质结。相似地,借助气体、液体或固体的扩散作为杂质扩散源,可在原位热处理的条件下形成p-n异质结。离子注入则是赋予杂质离子较高的能量,直接注入半导体基质,这种方法相比传统扩散具有诸多优势。但上述方法往往因其较低的合成效率、高能耗而受限。此外,溶胶-凝胶法和化学气相沉积在制备p-n异质结方面展现出优越的发展前景。其中,化学气相沉积(CVD)主要采用典型的气态化合物或混合物在基板的加热表面进行化学反应,然后生长出非挥发性涂层。该技术可以实现物质与杂质之间的有效结合;同时,它可以复制并应用于许多平面材料的合成。因此,在实际应用中,使用CVD可以巧妙地设计和制造许多新颖的p-n异质结或异质结构,这在各个领域都具有很大的潜力,如气体传感器、催化、储能等领域。

半导体金属氧化物具有丰富的物理化学特性,包括电学、光学和磁学性质[95,97],基于其本征电子特性,可实现材料的可控设计与性能调控。这些特性将受到材料其他特性,如晶体结构、缺陷、能带和p-n异质结的影响。因此,通过调节上述基本物理参数,可以设计出应用于不同学科或领域的SMO材料。

5. 其他潜在特性

利用丰富的表面态密度,可以获得相对丰富的表面费米能级,从而可以改变半导体氧化物敏感材料的表面电势并形成不同的半导体势垒。该半导体势垒在增强敏感特性方面起重要作用,最具代表性的半导体势垒是肖特基势垒。电子肖特基势垒高度可表示为$U_s = K(W_{Me} - W_s)$,其中W_{Me}和W_s分别是金属和半导体的电子函数。此外,半导体金属氧化物的高化学活性更有利于表面氧化还原反应的进行,提高传感器的灵敏度[111,127]。响应灵敏性材料还应具有稳定的化学活性,以提高传感器的抗干扰性能,特别是在腐蚀性气体环境和有毒传感器中。半导体气体传感器的耐湿性差主要是由于半导体氧化物表面水分子化学吸附形成的羟基化显著影响了半导体氧化物的敏感特性,合成对水分子具有低化学活性的半导体氧化物可以增强传感器的防潮性。此外,气体传感器的敏感材料在较高工作温度下必须具有高热稳定性,敏感材料的热稳定性越高,传感器基极电阻的漂移越小。通常,半导体氧化物材料的良好热稳定性和高灵敏度不能同时获得,因此该材料的传感器在高温运行的长时间内具有良好的长期稳定性和器件可靠性。

在半导体氧化物中,氧空穴的形成能导致半导体氧化物中的缺陷。同时,氧空穴的存在使得半导体氧化物在不同的氧分压下具有不同的导电性。在表面氧化还原反应过程中,氧空穴可以从半导体氧化物内部移动到氧化物晶界的边缘,并与晶界表面的吸附氧相互作用,达到平衡状态。氧空穴扩散与半导体氧化物气敏特性之间的关系以及反应过程中氧空穴扩散的机理仍有待进一步研究。当半导体氧化物暴露在被测大气中时,敏感材料表面会出现两种情况:气体吸附和解吸[128-131]。半导体氧化物和气体靶分子之间存在两种主要的吸附反应:物理吸附和化学吸附。物理吸附是中性的,不会形成新的物种;化学吸附则伴随着吸附剂和半导体氧化物之间的电荷交换。建立合理的吸附-解吸动力学模型并阐明吸附-解吸特性与传感器敏感特性之间的关系,有助于开发高性能半导体金属氧化物气体传感器。

此外,半导体的一些固有特征,包括晶粒尺寸、形态、暴露的晶面和孔隙率,特别是孔隙率,也可以影响半导体在许多领域的宏观应用。例如,在气体传感方面,在半导体氧化物内部,纳米粒子的孔隙率通常需要很小,这可以赋予传感器敏感受体更小的晶界势垒和更好的电子传输特性,对提高传感器的气体敏感性有很大帮助[56,132]。而在半导体氧化物表面,纳米粒子的孔隙率需要很大,大的孔隙率有利于气体分子在传感器的敏感受体内扩散,提高传感器的利用率和敏感特性。当外部纳米颗粒的孔径减小到几纳米甚至亚纳米时，气体分子从表面向内部颗粒的扩散将会受到阻碍。这也可能是具有三维分层结构的半导体氧化物具有良好的敏感特性的原因，因为它们具有更大的外表面孔隙率、更小的内部孔隙率和更好的晶界接触。

参考文献

[1] Heiland G, Mollwo E, Stöckmann F. Electronic processes in zinc oxide[J]. Solid State Phys., 1959, 8: 191-323.

[2] Heiland G. Zum Einfluß von adsorbiertem Sauerstoff auf die elektrische Leitfähigkeit von Zinkoxydkristallen[J]. Zeit. Phys., 1954, 138: 459-464.

[3] Zhu L, Ou L, Mao L, et al. Advances in noble metal-decorated metal oxide nanomaterials for chemiresistive gas sensors: overview[J]. Nano-Micro Lett., 2023, 15: 89-164.

[4] Bielański A, Dereń J, Haber J. Electric conductivity and catalytic activity of semiconducting oxide catalysts[J]. Nature, 1957, 179: 668-669.

[5] Myasnikov I A. The relation between the electric conductance and the adsorptive and sensitizing properties of zinc oxide. I. Electron phenomena in zinc oxide during adsorption of oxygen[J]. Zhurnal Fiz. Khimii., 1957, 31: 1721-1730.

[6] Yamazoe N, Sakai G, Shimanoe K. Oxide semiconductor gas sensors[J]. Catal. Surv. Asia, 2003, 7: 63-75.

[7] Seiyama T, Kato A, Fujiishi K, et al. A new detector for gaseous components using semiconductive thin films[J]. Anal. Chem., 1962, 34: 1502-1503.

[8] Seiyama T, Kato A, Fujiishi K, et al. A new detector for gaseous components using semiconductive thin films[J]. Anal. Chem., 1962, 34, 11, 1502-1503.

[9] Eranna G, Joshi B, Runthala D, et al. Oxide materials for development of integrated gas sensors — a comprehensive review[J]. Crit. Rev. Solid State Mater. Sci., 2004, 29: 111-188.

[10] Yamazoe N. Toward innovations of gas sensor technology[J]. Sens. Actuators B-Chem., 2005, 108: 2-14.

[11] Zou X, Wang J, Liu X, et al. Rational design of sub-parts per million specific gas sensors array based on metal nanoparticles decorated nanowire enhancement-mode transistors[J]. Nano Lett., 2013, 13: 3287-3292.

[12] Mizsei J. How can sensitive and selective semiconductor gas sensors be made? [J]. Sens. Actuators B-Chem., 1995, 23: 173-176.

[13] Korotcenkov G, Cho B. Bulk doping influence on the response of conductometric SnO_2 gas sensors: Understanding through cathodoluminescence study[J]. Sens. Actuators B-Chem., 2014, 196: 80-98.

[14] Barsan N, Koziej D, Weimar U. Metal oxide-based gas sensor research: How to? [J]. Sens. Actuators B-Chem., 2006, 121: 18-35.

[15] Korotcenkov G. Gas response control through structural and chemical modification of metal oxide films: state of the art and approaches[J]. Sens. Actuators B-Chem., 2005, 107:

209-232.

[16] Jin H, Kim S, Shiratori S. Fabrication of nanoporous and hetero structure thin film via a layer-by-layer self assembly method for a gas sensor[J]. Sens. Actuators B-Chem., 2004, 102: 241-247.

[17] Yamazoe N. New approaches for improving semiconductor gas sensors[J]. Sens. Actuators B-Chem., 1991, 5: 7-19.

[18] Arunkumar S, Hou T, Kim Y, et al. Au Decorated ZnO hierarchical architectures: Facile synthesis, tunable morphology and enhanced CO detection at room temperature[J]. Sens. Actuators B-Chem., 2017, 243: 990-1001.

[19] Campbell J. The surface science of metal oxides[J]. Int. Mater. Rev., 1995, 39: 125.

[20] Nowotny J. Surface segregation of defects in oxide ceramic materials[J]. Solid State Ion., 1988, 28-30: 1235-1243.

[21] Yamazoe N, Fuchigami J, Kishikawa M, et al. Interactions of tin oxide surface with O_2, H_2O and H_2[J]. Surf. Sci., 1978, 86: 335-344.

[22] Chang S. Oxygen chemisorption on tin oxide: correlation between electrical conductivity and EPR measurements[J]. J. Vac. Sci. Technol., 1980, 17: 366.

[23] Itoh T, Maeda T, Kasuya A. In situ surface-enhanced Raman scattering spectroelectrochemistry of oxygen species[J]. Faraday Dis., 2006, 132: 95-109.

[24] Amalric-Popescu D, Herrmann J, Ensuque A, et al. Nanosized tin dioxide: Spectroscopic (UV-VIS, NIR, EPR) and electrical conductivity studies[J]. Phys. Chem. Chem. Phys., 2001, 3: 2522-2530.

[25] Williams D. Semiconducting oxides as gas-sensitive resistors[J]. Sens. Actuators B-Chem., 1999, 57: 1-16.

[26] Bârsan N, Weimar U. Understanding the fundamental principles of metal oxide based gas sensors: the example of CO sensing with SnO_2 sensors in the presence of humidity[J]. J. Phys. Condens. Matter., 2003, 15, R813-R839.

[27] Shin J, Choi S J, Lee I, et al. Thin-wall assembled SnO_2 fibers functionalized by catalytic Pt nanoparticles and their superior exhaled-breath-sensing properties for the diagnosis of diabetes[J]. Adv. Funct. Mater., 2013, 23: 2357-2367.

[28] Kim H, Lee J. Highly sensitive and selective gas sensors using p-type oxide semiconductors: Overview[J]. Sens. Actuators B-Chem., 2014, 192: 607-627.

[29] Fergus J. Perovskite oxides for semiconductor-based gas sensors[J]. Sens. Actuators B-Chem., 2007, 123: 1169-1179.

[30] Jeong S, Moon Y, Wang J, et al. Exclusive detection of volatile aromatic hydrocarbons using bilayer oxide chemiresistors with catalytic overlayers[J]. Nat Commun., 2023, 14: 233-246.

[31] Choi S, Park J, Kim S. Synthesis of SnO_2-ZnO core-shell nanofibers via a novel two-step process and their gas sensing properties[J]. Nanotechnology, 2009, 20: 465597-465603.

[32] Korotcenkov G, Cho B. Metal oxide composites in conductometric gas sensors: Achievements and challenges[J]. Sens. Actuators B-Chem., 2017, 244: 182-210.

[33] Gurlo A. Nanosensors: towards morphological control of gas sensing activity. SnO_2,

In_2O_3, ZnO and WO_3 case studies[J]. Nanoscale, 2011, 3: 154-165.

[34] Ushio Y, Miyayama M, Yanagida H. Effects of interface states on gas-sensing properties of a CuO/ZnO thin-film heterojunction[J]. Sens. Actuators B-Chem., 1994, 17: 221-226.

[35] Muller S, Degler D, Feldmann C, et al. Exploiting synergies in catalysis and gas sensing using noble metal-loaded oxide composites[J]. ChemCatChem, 2017, 10: 864-880.

[36] Heidari E K, Zamani C, Marzbanrad E, et al. WO_3-based NO_2 sensors fabricated through low frequency AC electrophoretic deposition[J]. Sens. Actuators B-Chem., 2010, 146: 165-170.

[37] Wetchakun K, Samerjai T, Tamaekong N, et al. Semiconducting metal oxides as sensors for environmentally hazardous gases[J]. Sens. Actuators B-Chem., 2011, 160: 580-591.

[38] Song Z, Tang W, Chen Z, et al. Temperature-modulated selective detection of part-per-trillion NO_2 using platinum nanocluster sensitized 3D metal oxide nanotube arrays[J]. Small, 2022, 18: 2203212.

[39] Schütze A, Baur T, Leidinger M, et al. Highly sensitive and selective VOC sensor systems based on semiconductor gas sensors: how to?[J]. Environments, 2017, 4: 20-32.

[40] Wang D, Wan K, Zhang M, et al. Constructing hierarchical SnO_2 nanofiber/nanosheets for efficient formaldehyde detection[J]. Sens. Actuators B-Chem., 2019, 283: 714-723.

[41] Cai Y, Luo S, Chen R, et al. Fabrication of ZnO/Pd@ZIF-8/Pt hybrid for selective methane detection in the presence of ethanol and NO_2[J]. Sens. Actuators B-Chem., 2023, 375: 132867.

[42] Kim S, Choi S J, Jang J, et al. Exceptional high-performance of Pt-based bimetallic catalysts for exclusive detection of exhaled biomarkers[J]. Adv. Mater., 2017, 12: 1700737.

[43] Feng D, Du L, Xing X, et al. Highly sensitive and selective NiO/WO_3 composite nanoparticles in detecting H_2S biomarker of halitosis[J]. ACS Sens., 2021, 6: 733-741.

[44] Du L, Feng D, Xing X, et al. Nanocomposite-decorated filter paper as a twistable and water-tolerant sensor for selective detection of 5 ppb-60 v/v% ammonia[J]. ACS Sens., 2022, 7: 874-883.

[45] Liu Y, Li Y, Gao M, et al. Interfacial catalysis enabled acetone sensors based on rationally designed mesoporous metal oxides with erbium-doped WO_3 framework[J]. Adv. Mater. Interfaces., 2022, 9: 2200802.

[46] Wang Y, Li Y, Yang J, et al. Microbial volatile organic compounds and their application in microorganism identification in foodstuff[J]. TrAC-Trends Anal. Chem., 2016, 78: 1-16.

[47] Zhu Y, Zhao Y, Ma J, et al. Mesoporous tungsten oxides with crystalline framework for highly sensitive and selective detection of foodborne pathogens[J]. J. Am. Chem. Soc., 2017, 139: 10365-10373.

[48] Wang D, Deng L, Cai H, et al. Bimetallic PtCu nanocrystal sensitization WO_3 hollow spheres for highly efficient 3-hydroxy-2-butanone biomarker detection[J]. ACS Appl. Mater. Interfaces, 2020, 12: 18904-18912.

[49] Zhao C, Shen J, Xu S, et al. Ultra-efficient trimethylamine gas sensor based on Au nanoparticles sensitized WO_3 nanosheets for rapid assessment of seafood freshness[J]. Food

Chem., 2022, 392: 133318.

[50] Yang X, Shu L, Chen J, et al. A survey on smart agriculture: development modes, technologies, and security and privacy challenges[J]. IEEE/CAA J. Automatica Sin., 2021, 8: 273-302.

[51] Gao X, Li Y. Monitoring gases content in modern agriculture: a density functional theory study of the adsorption behavior and sensing properties of CO_2 on MoS_2 doped GeSe monolayer[J]. Sensors, 2022, 22: 3860-3870.

[52] Li X, Xu J, Jiang Y, et al. Toward agricultural ammonia volatilization monitoring: A flexible polyaniline/$Ti_3C_2T_x$ hybrid sensitive films based gas sensor[J]. Sens. Actuators B-Chem., 2020, 316: 128144.

[53] Smith A F, Liu X, Woodard T, et al. Bioelectronic protein nanowire sensors for ammonia detection[J]. Nano Res., 2020, 13: 1479-1484.

[54] Chen Y, Wu J, Xu Z, et al. Computational assisted tuning of Co-doped TiO_2 nanoparticles for ammonia detection at room temperatures[J]. Appl. Surf. Sci., 2022, 601: 154214.

[55] Xu J, Zhang C. Oxygen vacancy engineering on cerium oxide nanowires for room-temperature linalool detection in rice aging[J]. Adv. Ceram., 2022, 11: 1559-1570.

[56] Korotcenkov G, Cho B. Metal oxide composites in conductometric gas sensors: Achievements and challenges[J]. Sens Actuators B-Chem., 2017, 244: 182-210.

[57] Fine G, Cavanagh L, Afonja A, et al. Metal oxide semi-conductor gas sensors in environmental monitoring[J]. Sensors, 2010, 10: 5469-5502.

[58] Sun Y, Liu S, Meng F, et al. Metal oxide nanostructures and their gas sensing properties: a review[J]. Sensors, 2012, 12: 2610-2631.

[59] Kanan S, El-Kadri O, Abu-Yousef I, et al. Semiconducting metal oxide based sensors for selective gas pollutant detection[J]. Sensors, 2009, 9: 8158-8196.

[60] Arafat M, Dinan B, Akbar S, et al. Gas sensors based on one dimensional nanostructured metal-oxides: a review[J]. Sensors, 2012, 12: 7207-7258.

[61] Wetchakun K, Samerjai T, Tamaekong N, et al. Semiconducting metal oxides as sensors for environmentally hazardous gases[J]. Sens Actuators B-Chem., 2011, 160: 580-591.

[62] Tomchenko A, Harmer G, Marquis B, et al. Semiconducting metal oxide sensor array for the selective detection of combustion gases[J]. Sens Actuators B-Chem., 2003, 93: 126-134.

[63] Afzal A, Cioffi N, Sabbatini L, et al. NO_x sensors based on semiconducting metal oxide nanostructures: progress and perspectives[J]. Sens Actuators B-Chem., 2012, 171-172: 25-42.

[64] Huang J, Wan Q. Gas sensors based on semiconducting metal oxide one-dimensional nanostructures[J]. Sensors., 2009, 9: 9903-9924.

[65] Pinna N, Neri G, Antonietti M, et al. Nonaqueous synthesis of nanocrystalline semiconducting metal oxides for gas sensing[J]. Angew. Chem. Int. Ed., 2004, 43: 4345-4349.

[66] Concina I, Ibupoto Z, Vomiero A. Semiconducting metal oxide nanostructures for water splitting and photovoltaics[J]. Adv. Energy Mater., 2017, 7: 1700706.

[67] Franke M, Koplin T, Simon U. Metal and metal oxide nanoparticles in chemiresistors: does the nanoscale matter? [J]. Small, 2006, 2: 36-50.

[68] Artzi-Gerlitz R, Benkstein K D, Lahr D L, et al. Fabrication and gas sensing performance of parallel assemblies of metal oxide nanotubes supported by porous aluminum oxide membranes[J]. Sens Actuators B-Chem., 2009, 136: 257-264.

[69] Chen X, Sun K, Zhang E, et al. 3D porous micro/nanostructured interconnected metal/metal oxide electrodes for high-rate lithium storage[J]. RSC Adv., 2013, 3: 432-437.

[70] Ming J, Wu Y, Park J, et al. Assembling metal oxide nanocrystals into dense, hollow, porous nanoparticles for lithium-ion and lithium-oxygen battery application[J]. Nanoscale, 2013, 5: 10390-10396.

[71] Zhou X, Cheng X, Zhu Y, et al. Ordered porous metal oxide semiconductors for gas sensing[J]. Chin. Chem. Lett., 2018, 29: 405-416.

[72] Delaney P, McManamon C, Hanrahan J, et al. Development of chemically engineered porous metal oxides for phosphate removal[J]. J. Hazard Mater., 2011, 185: 382-391.

[73] Ren Y, Ma Z, Bruce P G. Ordered mesoporous metal oxides: synthesis and applications [J]. Chem. Soc. Rev., 2012, 41: 4909-4927.

[74] Wang C, Yin L, Zhang L, et al. Metal oxide gas sensors: sensitivity and influencing factors[J]. Sensors, 2010, 10: 2088-2106.

[75] Yoo K S, Park S H, Kang J H. Nano-grained thin-film indium tin oxide gas sensors for H_2 detection[J]. Sens Actuators B-Chem., 2005, 108: 159-164.

[76] Hübner M, Simion C E, Tomescu-Stănoiu A, et al. Influence of humidity on CO sensing with p-type CuO thick film gas sensors[J]. Sens Actuators B-Chem., 2011, 153: 347-353.

[77] Lupan O, Ursaki V V, Chai G, et al. Selective hydrogen gas nanosensor using individual ZnO nanowire with fast response at room temperature[J]. Sens Actuators B-Chem., 2010, 144: 56-66.

[78] Wagner T, Waitz T, Roggenbuck J, et al. Ordered mesoporous ZnO for gas sensing[J]. Thin Solid Films, 2007, 515: 8360-8363.

[79] Szilágyi I, Saukko S, Mizsei J, et al. Gas sensing selectivity of hexagonal and monoclinic WO_3 to H_2S[J]. Solid State Sci., 2010, 12: 1857-1860.

[80] Brezesinski T, Rohlfing D, Sallard S, et al. Highly crystalline WO_3 thin films with ordered 3D mesoporosity and improved electrochromic performance[J]. Small, 2006, 2: 1203-1211.

[81] Rothschild A, Komem Y. The effect of grain size on the sensitivity of nanocrystalline metal-oxide gas sensors[J]. J. Appl. Phys., 2004, 95: 6374-6380.

[82] Cheng J, Wang J, Li Q, et al. A review of recent developments in tin dioxide composites for gas sensing application[J]. J. Ind. Eng. Chem., 2016, 44: 1-22.

[83] Cheng J P, Liu L, Zhang J, et al. Influences of anion exchange and phase transformation on the supercapacitive properties of α-Co(OH)$_2$[J]. J. Electroanal. Chem., 2014, 722-723: 23-31.

[84] Yang X, Cao C, Hohn K, et al. Highly visible-light active C-and V-doped TiO_2 for

degradation of acetaldehyde[J]. J. Catal., 2007, 252: 296-302.

[85] Waitz T, Becker B, Wagner T, et al. Ordered nanoporous SnO_2 gas sensors with high thermal stability[J]. Sens. Actuators B-Chem., 2010, 150: 788-793.

[86] Zhou X, Cao Q, Huang H, et al. Study on sensing mechanism of $CuO-SnO_2$ gas sensors [J]. Mater. Sci. Eng., 2003, 99: 44-47.

[87] Choi K S, Park S, Chang S P. Enhanced ethanol sensing properties based on SnO_2 nanowires coated with Fe_2O_3 nanoparticles[J]. Sens. Actuators B-Chem., 2017, 238: 871-879.

[88] Liu H, Chen S, Wang G, et al. ordered mesoporous core/shell SnO_2/C nanocomposite as high-capacity anode material for lithium-ion batteries[J]. Chem. Eur. J., 2013, 19: 16897-16901.

[89] Comini E, Baratto C, Faglia G, et al. Quasi-one dimensional metal oxide semiconductors: Preparation, characterization and application as chemical sensors[J]. Prog. Mater. Sci., 2009, 54: 1-67.

[90] Batzill M, Diebold U. Surface studies of gas sensing metal oxides[J]. Phys. Chem. Chem. Phys., 2007, 9: 2307-2318.

[91] Yang J, Hidajat K, Kawi S. Synthesis of nano-SnO_2/SBA-15 composite as a highly sensitive semiconductor oxide gas sensor[J]. Mater. Lett., 2008, 62: 1441-1443.

[92] Zhao X, Zhou R, Hua Q, et al. Recent progress in ohmic/schottky-contacted ZnO nanowire sensors[J]. J. Nanomater., 2015: 1-20.

[93] Zhou X, Lee S, Xu Z, et al. Recent progress on the development of chemosensors for gases [J]. Chem. Rev., 2015, 115: 7944-8000.

[94] Zakrzewska K. Gas sensing mechanism of TiO_2-based thin films[J]. Vacuum, 2004, 74: 335-338.

[95] Jiménez I, Arbiol J, Dezanneau G, et al. Crystalline structure, defects and gas sensor response to NO_2 and H_2S of tungsten trioxide nanopowders[J]. Sens. Actuators B-Chem., 2003, 93: 475-485.

[96] Zhang Y H, Chen Y B, Zhou K G, et al. Improving gas sensing properties of graphene by introducing dopants and defects: a first-principles study[J]. Nanotechnology, 2009, 20: 185504-185511.

[97] Fergus J. Perovskite oxides for semiconductor-based gas sensors[J]. Sens. Actuators B-Chem., 2007, 123: 1169-1179.

[98] Schmidt-Mende L, MacManus-Driscoll J. ZnO - nanostructures, defects, and devices[J]. Mater. Today, 2007, 10: 40-48.

[99] Adepalli K, Kelsch M, Merkle R, et al. Influence of line defects on the electrical properties of single crystal TiO_2[J]. Adv. Funct. Mater., 2013, 23: 1798-1806.

[100] Nisar J, Topalian Z, De Sarkar A, et al. TiO_2-based gas sensor: a possible application to SO_2[J]. ACS Appl. Mater. Interfaces, 2013, 5: 8516-8522.

[101] Kim K, Lee H, Johnson R, et al. Selective metal deposition at graphene line defects by atomic layer deposition[J]. Nat. Commun., 2014, 5: 4781-4789.

[102] Ahn M, Park K, Heo J, et al. Gas sensing properties of defect-controlled ZnO-nanowire

gas sensor[J]. Appl. Phys. Lett., 2008, 93: 263103 - 263106.

[103] Rothschild A, Litzelman S, Tuller H, et al. Temperature-independent resistive oxygen sensors based on $SrTi_{1-x}Fe_xO_{3-\delta}$ solid solutions[J]. Sens Actuators B-Chem., 2005, 108: 223 - 230.

[104] Zaleska A. Doped-TiO_2: a review[J]. Recent Pat. Eng., 2008, 2: 157 - 164.

[105] Li S, Xia J. Electronic states of a hydrogenic donor impurity in semiconductor nanostructures[J]. Phys. Lett. A, 2007, 366, 120 - 123.

[106] Waldrop J, Grant R. Semiconductor heterojunction interfaces: nontransitivity of energy-band discontiuities[J]. Phys. Rev. Lett., 1979, 43: 1686 - 1689.

[107] Anothainart K, Burgmair M, Karthigeyan A, et al. Light enhanced NO_2 gas sensing with tin oxide at room temperature: conductance and work function measurements[J]. Sens Actuators B-Chem., 2003, 93: 580 - 584.

[108] Heyd J, Peralta J E, Scuseria G E, et al. Energy band gaps and lattice parameters evaluated with the Heyd-Scuseria-Ernzerhof screened hybrid functional[J]. J. Chem. Phys., 2005, 123: 174101.

[109] Sadeghi E. Impurity binding energy of excited states in spherical quantum dot[J]. Phys. E, 2009, 41: 1319 - 1322.

[110] Zhuravlev M, Tsymbal E, Vedyayev A. Impurity-assisted interlayer exchange coupling across a tunnel barrier[J]. Phys. Rev. Lett., 2005, 94: 026806.

[111] Wehling T, Katsnelson M, Lichtenstein A. Adsorbates on graphene: Impurity states and electron scattering[J]. Chem. Phys. Lett., 2009, 476: 125 - 134.

[112] White S R, Sham L. Electronic properties of flat-band semiconductor heterostructures[J]. J. Phys. Rev. Lett., 1981, 47: 879 - 882.

[113] Langer J M, Heinrich H. Deep-level impurities: a possible guide to prediction of band-edge discontinuities in semiconductor heterojunctions[J]. Phys. Rev. Lett., 1985, 55: 1414 - 1417.

[114] Basu S, Bhattacharyya P. Recent developments on graphene and graphene oxide based solid state gas sensors[J]. Actuators B-Chem., 2012, 173: 1 - 21.

[115] Bittencourt C, Felten A, Espinosa E, et al. WO_3 films modified with functionalised multi-wall carbon nanotubes: morphological, compositional and gas response studies[J]. Sens. Actuators B-Chem., 2006, 115: 33 - 41.

[116] Cui S, Pu H, Wells S A, et al. Ultrahigh sensitivity and layer-dependent sensing performance of phosphorene-based gas sensors[J]. Nat. Commun., 2015, 6: 8632 - 8640.

[117] Costello B, Ledochowski M, Ratcliffe N. The importance of methane breath testing: a review[J]. J. Breath Res., 2013, 7: 024001.

[118] Dong C, Liu X, Han B, et al. Nonaqueous synthesis of Ag-functionalized In_2O_3/ZnO nanocomposites for highly sensitive formaldehyde sensor[J]. Sens. Actuators B-Chem., 2016, 224: 193 - 200.

[119] Comini E. Metal oxide nano-crystals for gas sensing[J]. Anal. Chim. Acta., 2016, 568: 28 - 40.

[120] Silva L, M'Peko J, Catto A, et al. UV-enhanced ozone gas sensing response of ZnO -

SnO$_2$ heterojunctions at room temperature[J]. Sens. Actuators B-Chem., 2017, 240: 573 - 579.

[121] Dandeneau C, Jeon Y, Shelton C, et al. Thin film chemical sensors based on p - CuO/n - ZnO heterocontacts[J]. J. Thin Solid Films, 2009, 517: 4448 - 4454.

[122] Dhawale D, Salunkhe R, Patil U, et al. Room temperature liquefied petroleum gas (LPG) sensor based on p-polyaniline/n - TiO$_2$ heterojunction[J]. Sens. Actuators B-Chem., 2008, 134: 988 - 992.

[123] Huang H, Gong H, Chow C, et al. Low-temperature growth of SnO$_2$ nanorod arrays and tunable n - p - n sensing response of a ZnO/SnO$_2$ heterojunction for exclusive hydrogen sensors[J]. Adv. Funct. Mater., 2011, 21: 2680 - 2686.

[124] Ju D, Xu H, Xu Q, et al. High triethylamine-sensing properties of NiO/SnO$_2$ hollow sphere P - N heterojunction sensors[J]. Sens. Actuators B-Chem., 2015, 215: 39 - 44.

[125] Ma L, Fan H, Tian H, et al. The n - ZnO/n - In$_2$O$_3$ heterojunction formed by a surface-modification and their potential barrier-control in methanal gas sensing[J]. Sens. Actuators B-Chem., 2016, 222: 508 - 516.

[126] Miller D, Akbar S, Morris P. Nanoscale metal oxide-based heterojunctions for gas sensing: a review[J]. Sens. Actuators B-Chem., 2014, 204: 250 - 272.

[127] O'Donnell K P, Chen X. Temperature dependence of semiconductor band gaps[J]. Appl. Phys. Lett., 1991, 58: 2924 - 2926.

[128] Han D, Zhai L, Gu F, et al. Highly sensitive NO$_2$ gas sensor of ppb-level detection based on In$_2$O$_3$ nanobricks at low temperature[J]. Sens. Actuators B-Chem., 2018, 262: 655 - 663.

[129] Xing X, Xiao X, Wang L, et al. Highly sensitive formaldehyde gas sensor based on hierarchically porous Ag-loaded ZnO heterojunction nanocomposites[J]. Sens. Actuators B-Chem., 2017, 247: 797 - 806.

[130] Shendage S, Patil V, Vanalakar S, et al. Sensitive and selective NO$_2$ gas sensor based on WO$_3$ nanoplates[J]. Sens. Actuators B-Chem., 2017, 240: 426 - 433.

[131] Liu J, Wang T, Wang B, et al. Highly sensitive and low detection limit of ethanol gas sensor based on hollow ZnO/SnO$_2$ spheres composite material[J]. Sens Actuators B-Chem., 2017, 245: 551 - 559.

[132] Li Y, Chen N, Deng D, et al. Formaldehyde detection: SnO$_2$ microspheres for formaldehyde gas sensor with high sensitivity, fast response/recovery and good selectivity[J]. Sens. Actuators B-Chem., 2017, 238: 264 - 273.

第2章
半导体金属氧化物气体传感器的传感机理

金属氧化物半导体的气体传感机理主要基于传感材料在目标气体中电阻的变化。例如,对于 n 型半导体,传感器的气体响应定义为 $S=R_a/R_g$(还原气体)或 $S=R_g/R_a$(氧化气体),其中 R_a 和 R_g 分别是在空气和目标气体中的电阻。通常,响应或恢复时间定义为在接触气体或释放气体后,传感器输出电阻达到或恢复 90% 初始值所用的时间[1-2]。

半导体金属氧化物气体传感器的敏感机理主要分为两类。

(1) 表面控制型。氧气的化学吸附改变了金属氧化物的电阻,金属氧化物表面气体的吸附和反应也相应地改变了电阻。

$$O_2(g) + xe^- \rightleftharpoons O_2^{x-}(ad) \tag{2.1}$$

$$2H_2(g) + O_2^{x-}(ad) \rightleftharpoons 2H_2O(g) + xe^- \tag{2.2}$$

(2) 体电阻控制型。该机理适用于 $\gamma\text{-}Fe_2O_3$ 和 ABO_3 型气敏材料。下面给出了 $\gamma\text{-}Fe_2O_3$ 的一个例子,$\gamma\text{-}Fe_2O_3$ 是一种亚稳态结构,可在高温下转化为稳定的 $\alpha\text{-}Fe_2O_3$,在还原气作用下转化为类似的 Fe_3O_4 结构。几种氧化铁的转化关系如下:

$$Fe_3O_4 \xleftarrow{\text{还原}} \gamma\text{-}Fe_2O_3 \xrightarrow{\text{还原}} \alpha\text{-}Fe_2O_3 \tag{2.3}$$

对于大多数金属氧化物半导体(如 n 型和 p 型半导体)传感材料,传感机制多为表面控制型,即取决于金属氧化物电子耗尽层(electron depletion layer, EDL)载流子浓度的变化。

2.1 纯金属氧化物半导体

金属氧化物半导体的气敏特性基于氧吸附可形成电子核壳结构(见图2.1)。在高温(>100℃)下,氧分子被吸附到 n 型金属氧化物半导体(如 SnO_2 和 ZnO)的表面上,并通过从半导体表面获取电子而电离成氧物种,如 O^{2-}、O^- 和 O_2^-。一般来说,在<150℃、150~400℃ 和>400℃ 时,O^{2-}、O^- 和 O_2^- 离子分别占优势,这导

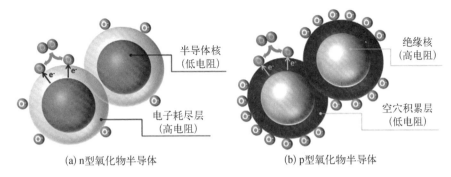

图 2.1　n 型和 p 型氧化物半导体中电子核壳结构的形成[2]

致了电子核壳结构[见图 2.1(a)];也就是核的 n 型半导体区域和粒子外壳的电阻电子耗尽层的形成。同样地,p 型金属氧化物半导体中氧的吸附作用在材料表面形成空穴积累层(HAL),这是由于电荷种类之间的静电相互作用[见图 2.1(b)],这再次建立了电子核壳结构,即在粒子表面的核和半导体 HALs。

金属氧化物半导体的气敏活性取决于金属氧化物在目标气体中的电阻变化。半导体的电阻会随着其物理性质和气体分析物的不同而增大或减小。例如,n 型半导体的传感机理如图 2.2(a)所示。对于 n 型半导体,载流体为电子(e^-),在环境中,氧分子吸附在氧化物表面并"抓住"表面的电子,形成氧负离子,如 O^{2-} 和 O^-,导致电子密度减小和电阻增加。半导体的电子分布只能在靠近表面的有限深度内受到不同吸附氧的影响。电子密度较小的受影响区域称为电子耗尽层,其离表面的深度称为德拜长度(L_D),通常为几纳米。半导体德拜长度的计算在式(2.4)中给出,其中,ε 是介电常数,k_B 是玻尔兹曼常数,T 是绝对温度,在开尔文温度中,q 是基本电荷,N_d 是掺杂剂(供体或受体)的密度。当 n 型氧化物暴露在诸如 CO、H_2、CH_4、C_2H_5OH 和丙酮等还原性气体中时,电子会通过负氧物质和还原性气体之间的表面反应回流到耗尽的氧化物中,从而降低了金属氧化物的电阻;而诸如 Cl_2、NO_x 和 SO_2 等氧化性气体则会加剧电子的脱失,引起电阻的增加。对于 p 型金属氧化物,电流载体是空穴(H^+),电阻对还原性气体和氧化性气体的变化与 n 型半导体完全相反。在电阻式传感器中,传感材料沉积在两个或多个电极上,这些电极测量暴露在目标气体中时氧化物的电阻会发生变化。传感测量的典型简化电路如图 2.2(b)所示。传感材料与包括一个串联负载电阻的分压电路连接。根据负载电阻的输出电压,可以计算出感测材料的电阻及其接触目标气体时的电阻变化。为了保证计算的准确性,需要一个合适的负载电阻(R_L),其电阻可与传感材料相当。通常,在早期的测量系统中人工选择 R_L。最近开发的测量系统引入了一种 R_L 自动开关,可以大大降低测量误差,提高实际应用的潜力[3-5]。

$$L_D = \sqrt{\frac{\varepsilon k_B T}{q^2 N_d}} \qquad (2.4)$$

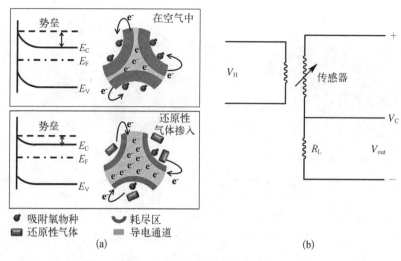

图 2.2 传感机理示意图[4]

(a) n 型金属氧化物半导体的传感机理,当接触到还原气体时,传导区域会膨胀;(b) 用于气体传感测量的典型电路,其中,R_L:负载电阻;V_C:电路电压;V_{out}:输出电压;V_H:加热电压

2.1.1 n 型金属氧化物

Zhu 等[6]合成了用于检测食源性病原体的介孔晶态氧化钨,介孔 WO_3 基化学电阻传感器检测 3-羟基-2-丁酮时具有快速响应、高灵敏度和高选择性的能力[6]。介孔 WO_3 基传感器的传感机理可以用表面耗竭模型来解释。如图 2.3 所示,当传感器暴露在空气中时,氧分子可以化学吸附在 WO_3 的表面上,从传导带捕获电子并形成吸附的氧阴离子(O^{2-}、O^- 和 O_2^-)。同时,在 WO_3 表面附近形成一层较厚的空间电荷层,以较高的电阻增加势垒(绿色标记)。相反,当 WO_3 基传感器暴露在还原性 3-羟基-2-丁酮气体中时,目标分子可以与氧负离子反应并释放自由电子,导致势垒厚度(绿色标记)和电阻减小。

Xiao 等[7]报道了介孔 SnO_2 用于制备具有良好传感性能的硫化氢气体传感纳米器件,该器件显示出较高的灵敏度(R_a/R_g=170,50 ppm)和稳定性,其化学机理研究表明,在 SnO_2 基传感器的气体传感过程中,会同时产生 SO_2 和 SnS_2。如图 2.4 所示,在介孔 SnO_2 材料的结晶孔壁中,两个相邻纳米颗粒之间形成了丰富的同质连接。在空气中,氧分子可以通过纳米颗粒的介孔和空隙扩散,从而完全覆盖纳米颗粒的表面。吸附氧物种能从纳米颗粒中提取电子,因此在 SnO_2 纳米颗粒表面形成电子耗尽层,导致边界上形成势垒。势垒的存在有助于限制电子通过边界的流动。当接触到硫化氢时,负载电阻的电压迅速增加,因为表面反应将电子从硫化氢返回到 SnO_2,导致 SnO_2 传感器的电阻降低。

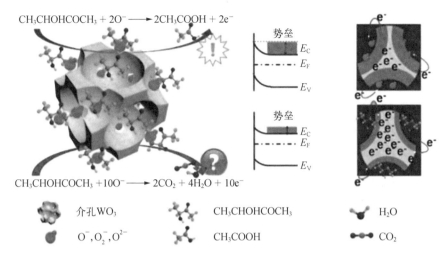

图 2.3 检测暴露于空气和目标气体-空气混合物中的 3-羟基-2-丁酮的
介孔 WO_3 基传感器的传感机理示意图(后附彩图)[6]

E_V：价带边；E_C：导带边；E_F：费米能级

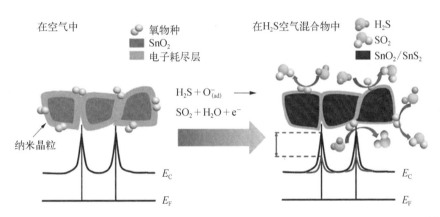

图 2.4 介孔 SnO_2 基传感器暴露在空气和 H_2S 空气混合物中的传感机理(后附彩图)[7]

E_C：传导带边缘；E_F：费米能级

2.1.2 p 型金属氧化物

Wang 等[8]合成了具有大介孔和石墨化孔壁的有序介孔碳/氧化钴纳米复合材料。由于具有大孔的石墨化 OMC 与均匀活性的 p 型 CoO_x 纳米颗粒具有很强的协同作用，制备的介孔纳米复合材料在氢传感方面表现出优异的性能。CoO_x 纳米颗粒作为活性物质，在气体传感过程中遵循氧吸附机理(见图 2.5)。首先，氧分子趋向于捕获并与 CoO_x/C 复合材料导带中的电子反应，并在 CoO_x 纳米颗粒表面形成含负氧离子(O^{2-}、O^-、O_2^-)的耗尽层。然后，当感测材料暴露于 H_2-空气混合物中时,还原性 H_2 会与氧阴离子反应形成 H_2O,电子流回 CoO_x 并与一定数量的空穴复合,导致电阻迅速增大。

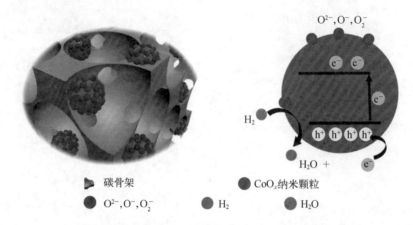

图 2.5　CoO_x/C 暴露在空气(左)和 H_2-空气混合物(右)中时的氢气感应机制示意图[8]

通常,CoO_x 是 p 型半导体材料,当其在一定温度下暴露于空气中时,吸附的氧从 CoO_x 中捕获电子,并在 CoO_x 纳米颗粒表面附近形成氧阴离子(O^{2-}、O^-、O_2^-);当暴露在 H_2-空气混合物中时,H_2 与氧负离子反应,电子返回到 CoO_x 纳米颗粒,与空穴结合,导致电阻增加

2.2　金属氧化物异质结

2.2.1　n-n 异质结

Han 等[9]成功地制备了一种新型的基于有序介孔 WO_3/ZnO(OM-WO_3/ZnO)的 n-n 型复合传感器,该传感器具有响应好、响应时间短、检测限低、对氮氧化合物气体选择性好等优点[9]。制备的 WO_3-ZnO 传感器的孔的内表面积比外表面积大得多,传感器与目标气体的反应主要发生在孔的表面。该过程可分为三个部分,如图 2.6 所示。从图 2.6(a)可以看出,WO_3 的费米能级高于 ZnO,WO_3 中的电子倾向于流入 ZnO 层,以实现并维持费米能级的平衡。在这种情况下,超流态电子分布在 ZnO 侧,超流态空穴分布在靠近 n-n 异质结层的 WO_3 侧。当电子被注入 n-n 异质结层时,电子将流向 WO_3 侧。因此,在外加电流作用下,n-n 异质结层降低了电子从 ZnO 层迁移到 WO_3 层的势垒。当 WO_3/ZnO 传感器暴露在空气中时,氧气被吸附在 WO_3 层的表面。如上所述,发生吸附氧反应,氧分子捕获 WO_3 层的电子,形成一系列氧物种(O^{2-}、O^-、O_2^-)。如图 2.6(b)所示,这是一个电子耗尽的过程,导致 WO_3 层中电子的浓度降低,并在 n-n 异质结的两侧形成可膨胀的电子浓度差。它为在外加电流作用下电子从 ZnO 层迁移到 WO_3 层提供了更大的驱动力。

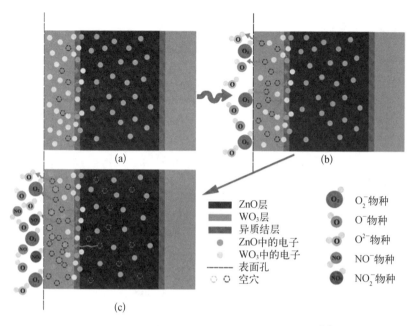

图 2.6 WO_3-ZnO 气体传感器传感机理示意图[9]

从图 2.6(c)可以看出,当氧化性氮氧化物注入测试室时,NO_2 和 NO 也被吸附在 WO_3 层的表面。氮氧化物捕获了 WO_3 的电子,WO_3 层的电子浓度依次降低,n-n 异质结两侧的电子浓度差进一步增大,导致 ZnO 层的电子在外加电流下通过 n-n 异质结向 WO_3 层移动。因此,传感器的电阻急剧下降。

Gao 等[10]称,Al_2O_3-In_2O_3 纳米纤维传感器对室温下 0.3~100 ppm 范围内的氮氧化物浓度表现出非常高的响应,独特的一维介孔管结构与 Al_2O_3 改性作用之间的协同作用,使其具有良好的气体传感性能。该模型可用于解释图 2.7 中的传感机制。

介孔-Al_2O_3/In_2O_3 中的电子传输和氮氧化物气体反应如下:① 当传感器[见图 2.7(a)]暴露在空气中时,氧气可能被吸附在管状表面的外层/内层以及整个介孔壁上[见图 2.7(b)中的 1]。② 氧分子可以很容易地从其导带或供体水平捕获自由电子,形成氧离子或化学吸附氧(O^{2-} 或 O^-)。如图 2.7(b)中的 2 所示,这些氧物种中的负电荷会导致耗尽层的出现。这种化学吸附氧被认为是一种电子供体,其在很大程度上取决于材料的导电性。许多氧物种(氧气或 O^-)的形成是因为它们在气体传感器中具有良好的催化活性。③ 当传感器暴露在氮氧化物气体中时,由于氮氧化物具有很高的电子亲和力,气体分子倾向于从中间 Al_2O_3/In_2O_3 的传导带或供体中吸引电子。这使得电子从介孔-Al_2O_3/In_2O_3 传感器转移到表面吸附氮氧化物,如图 2.7(b)中的 3 所示。同时,氧化铟与氮氧化物的化学反应以及缺

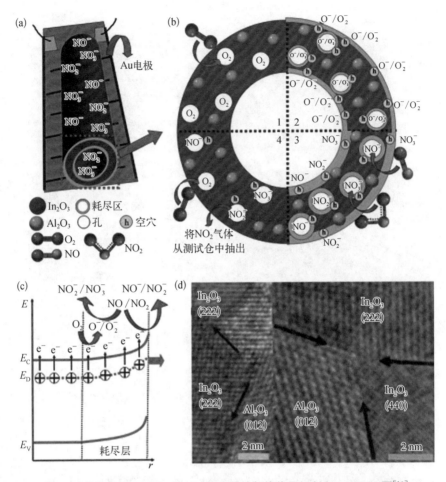

图 2.7 有序介孔-Al_2O_3/In_2O_3 传感器的气体传导机制和 HRTEM 图[10]

(a) 介孔-Al_2O_3/In_2O_3 传感器;(b) 传感器的气体传导过程:其中 1 为空气气氛,2 为氧分子俘获电子并形成化学吸附氧,3 为氮氧化物气体传感响应,4 为氮氧化物浓度降低;(c) 表面区域附近纳米管的能带图,E_C 为导带,E_D 为施主能级,E_V 为价带;(d) 有序介孔 Al_2O_3/In_2O_3 的 HRTEM 图,显示其缺陷位点和异质结构,箭头指示线性和平面缺陷

陷浓度的变化也可能发生,这一过程从介孔 Al_2O_3/In_2O_3 的传导带或供体能级捕获电子,最终导致电子密度降低。NO_2 和 NO 在介孔-Al_2O_3/In_2O_3 上的吸附形成 NO_2^- 和 NO^-,吸附在这些 NO/NO_2 中的负电荷导致耗尽区厚度和电阻增大。另外,目标气体分子 NO_2 和 NO 直接吸附在介孔 Al_2O_3/In_2O_3 上,然后与 O^{2-}/O^- 反应生成双齿 NO_3^-(S),进而生成 NO_2^-。④ 用空气吹扫气室,以恢复中间 Al_2O_3/In_2O_3 传感器的电阻[见图 2.7(b)中的 4]。在图 2.7(c)中,位于导带下方的供体能级可能是由 Al_2O_3 添加剂导致的氧空位/缺陷。图 2.7(d)显示了 Al_2O_3 和 In_2O_3 晶粒之间存在许多缺陷。

Sun 等[11]报道,4 mol%① MoO_3/WO_3 复合纳米结构表现出较纯 WO_3 增强的气体传感性能,具有较低的检测限(500 ppb),在 320℃的操作温度下,对 100 ppm 乙醇和丙酮的响应值很高,分别为 28 和 18,分别是纯 WO_3 的 2.3 倍和 1.7 倍。对于纯 WO_3,当气体传感器暴露在空气中时,由于氧分子的电离作用,电子耗尽层的厚度会增加。对于纯 WO_3,当气体传感器暴露在空气中时,由于氧分子的电离作用,导致导带中自由电子浓度降低,电子耗尽层厚度增加,因此 WO_3 的电阻升高。[见图 2.8(a)]。当气体传感器暴露在还原气体中时,氧物种将与目标气体发生反应,并将电子释放到传导带中。因此,界面处的耗尽层厚度将减小,进而导致 WO_3 气体传感器的测量电阻减小[见图 2.8(b)]。与纯 WO_3 相比,MoO_3/WO_3 气体传感器气敏性能的提高主要是由于 WO_3 和 MoO_3 的协同效应和异质结。首先,MoO_3 和 WO_3 都是重要的传感材料。不同的气敏材料都具有协同效应,在其他的层状复合材料中也观察到了这种效应。其次,它可以归因于形成于 MoO_3 和 WO_3 界面上的异质结。不同的功函数将导致带负电荷的载流子从 WO_3 移动到 MoO_3,直到它们的费米能级对齐,在界面上形成一个更厚的电子耗尽层[见图 2.8(c)、图 2.8(d)]。结果表明,与纯 WO_3 相比,MoO_3/WO_3 气体传感器对乙醇和丙酮的传感性能增强。在其他复合材料中也发现了由电子耗尽层厚度变化控制的传感机理。然而,过量的钼元素会降低样品的传感性能,因为过量的掺杂会减少 WO_3 和目标气体之间的有效吸附位点。

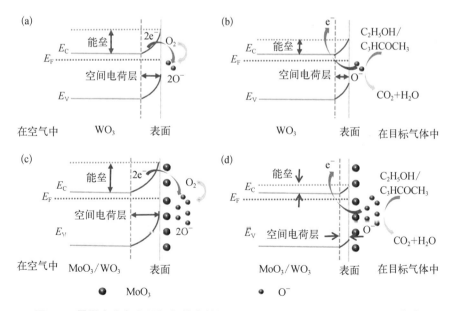

图 2.8 暴露在空气和目标气体中的纯 WO_3 和 MoO_3/WO_3 的带状示意图[11]

① 指 MoO_3 的摩尔分数为 4%。

2.2.2 p-p 异质结

Alali 等[12]通过静电纺丝技术构建了 p-p 异质结 $CuO/CuCo_2O_4$ 纳米管,用于在室温下检测正丙醇气体。复合 $CuO/CuCo_2O_4$ 纳米管是典型的 p-p 半导体金属氧化物,以空穴作为载流子。为了进一步了解测试气体分子和敏感材料之间的反应,图 2.9 描绘了空气环境和正丙醇环境中反应过程随带隙能量变化的示意图。在第一种情况下,当传感器处于空气环境中时,氧分子(O_2,称为自由态氧)将被吸附在敏感材料的表面,并扩散到敏感材料的晶体结构中,该过程如图 2.9(a)所示。被吸附的氧从敏感材料的导带(CB)捕获电子(e^-)产生离子氧,同时在 CB 中产生许多空穴(H^+)。根据工作温度的不同,金属氧化物结构中的氧离子可以处于不同的离子状态:在低温状态下,主要以 O^{2-} 的状态存在;在高温状态下,氧离子可以同时存在 O^- 和 O^{2-} 态。在这种情况下,CuO 和 $CuCo_2O_4$ 晶体表层中的空穴浓度会增加,从而在表面形成一层较厚的空穴积累层,导致敏感材料的导电性增加。当传感器转移到正丙醇环境中时,正丙醇分子与传感材料表面吸附的氧物种发生反应,所生成的反应产物(二氧化碳和水)将释放到大气中。该过程如图 2.9(b)所

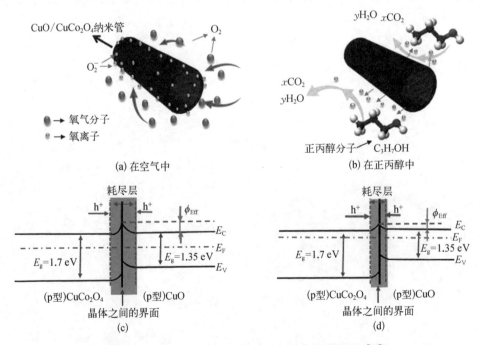

图 2.9 复合 $CuO/CuCo_2O_4$ 纳米管对正丙醇传感机理[12]

(a)、(b) 空气环境和正丙醇环境中复合 $CuO/CuCo_2O_4$ 纳米管的传感机理示意图;
(c)、(d) 空气和正丙醇环境中带隙能量的示意图

示。空气环境中带隙能量结构的示意图如图 2.9(c) 所示。正丙醇分子与吸附氧之间的反应可以描述为

$$2C_3H_7OH + 9O_2^- + 18h^+ \rightarrow 6CO_2 + 8H_2O \qquad (2.5)$$

反应后,被捕获的电子被释放到 CuO 和 CuCo$_2$O$_4$ 晶体的 CB 中,中和空穴直至达到饱和状态,从而减小积累层的厚度。暴露于正丙醇后的带隙结构变化如图 2.9(d) 所示,其中 Φ_{eff} 为有效结势垒高度,E_C 为传导能带隙有效结的较低能级,E_V 为传导能带隙有效结的较高能级,E_F 为费米能级。在这种状态下,敏感材料表面层中的空穴浓度较低,并且晶体之间的能垒高度也会降低,因此,传感器电阻增大。当传感器被移出正丙醇环境时,由于空气分子与敏感材料的反应,导电率增加到初始值。传感器在空气和目标气体中的每一个变化都会重复这些过程。综上,基于有序结构复合材料的气敏材料在响应时间和响应恢复时间上具有优势。

2.2.3　p-n 异质结

在 p-n 异质结中,高能电子可以通过氧化物界面转移到未被占用的低能态,以平衡费米能级,从而导致"带弯曲"。这种能量转换可以改变 p 侧和 n 侧的能量结构,从而获得更高的灵敏度。以 p-NiO/n-SnO$_2$ 异质结为例,p-NiO/n-SnO$_2$ 的响应高于原始 SnO$_2$ 或 NiO[13]。能带结构的可能解释如图 2.10 所示:电子从 SnO$_2$ 转移到 NiO,而空穴从 NiO 转移到 SnO$_2$,直到系统的费米能级平衡,导致氧化物界面处的耗尽区域变宽和电阻增加。当传感器暴露在还原性的 C$_2$H$_5$OH 气体中时,C$_2$H$_5$OH 与吸附的氧物种反应,并将电子释放回本体,提高材料的导电性。此外,C$_2$H$_5$OH 还将电子释放到 p 型 NiO 中,导致电子和空穴复合,从而降低空穴的浓度。NiO 中空穴的减少会导致电子的增加,因此,p-n 异质结两侧相同

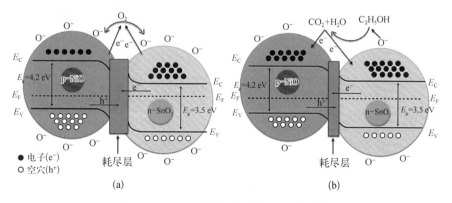

图 2.10　p-n 异质结传感机理示意图[13]

(a)、(b) 空气和乙醇气体包围时 p-NiO/n-SnO$_2$ 异质结的带结构模型,其中 E_C:导带;E_F:费米能级;E_V:价带

载流子的浓度梯度减小,载流子的扩散大大减小,导致界面处的耗尽层变薄,从而进一步减小了 SnO_2/NiO 复合材料在乙醇中的电阻。一般来说,与纯 SnO_2 传感器相比,SnO_2 传感器与 NiO 传感器之间形成 p-n 异质结大大增加了空气中的电阻,降低了乙醇气体中的电阻。该理论模型也可用于解释其他异质结材料体系,如 CuO/ZnO 和 CuO/SnO_2。

Kim 等[14]将 Sn 前驱体均匀地涂在 Ni 球上,加热 Ni 球使其部分氧化,并将 Sn 前驱体转化为 SnO_2,随后通过加热将复合微球的 Ni 金属芯溶解得到 SnO_2 空心球,并且在其内壁修饰了 NiO 纳米粒子。NiO 修饰的 SnO_2 空心球显示出超快速的响应-恢复速率(<5 s)(见图 2.11)。对比商业生产的 SnO_2 纳米粒子,将其暴露于 20~100 ppm 乙醇中,其要在很长时间(124~708 s)后才能恢复初始电阻。此外,将 NiO 负载到商业 SnO_2 微球上,能够将其恢复时间缩短到小于 30 s。Wang 等[15]还报道了 NiO-SnO_2 空心球在 300 ℃下对 20 ppm NH_3 的传感中显示出非常快速的恢复速率(4 s)。由于 NiO 能够吸附大量的氧气,氧阴离子或电子从 NiO 转移到相邻的 SnO_2,这可能是 NiO 修饰的 SnO_2 空心球具有快速恢复速率的原因。值得注意的是,吸附在 p 型氧化物半导体上的氧浓度通常高于 n 型氧化物半导体,因此将 p 型氧化物材料负载到 n 型氧化物半导体气体传感器上,可以将吸附的氧物种或电子从 p 型氧化物半导体转移到 n 型氧化物半导体,从而增强恢复动力学。

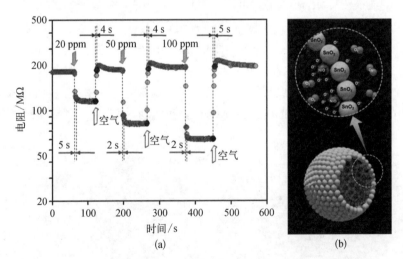

图 2.11　450 ℃下测得的 NiO 功能化 SnO_2 空心球的动态乙醇传感曲线[14]

2.3　掺杂金属氧化物半导体

将氧化铟与氮掺杂是一种能够将其光响应范围扩展到可见光区域并改变其能

带结构的策略,使该材料在光催化、光电传感和气体传感中具有潜在的应用前景。Gai 等[16]报道了一种简单的溶剂热合成方法,用于制备含氮氧化铟纳米晶体,该纳米晶体由层状结构组成,通过控制氮掺杂含量可以调节氧化铟的光学性质和气体传感性能,如图 2.12 所示。此外,由于掺杂的氮物种类似于表面改性剂,故氮掺杂可以提高氧化铟对乙醇的响应。

如图 2.13 所示,在 200℃以上的温度下,由于水分子通过化学吸附释放,表面氧阴离子和氮化物离子占主导地位,氮化物的表面修饰作用不仅影响受体功能,而且增加了氧化铟的德拜长度。这使得由氮掺杂材料构筑的传感器具有更高的灵敏度。此外,与未掺杂样品相比,氮掺杂样品具有更大的比表面积和更多从晶界获得的活性位点,也有助于增强传感器响应。

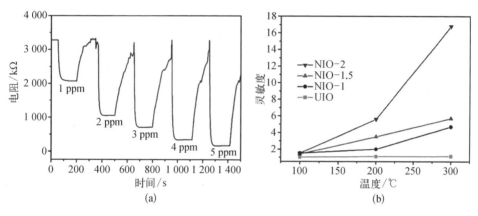

图 2.12 NiO 对乙醇的传感性能[16]

(a) 在 300℃条件下,基于 NiO-2 用 1~5 ppm 乙醇构建的传感器的响应和灵敏度变化;
(b) 传感器对 5 ppm 乙醇的灵敏度随温度的变化

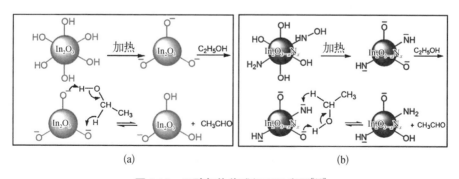

图 2.13 乙醇气体传感机理示意图[16]

(a) 未掺杂样品;(b) 掺氮样品

Kim 等[17]报道,将 p 型 NiO 与铁掺杂可显著增强 p 型 NiO 的气体响应。他们首先在 Ni 模板球上涂上一层薄薄的铁前驱体,将外层镍球部分氧化成 NiO,然后将金属镍核溶解在稀盐酸中以去除 Ni 模板,最后在 600℃加热空心球,制备了 0.3%Fe 掺杂的 NiO 空心球,并研究了 Fe 掺杂对 NiO 气体传感特性的影响。该团队还通过对 Ni 前驱体进行类似的涂层、加热以及随后的去除 Ni 芯模板的工艺,合成了没有 Fe 掺杂的 Ni 空心球,并与 Fe 掺杂的 Ni 空心球进行气敏性能比较。在 350℃下,C_2H_5OH 浓度从 5 ppm 增加到 100 ppm 时,Fe 掺杂 NiO 传感器的响应(R_g/R_a)可从 5.5 增加到 172.5(见图 2.14)。所产生的电子以补偿形式将 Fe^{3+} 代入 Ni^{2+} 位点,从而通过复合反应降低 NiO 中空穴的浓度,如下所示:

$$Fe_2O_3 \xrightarrow{NiO} Fe_{Ni}^{\cdot} + 2O_O^X + \frac{1}{2}O_2(g) + 2e^- \tag{2.6}$$

$$Fe_2O_3 \xrightarrow{3NiO} 2Fe_{Ni}^{\cdot} + 3O_O^X + V_{Ni}'' \tag{2.7}$$

以上掺入反应清楚地解释了随着 Fe 掺杂剂浓度的增加,传感器电阻增加的机制。这一发现强烈表明传感器的气体响应与材料中载流子的浓度有关。也就是说,Fe 掺杂的 NiO(在 HAL 中含有较低浓度的空穴)比未掺杂的 NiO(在 HAL 中含有较高浓度的空穴)对与分析物气体反应注入传感器相同数量的电子更敏感。

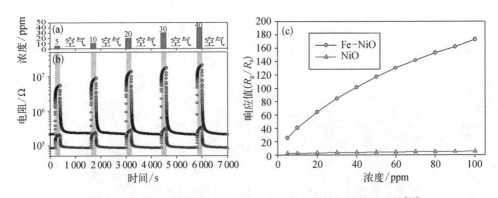

图 2.14 350℃下未掺杂 NiO 和 Fe‐NiO 空心球的气体传感特性[17]
(a) 气体浓度曲线;(b) 对 5~10 ppm C_2H_5OH 的感应瞬变;
(c) 对 5~100 ppm C_2H_5OH 的气体响应

稀土元素已被证明在微观结构细化和改善 SMO 基传感材料的表面化学特性方面发挥着不可替代的作用[18-20]。特别是这些具有多价态的元素,如 Ce、Pr

和 Tb,可以赋予 SMO 传感器独特的物理和化学特性,包括极端的表面碱性、改进的催化活性、快速的氧离子流动性和较强的耐湿性,都有助于大幅提升气敏性能[21-22]。Liu 等[23]报道了一种新型的 Ce 掺杂介孔 WO_3(Ce-2/mWO_3,其中 2 表示 CeO_2/WO_3 的掺杂质量分数为 2%),其在低工作温度(150℃)下具有优异的 H_2S 传感能力,如超高灵敏度(R_a/R_g=381@50 ppm H_2S)、快速响应速度(6 s)、高选择性,并且具有优异的长期稳定性。在介孔 WO_3 孔壁中引入了 Ce 原子,氧空位(O_v)显著增加,并产生 $W^{\delta+}$-O_v 位点,是其具有优异的气敏性能的原因之一。在敏感层的吸附催化反应过程中,表面吸附的 H_2S 分子明显转化为 SO_x 和 SO_x^{2-} 物种,是其气敏性能显著改善的另一个原因[见图 2.15(a)]。同时该团队还进行了 DFT 计算,以深入阐明气敏机理[见图 2.15(c)、图 2.15(f)]。结果表明,Ce^{4+} 被插入 WO_3 的晶格中而不是原子置换,并且嵌入 WO_3 的 Ce^{4+} 和 S 原子之间具有较强的相互作用,表现出更强的 H_2S 吸附和电子转移能力,从而导致 Ce-2/mWO_3 传感器具有优异的 H_2S 传感性能。此外,特别值得一提的是,镧系元素的三价/四价态共存对于防止水中毒至关重要[24],这使得 MOS 气敏材料上的多价镧系元素改性成为设计具有抗湿度特性的可靠气体传感器的新兴方法。Kim 等[18]制备了一种掺杂 Pr 的 In_2O_3 大孔球体,该球体具有强耐湿性的气敏特性,并在 450℃ 的宽相对湿度范围(0~80%)内对丙酮具有高响应,对干扰气体的交叉响应可忽略不计(见图 2.16)。这项研究给出了优良的抗湿度机理,其描述如下:n 型 MOS 的常规水中毒机制可以反映为式(2.8),即表面吸附的水分子与活性氧阴离子反应,引起与目标气体的竞争反应。水蒸气和活性氧物种之间的反应可以形成羟基和电子回到敏感层,导致传感器响应和电阻的显著变化。Pr 掺杂 In_2O_3 传感器的高抗湿干扰性能主要取决于其在 Pr^{3+}/Pr^{4+} 对之间的氧化还原循环反应,如式(2.9)和式(2.10)所示。Pr^{3+} 作为羟基的清除剂起着重要作用,Pr^{4+} 可以通过捕获 In_2O_3 提供的电子而容易地还原为 Pr^{3+},这表明 Pr^{3+} 是抑制水中毒效应的先决条件。此外,额外的电子可以被清除,吸附的氧也可以通过循环氧化还原反应[见式(2.9)和式(2.10)]再生,促进水中毒反应[见式(2.8)]的相反过程。因此,表面吸附的水分子和氧阴离子之间的竞争反应产生的所有羟基和电子都可以通过 Pr^{3+}/Pr^{4+} 对去除,因此以 Pr 作为掺杂剂的传感器电阻和气体响应几乎不受在潮湿气氛中反应[见式(2.8)]的影响。

$$H_2O + O_{ad}^- (或 O_{ad}^{2-}) \longleftrightarrow 2OH + e^- (2e^-) \tag{2.8}$$

$$Pr^{3+} + 2OH \longrightarrow Pr^{4+} + H_2O + O_{ad}^- \tag{2.9}$$

$$Pr^{4+} + e^- \longrightarrow Pr^{3+} \tag{2.10}$$

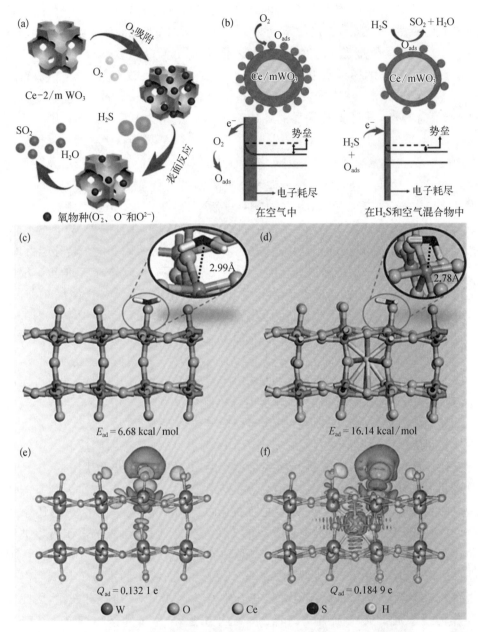

图 2.15 基于 Ce 掺杂有序介孔 WO₃ 的 H₂S 传感器机理示意图和 DFT 计算(后附彩图)[23]

(a) 用于 H₂S 检测的 Ce-2/mWO₃ 基气体传感器的传感机理；(b) 暴露于空气和 H₂S 空气混合物中的 Ce-2/mWO₃ 敏感材料的能带结构和电子转移过程；掺杂 Ce 的 WO₃ 传感器的 H₂S 吸附能和电荷转移行为的 DFT 计算：(c)(d) H₂S 分子在 WO₃、嵌入 Ce 的 WO₃ 的最佳结构上的吸附构型和能量；(e)(f) WO₃、Ce⁴⁺ 嵌入 WO₃ 的 H₂S 分子吸附的微分电荷密度以及电子密度，青色和黄色分别表示增加和减少

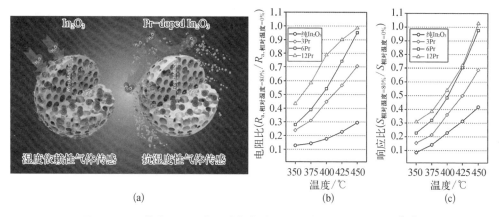

图 2.16　Pr 掺杂 In_2O_3 大孔球气体传感器示意图和抗湿度特性[18]

(a) 高相对湿度下用于丙酮检测的 In_2O_3 和 Pr 掺杂 In_2O_3 大孔球气体传感器示意图；(b)、(c) 3%～12%① Pr 掺杂的 In_2O_3 大孔球在丙酮浓度为 20 ppm、350～450℃ 条件下测量所得纯的电阻比和响应比

2.4　贵金属敏化金属氧化物

要进一步提高气体传感性能，可利用贵金属（如 Au、Pd、Pt 和 Ag）的催化和敏化作用，使金属氧化物传感材料功能化。近年来，研究人员已经做了很多工作，将纳米多孔半导体金属氧化物传感材料和贵金属纳米催化剂集成到高性能气体传感器中。

Ma 等[25]发现，Pt 敏化介孔 WO_3 在低工作温度下对低浓度 CO 具有良好的催化传感响应，灵敏度高，响应恢复时间短（16 s/1 s），选择性高。由于 WO_3 和 Pt 的带隙和功函数不同，可以在两者之间的界面形成肖特基结。铂纳米颗粒作为一种电子敏化剂，在气体吸附和解吸过程中可以改变氧化状态。为了研究铂纳米颗粒在气敏测量中的氧化状态，该团队对 WO_3/Pt 进行了原位 XPS 分析。先在 125℃ 下 100 ppm CO 中测量，再将两个 WO_3/Pt 传感器分别暴露在空气中和 100 ppm CO 中冷却至 25℃ 进行 XPS 分析。图 2.17(a) 显示，暴露在空气中后，检测到氧化铂物种（即 Pt^{2+}），这可归因于带隙为 0.7 eV 的 p 型半导体 PtO。相比之下，暴露在一氧化碳中后，XPS 结果显示只有铂（Pt^0）的还原状态[见图 2.17(b)]。当 WO_3/Pt 材料暴露于空气中时，在 PtO_x 和 WO_3 的界面处形成的 p-n 异质结导致 WO_3/Pt 复合材料中的耗尽区比 WO_3 进一步扩大。测量得到 WO_3/Pt 的基极电阻为 16.0 MΩ，远高于 WO_3（1.58 MΩ）[见图 2.17(c)、图 2.17(d)]，进一步证实存在 Pt/PtO_x NP 可显著增加基线电阻。在空气中 WO_3 表面吸附的 O^-、O_2^- 等

---

① 此处指原子百分数。

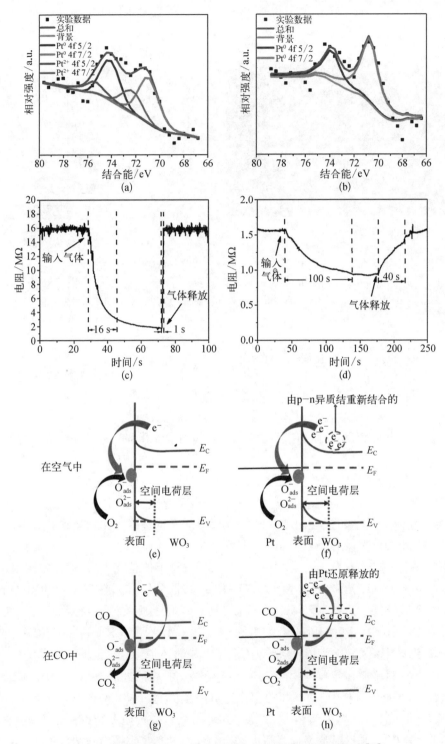

图 2.17 WO$_3$/Pt 的 CO 传感机理研究和示意图(后附彩图)[25]

(a)、(b) 分别为在 125℃ 及 57% 相对湿度下对 100 ppm CO 进行 5 次传感测试后,在空气中和在 CO 中冷却至室温后 WO$_3$/Pt 的原位 Pt 4f XPS 光谱;(c)、(d) WO$_3$/Pt、WO$_3$ 在 125℃ 及 57% 相对湿度下向 100 ppm CO 的动态电阻过渡特性;(e)~(h) 介孔 WO$_3$/Pt 和介孔 WO$_3$ 传感机理示意图

带电物质也通过捕获传感材料中的自由电子而导致电子耗尽[见图 2.17(e)、图 2.17(f)]。当传感过程暴露于 CO 中时，氧化的铂纳米颗粒被还原为金属铂，从而导致 PtO_x 和 WO_3 之间的 p-n 异质结消失，并向 WO_3 基体提供电子，从而导致表面耗尽层的有效调制。另外，随着 WO_3/Pt 表面吸附氧的种类增多，CO 与化学吸附氧之间的表面氧化还原反应增多。这些增多的表面反应产生的额外电子导致 CO 中基于 WO_3/Pt 的传感器电阻显著降低[见图 2.17(g)、图 2.17(h)]。因此，基于 WO_3/Pt 的传感器可以提高 CO 的传感性能。

Zhang 等[26]证明，金纳米粒子修饰的单晶 MoO_3 纳米带对三甲胺（TMA）具有增强的传感性能。作为一种催化剂，金的增强作用可以用电子敏化来解释。当引入贵金属时，Au/MoO_3 纳米复合材料中均匀分布的金纳米粒子能促进活性氧作为化学敏化剂，最可能的提高传感器响应的机制是通过溢出效应实现化学敏化。可以通过改变金纳米粒子的电荷状态来改变其表面势垒高度，进而导致 MoO_3 的电导发生变化。这一机制大大促进了金纳米粒子和 MoO_3 之间的直接电子交换。Au 催化剂使氧分子离解，许多氧离子会溢出到 MoO_3 的表面。在这一过程中，吸附氧分子向氧离子的转化率加快，从而导致电子耗尽层厚度和电阻的增加。因此，化学敏化机制被认为是通过添加金纳米粒子来增强气体传感性能的主要机制。

此外，Au 纳米粒子（5.1 eV）的功函数高于 MoO_3 纳米带（2.9 eV），因此当 Au 纳米粒子与 MoO_3 接触时，电子很容易从 MoO_3 纳米带流向 Au 纳米粒子，在 Au 和 MoO_3 之间形成电子耗尽区或肖特基势垒。TMA 气体与吸附氧离子的相互作用将电子释放回 MoO_3 纳米带，并显著降低肖特基势垒的高度，导致传感器电阻降低。因此，Au/MoO_3 纳米复合材料比纯 MoO_3 纳米带具有更好的传感性能。气敏机理示意图如图 2.18 所示。

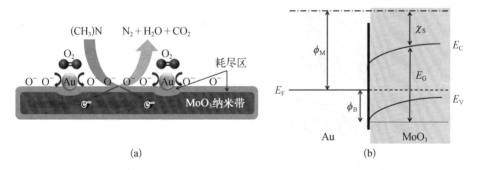

图 2.18　基于 Au/MoO_3 纳米带的三甲胺传感机理研究[26]

(a) 用 Au 纳米粒子修饰的 MoO_3 纳米带的气敏机理示意图；
(b) Au 和 MoO_3 界面处形成的肖特基势垒的能带图

Liu 等[27]报道了 4.0% Ag-α-Fe$_2$O$_3$ 表现出优异的丙酮传感性能。具有三维海胆状结构的 Ag-α-Fe$_2$O$_3$ 较 Fe$_2$O$_3$ 纳米管表现出增强的气体传感性能的原因可以总结如下：① 与 Fe$_2$O$_3$ 纳米管相比，海胆状分层结构提供了较大的比表面积；此外，这些三维结构与纳米分支组装产生了更高的多孔性，有助于分析物和背景气体进入传感单元中所含颗粒的所有表面。② 由于化学敏化和电子敏化作用，Ag 的负载能显著改善气固界面的气敏催化反应。化学敏化是由溢出效应介导的，而电子敏化则是由半导体和贵金属添加剂之间的电子直接交换介导的。如图 2.19 所示，一方面，Ag 可以改善表面反应，包括氧的吸附、解离和电离，然后大量的化学吸附氧和气体分子溢出到 Fe$_2$O$_3$ 载体的表面；另一方面，在气敏恢复过程中，Ag 表面丰富的带负电荷的表面氧可以促进电子或负电荷的吸附氧转移到 Fe$_2$O$_3$ 表面。因此，研究者认为负载的 Ag 不涉及，而是加速传感器和测试气体之间的电子交换。此外，Ag 的高度分散导致了较高的 BET 表面积，从而提供了更多的活性位点。因此，与 Fe$_2$O$_3$ 纳米粒子传感器相比，具有 3D 海胆结构的 Ag-α-Fe$_2$O$_3$ 传感器表现出更优异的气敏性能。例如，4.0% Ag-α-Fe$_2$O$_3$ 对 200 ppm 丙酮的灵敏度比具有 3D 海胆状的纯 Fe$_2$O$_3$ 高 2.2 倍，比 Fe$_2$O$_3$ 纳米粒子高 5.7 倍。

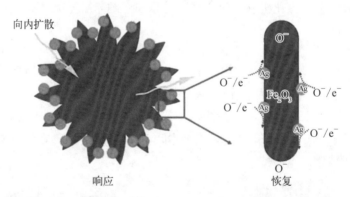

图 2.19　三维类海胆 Ag-α-Fe$_2$O$_3$ 纳米复合微球气敏机理示意图[27]

Cui 等[28]提出了一种静电纺丝技术和退火工艺，获得了掺杂 Pt 纳米颗粒的超细 SnO$_2$ 介孔纳米纤维。如图 2.20 所示，Pt 和超细 SnO$_2$ 介孔纳米纤维结构之间的强协同效应促进了氧气分子的吸附，导致电阻变化更加显著。超细纳米纤维结构和 SnO$_2$ 的一维互连有利于提高气体可及性和快速恢复。结果表明，Pt/SnO$_2$ 传感器具有较高的响应值（R_a/R_g=31.2@2 ppm 丙酮）、较短的响应恢复时间、较低的实际检测限（100 ppb）和优异的丙酮选择性。因此，基于 Pt 掺杂的超细 SnO$_2$ 介孔纳米纤维可开发一种能够检测呼气中微量疾病生物标志物的高灵敏度半导体氧化物传感器，有望实现无创、实时糖尿病诊断。

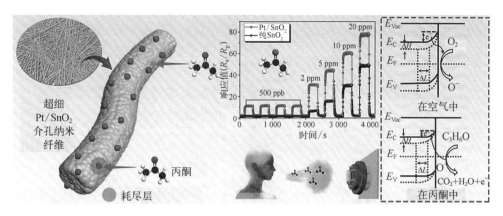

图 2.20 用于增强丙酮传感的超细 Pt 掺杂 SnO_2 介孔纳米纤维基气体传感器[28]

与单一贵金属敏化相比,双金属合金的协同效应和强相互作用可以放大贵金属的优异功能,得到远远超出简单添加的效果[29-30]。Li 等[32]发现,具有富 Pd 壳和富 Pt 核的 PdPt 双金属官能化 SnO_2 在 100 ℃ 和 320 ℃ 对 CO 和 CH_4 显示出温度依赖的双重选择性,具有优异的长期稳定性和耐湿能力。由于 Pt-Pd 纳米颗粒对 CO 和 CH_4 的催化活化能力存在明显差异,因此他们提出了双选择性机理,即 Pt-Pd 纳米颗粒在低温下对 CO 氧化显示出很强的催化性能,而由于 CH_4 的非极性和强稳定性,CH_4 催化氧化则通常需要较高的温度(超过 350 ℃)。这项研究还提出扩散深度是实现 CO 和 CH_4 双重选择性的主要机制(见图 2.21)。一般来说,感测层的利用与气体的扩散深度有很强的关联关系,从而影响感测性能。以 CO 为代表,在低温(40~80 ℃)下,由于 CO 与表面吸附氧物种之间的催化氧化反应速率较慢,CO 在传感层中的扩散深度相对较大,并且传感层几乎完全被 CO 氧化所消耗 [见图 2.21(a)],这涉及大量的电子转移,从而导致电阻的显著变化。在高温 (100~160 ℃)下,由于感测材料表面上的 CO 氧化反应速率较快,CO 氧化仅停留在感测层表面[见图 2.21(b)],因此感测层中的 CO 扩散无法有效运行,从而导致低响应。上述解释合理地揭示了 CO 灵敏度随着工作温度的增加而逐渐降低的原因。然而,对于 CH_4[见图 2.21(e)、图 2.21(f)],尽管在低温(低于 350 ℃)下 CH_4 在感测层中深度扩散,但由于 CH_4 的低温氧化反应性差,其灵敏度仍然较低;在高工作温度(高于 350 ℃)下,与 CO 一样,CH_4 氧化仅发生在敏感层的表面,因此导致低响应。

已知单原子催化剂(SAC)在多相催化领域具有远优于贵金属纳米颗粒(NPs)催化剂的催化能力,因为其具有最大化的原子利用效率、强大的金属载体界面相互作用、均匀的活性位点和活性原子的不饱和配位环境[33-35]。气敏机理通常涉及目标气体和气敏材料之间的相互作用,以及气体分析物和活性氧物种之间的催化反

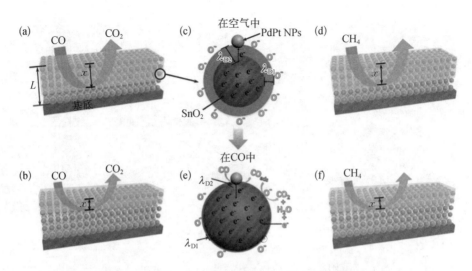

图 2.21 基于 PdPt/SnO$_2$ 微球的 CO 传感机理研究[32]

(a)、(b) 在低于 100℃、高于 100℃时 CO 扩散到由 PdPt/SnO$_2$ 组成的传感层的示意图；(c)~(f) 耗尽层 λ_D(灰色区域)在低于 350℃、高于 350℃、在空气中、在 CO 中 CH$_4$ 扩散到传感层中发生的变化；L 为传感层的厚度，λ_D 为电子耗尽区域，x 为气体在传感层中的扩散路径

应[36]。因此，必须综合考虑多相催化原理，才能设计和构建具有相当最佳总体性能的气体传感器。事实上，将 SACs 策略应用于气体传感最近已成为一种提高灵敏度和选择性的先进技术[37-39]。Ye 等[35]提出，钯单原子负载的 TiO$_2$(Pd$_1$-TiO$_2$)在室温下表现出高灵敏度、极低的检测限和出色的 CO 传感选择性[见图 2.22(a)和图 2.22(b)]，并揭示了改善 CO 传感性能的机理。如图 2.22(c)所示，Pd$_1$-TiO$_2$ 具有以空穴为载体的 p 型半导体的典型性质。类似于上述 p 型金属氧化物的传统感测过程，当电子暴露于空气中时，敏感材料表面吸附的氧分子抓住电子可以产生活性氧阴离子。该过程增加了敏感材料中空穴的浓度，使其处于高导电状态。当 Pd$_1$-TiO$_2$ 处于 CO(还原气体)气氛中时，CO 与氧阴离子反应，释放与空穴复合的电子，降低载流子密度，并导致其处于较低的导电状态。值得注意的是，与纯 TiO$_2$ 和 Pd NPs-TiO$_2$(Pd 纳米颗粒负载 TiO$_2$)相比，Pd 单原子的负载诱导了更多的表面吸附氧物种(OAD)，这意味着在 CO 传感期间有更多的电子转移，极大地提高了它们的灵敏度。此外，Pd 单原子的引入赋予了 Pd$_1$-TiO$_2$ 比原始 TiO$_2$ 和 Pd NPs-TiO$_2$ 更强的 CO 吸附能力[见图 2.22(e)]，进一步证明了 Pd$_1$-TiO$_2$ 在室温下具有超高 CO 敏感性的机理。通过 DFT 计算发现，与六种干扰气体相比，Pd$_1$-TiO$_2$ 的 CO 吸附能最低，这意味着由于单活性中心的特性，Pd$_1$-TiO$_2$ 对 CO 的吸附最强，从而深入解释了其优良的气体选择性。

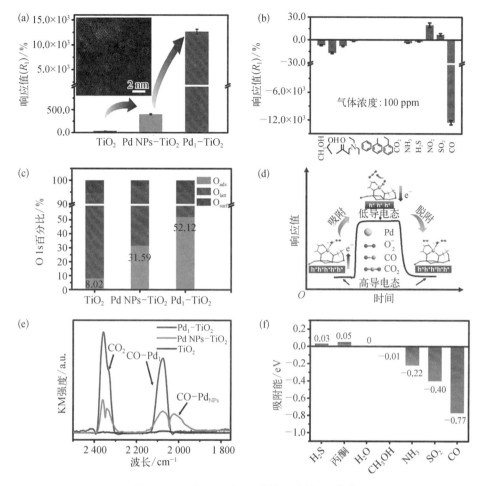

图 2.22 Pd/TiO$_2$ 对 CO 的传感特性研究[35]

(a) 原始 TiO$_2$、Pd NPs-TiO$_2$ 和 Pd$_1$-TiO$_2$ 对 100 ppm CO 的响应柱状图,插图是 Pd$_1$-TiO$_2$ 的 AC-HAADF-STEM 图;(b) Pd$_1$-TiO$_2$ 传感器对不同 100 ppm 气体的响应;(c) 纯 TiO$_2$、Pd NPs-TiO$_2$ 和 Pd-TiO$_2$ 的 O 1s 光谱中 O 物种的百分比;(d) Pd$_1$-TiO$_2$ 的模拟 CO 传感机制;(e) 纯 TiO$_2$、Pd NPs-TiO$_2$ 和 Pd$_1$-TiO$_2$ 的 DRIFT 光谱;(f) 选定气体在 Pd$_1$-TiO$_2$ 上的吸附能

2.5 晶粒尺寸的影响

晶粒尺寸对电阻变化起着重要作用,并影响气体传感性能。如上所述,电子密度变化只发生在氧化物颗粒的外层,因此,耗尽区的内部区域对电阻变化没有贡献[38-41]。理论上,粒径越大,超过耗尽层宽度(L)的 2 倍,越不利于材料电导的整体变化,如图 2.23 所示。对于大颗粒($D \gg 2L$),当其接触氧化气体和还原气体时,高导电率核心区几乎不参与电子的获取和释放过程,从而表现出不良的响应。对

于 D 值接近但仍大于 $2L$ ($D>2L$) 的晶粒,其在每个颈部周围的耗尽区形成一个狭窄的通道,因此,电导率取决于晶界屏障和横截面面积,因此灵敏度增强,并与晶粒尺寸相关。当 $D<2L$ 时,晶体几乎完全耗尽,这意味着整个晶粒参与了与气体分子之间的电荷转移相互作用。这些相互连接的晶粒能带几乎是平的,因为晶界中不存在明显的电荷转移障碍。Xu 等[41]研究了多孔 SnO_2 传感器的晶粒尺寸效应,发现当 SnO_2 晶体尺寸(D)控制在 5~32 nm 范围内时,当 D 减小至约等于 $2L$(约 6 nm)或小于 $2L$(约 6 nm)时,H_2 和 CO 的响应随之增大,如图 2.24 所示。值得一提的是,尽管小晶体在理论上可以产生较高的灵敏度,但由于纳米颗粒在高工作温度下约导电率急剧下降且性质不稳定,因此超小金属氧化物颗粒的实际应用仍受到限制。

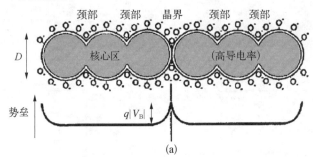

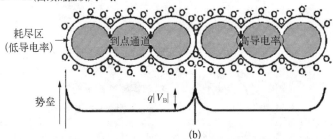

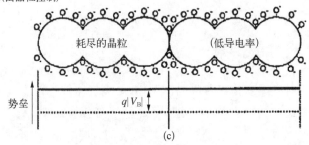

图 2.23 晶体尺寸对金属氧化物气体传感器灵敏度影响的示意模型[40]

(a) $D\gg2L$;(b) $D>2L$;(c) $D<2L$

Righettoni 等[42] 通过可缩放火焰气溶胶技术,制备了纯 WO_3 纳米颗粒和掺杂 SiO_2 的 WO_3 纳米微粒传感膜,并将其直接沉积和原位退火到叉指电极上。这里的一个独特创新是这些薄膜由 $\varepsilon\text{-}WO_3$ 组成,这是一种对丙酮具有高选择性的亚稳相。更重要的,这项研究详细探究了无毒硅掺杂对 $\varepsilon\text{-}WO_3$ 含量、晶体和晶粒尺寸的影响(见图 2.25),并解释了其与丙酮传感性能的相关性。结果显示,随着 SiO_2 掺杂量的增加,$\varepsilon\text{-}WO_3$ 含量先增加,随后达到稳定的水平。而 $\varepsilon\text{-}WO_3$ 的晶体尺寸则随 SiO_2 掺杂量的增加而逐渐减小。对其气敏性

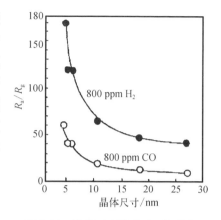

图 2.24 粒度对多孔 SnO_2 传感器响应的影响[41]

能的测试结果显示,摩尔分数为 10% 的 SiO_2 的掺杂水平使得丙酮传感器表现出最佳气敏性能,且对低至 20 ppb 的丙酮仍具有较高的灵敏度和选择性。

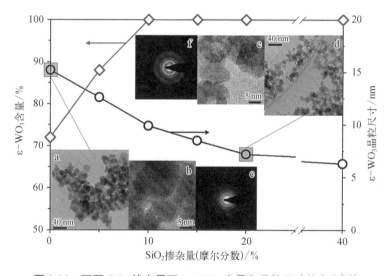

图 2.25 不同 SiO_2 掺杂量下 $\varepsilon\text{-}WO_3$ 含量和晶粒尺寸的变化[42]

2.6 气体传感器评价标准

作为物联网(IoT)的重要组成部分,气体传感器广泛应用于工业过程、制药、环境监测、安全等领域。为实现快速、可靠、全面的检测,根据实际应用需求,气体传感器须满足灵敏度、响应-恢复时间、选择性、稳定性等多种要求。在本节中,我们

将介绍几种最重要的对气体传感器评价标准的定义和理解。

2.6.1 灵敏度

灵敏度是最重要的指标之一，它描述了传感材料对目标分析物的响应活性。灵敏度(S)定义为输出变化($\mathrm{d}y$)与输入变化($\mathrm{d}x$)之比。

$$S = \frac{\mathrm{d}y}{\mathrm{d}x} \tag{2.11}$$

对于化学阻抗气体传感器，输出量是响应值(R)，输入量是目标气体的浓度。响应值 R 通常根据载气(通常为空气)中传感器的电阻(R_a)与暴露于某一浓度的目标分析物时的电阻(R_g)之比来计算。

$$R = \frac{R_a}{R_g} \tag{2.12}$$

有时灵敏度和响应这两个概念的界限并不清晰，因为它们都描述了电阻的变化程度。因此，灵敏度也存在其他表示方法：

$$S = \frac{|R_a - R_g|}{R_g} = \frac{\Delta R}{R_g} \tag{2.13}$$

理想情况下，S 仅与气体分压(p)和温度(T)相关[41]。

研究表明，描述基于 SnO_2 的传感器的电导(G)或电阻(R)对目标还原气体浓度的经验方程可表示为

$$G = \frac{1}{R} = Ac^\beta \tag{2.14}$$

式中，c 是目标气体的浓度；A 和 β 为孤立的经验常数[43-44]。

2.6.2 工作温度

众所周知，温度对半导体金属氧化物的化学和物理性质有很大影响。对于化学阻抗气体传感器来说，工作温度控制着反应动力学、电导率和电子迁移率。由于传感过程需要加热以使表面氧化还原反应克服活化能障碍，传统的 MOS 气体传感器通常需要在 200~500℃的温度范围内工作[45]。在大多数情况下，传感器对某些气体的响应会随着操作温度的升高而发生变化，从而产生如图 2.26 所示的火山

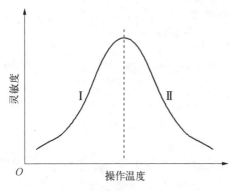

图 2.26 典型的操作温度-灵敏度曲线示意图

型曲线。温度对灵敏度的影响主要来自敏感材料表面吸附氧物种活性和气体吸附-脱附。当温度相对较低时(见图 2.26 区域Ⅰ),吸附氧物种的活性随温度升高而增加,灵敏度随之升高;同时,温度越高时材料表面吸附的反应物越容易解吸,导致目标气体与材料表面之间的电子转移减少,从而使灵敏度降低(见图 2.26 区域Ⅱ)。Pulkkinen 等[46]使用动力学蒙特卡罗方法对 SnO_2 表面氧物种吸附形态进行了模拟计算(见图 2.27)。结果表明,氧物种在较低温度时(<650 K)主要以 O_2^- 的形式吸附在 SnO_2 表面,而在较高温度下(>750 K)主要以 O^- 的形式存在。O_2 随温度升高的吸附状态变化可表示为

$$\frac{1}{2}O_{2(gas)} \rightarrow \frac{1}{2}O_{2(phys)} \rightarrow \frac{1}{2}O^-_{2(chem)} \rightarrow O^-_{(chem)} \rightarrow O^{2-}_{(chem)} \tag{2.15}$$

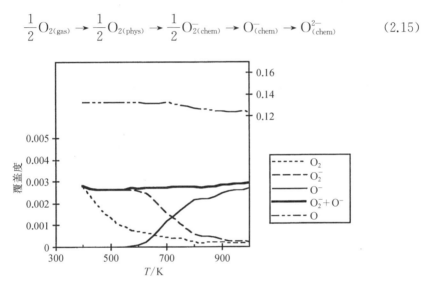

图 2.27 SnO_2 表面不同氧物种覆盖率随温度变化曲线图[46]

由于不同材料、不同气体的吸附-解吸行为和表面反应模型通常存在很大差异,材料表面吸附分子的状态也会有不同。为得到高灵敏度,大多数 MOS 气体传感器通常需要保持较高工作温度,这使得 MOS 传感器的应用受到限制,因为当活性材料暴露于易燃易爆气体时,高温通常会导致能量浪费并存在安全隐患。此外,高温操作还会导致传感器信号漂移,从而导致结果偏差或误报。因此,在保持可接受的灵敏度的同时降低工作温度已成为 MOS 气体传感器的发展趋势。

人们在开发能够在室温下高效工作的气体传感器方面做出了巨大的努力,目前,室温传感器仍然存在一些挑战,首先这类传感器的传感性能有限,许多传感器由于吸附能大,表面活性氧含量不足,导致灵敏度较低,响应/恢复时间很长。其次,相对湿度对于室温下工作的传感器是一个主要的干扰,在此温度下,H_2O 分子与氧分子发生表面竞争吸附,导致传感材料表面活性氧减少。此外,传感器的选择性也会受到很大的限制。因此,工业生产、环境监测和临床医学等都对高性能气体

传感器提出了迫切的需求。Cui 等[47]报道了一种基于二维碲纳米片的室温 NO_2 传感器,并结合实验和 DFT 计算揭示了碲传感器的气敏机制(见图 2.28)。这里的碲纳米片是通过简单的液相剥离制备的,其在室温下对 NO_2 具有良好的响应,对 25 ppb 和 150 ppb 的 NO_2 分别可达到 201.8% 和 264.3% 的响应值。另外,二维碲纳米片在室温下对 NO_2 传感表现出超低的理论检测限(0.214 ppb),对 ppb 级别的 NO_2 具有快速的响应-恢复速率,这与大多数报道的二维材料相比是更优异的。DFT 计算进一步从原子尺度上解释了碲的气敏机理。电荷转移分析表明吸附 NO_2 后碲电导率的提高主要是由于碲与 NO_2 界面的电子转移。吸附 NO_2 后,碲能带结构也发生明显分裂,自旋向上和自旋向下的能带隙都变窄。综上,因其优越的灵敏度、良好的选择性、超低的检测限、快速的响应恢复和高稳定性,碲传感器可以作为复杂大气环境中 NO_2 检测的候选者。

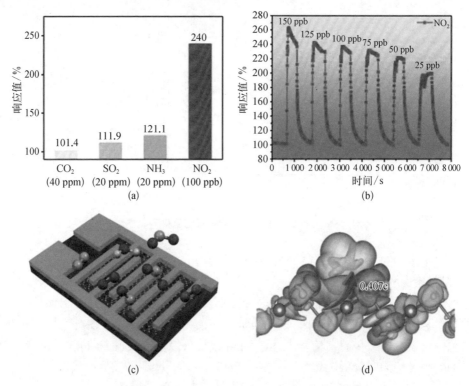

图 2.28 基于二维碲纳米片的室温 NO_2 传感器[47]

2.6.3 选择性

传感器的选择性 Q 描述了传感器在检测时受非目标气体干扰的程度,反映了传感器区分待检测的特定气体 x 和气体环境的其他组分 x' 的能力,可表示为

$$Q = \frac{\dfrac{\mathrm{d}y}{\mathrm{d}x'}}{\dfrac{\mathrm{d}y}{\mathrm{d}x}} \tag{2.16}$$

选择性低是很多MOS敏感材料的主要缺点,即难以区分检测物与干扰气体的信号。因为材料表面存在大范围的吸附位点,所以无法区分每种气体分子对总电信号的贡献[48]。而过渡金属及其氧化物的表面性质对材料的电化学性质有极大影响,因此最常用的提高其选择性的方法是对金属氧化物材料进行高度均匀的表面改性[49-51]。改性剂的种类与用量是提高材料选择性的最关键因素,还须考虑金属氧化物基底与改性剂的相互作用、改性剂与目标气体的相互作用以及复合材料对目标气体的增敏效果等。传感器的选择性通常以条形图的形式表现,如图2.29所示。贵金属是最为常用的选择性改性剂之一,如0.5%Pt的掺杂可以选择性提高WO_3对一氧化碳的灵敏度,从而提高对一氧化碳的选择性[25]。

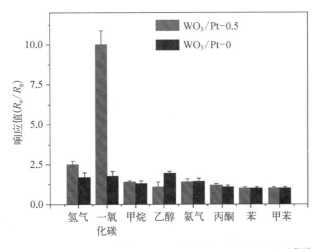

图2.29 WO_3和Pt掺杂的WO_3对各种目标气体的响应[25]

由于不同气体的响应机理存在差异,对同一材料,不同温度可能对不同目标气体产生差异化的灵敏度影响,一些材料在不同的操作温度下可以显示出不同的选择性分布。Jing和Zhan[52]报道的多孔ZnO纳米薄片作为多功能选择性气体传感材料,在200℃时对氯苯有最大灵敏度,对乙醇的敏感性则相对较低;而在380℃时,该材料对乙醇的响应远高于氯苯(见图2.30)。此传感器可以在较低工作温度(150℃<T<250℃)下用作氯苯传感器,在较高工作温度(250℃<T<450℃)下作为乙醇传感器使用。

Jo等[53]设计了一种能在室温下检测ppb级甲醛的化学电阻型TiO_2传感器,其在紫外线照射下显示出对甲醛和乙醇的高选择性检测。如图2.31所示,研究者

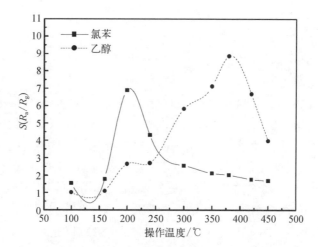

图 2.30 多孔 ZnO 纳米薄片传感器对 100 ppm 氯苯和乙醇的气体响应与操作温度关系图[52]

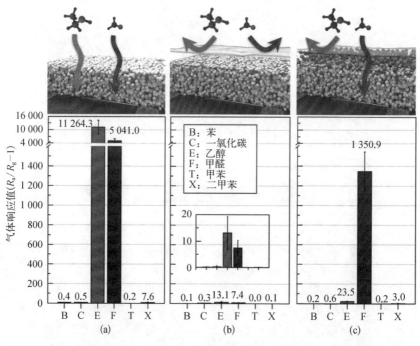

图 2.31 使用单片柔性化学电阻型传感器在室温下高选择性检测甲醛[53]

们由沸石咪唑骨架(ZIF-7)纳米粒子和聚合物组成的混合基质膜(MMM)涂层在 TiO_2 传感膜上,其可以起到分子筛作用以去除乙醇干扰,从而实现了室温下对 5 ppm 甲醛的超高选择性(响应比＞50)和响应(电阻比＞1 100)。此外,他们使用夹在柔性聚对苯二甲酸乙二醇酯基材和 MMM 覆盖层之间的 TiO_2 薄膜,成功地制作了单片柔性传感器。这项研究工作提供了一种可以提高气体选择性的分子筛策略,并展示了将柔性气体传感器应用于室内空气监测的潜力。

甲醇中毒会导致失明、器官衰竭,甚至导致死亡。通常情况下,在乙醇作为背景气时,化学传感器无法对甲醇和乙醇进行选择性检测,因此通过呼吸分析进行快速诊断甲醇中毒或筛选含酒精饮料仍存在着巨大挑战。Broek 等[54]介绍了一种廉价的手持式传感器,用于高选择性甲醇检测。如图 2.32 所示,它包括一个分离柱(Teflon 管中的 Tenax TA 颗粒)以及一个化学电阻式气体传感器。分离柱类似于气相色谱柱,可以将甲醇与乙醇、丙酮或氢等干扰物分离,气体传感器的敏感材料是 Pd 掺杂的 SnO_2 纳米颗粒,可以用于量化甲醇浓度。使用该传感器,可以在 2 分钟内实现 1~1 000 ppm 的甲醇传感,且不受高含量乙醇的干扰(最高可区分 62 000 ppm 乙醇)。为验证其实际应用性能,对加标呼气样品和白酒中的甲醇浓度进行了检测,从而证明了传感器在呼吸分析或空气质量监测等新兴应用中实现高选择性传感的潜力。

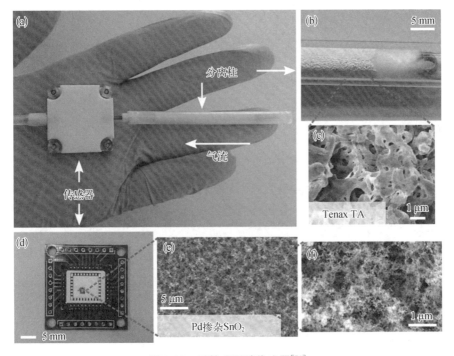

图 2.32　手持式甲醇传感器[54]

2.6.4　稳定性

由于金属氧化物具有较好的热稳定性和化学稳定性,MOS 气体传感器是应用最广泛的类型。稳定性描述了传感器材料在长时间工作时保持其输出信号准确性的能力,这种再现性可能受传感器层(热)老化以及传感器材料性质改变的影响。引起 MOS 传感器传感性能变化最常见的原因是在高工作温度下金属氧化物晶粒的生长[55],金属氧化物的结晶和晶粒生长可导致基线电阻偏移,并降低传感器的灵敏度。研究表明[56],每种尺寸的微晶都有其自身的临界温度,高于该临界温度则会出现晶粒尺寸增长的趋势。对于最常用的 SnO_2 材料来说,其边界阈值温度(T_{th},以℃为单位)与晶粒尺寸(t,以 nm 为单位)之间的关系可表示为

$$T_{th} = 420 \cdot (\lg t)^{\frac{3}{4}} \tag{2.17}$$

因此,具有较高结晶度的传感材料通常表现出比无定形或多晶金属氧化物更好的长期稳定性[44,57-58]。

电化学分子纳米传感器需要两种相互冲突的表面特性,即分子选择性的催化活性和长期数据收集的热稳定性。Zhang 等[59]发现使用强酸进行简单的表面处理会产生 WO_3 水合物纳米线的两种相互冲突的表面性质,从而增强醛类(壬醛,生物标记物)的电子分子传感。质谱测量表明,使用强酸进行的表面处理大大促进了壬醛的氧化以及壬酸产品在 50℃下从表面上的解吸,该温度低于未处理表面所需的300℃。光谱和结构测量结合数值模拟确定了表面羧基的两种不同吸附结构,其中分子直接与配位不饱和表面钨结合,优先进行催化氧化反应和随后的脱附过程。此外,该项研究还发现酸处理 WO_3 水合物纳米线表面上的催化活性的热耐久性(超过 10 年)高达 300℃,灵敏度可保持多年,这能够保证传感器的长期稳定操作。

2.6.5 响应-恢复时间

响应-恢复时间体现了传感器在接触到目标气体后发出信号的速度,以及脱离目标气体后恢复基态的速度。在气体传感器实际应用中,更短的响应-恢复时间有利于快速检测出目标物质。响应或恢复时间分别表示为在输入或脱离分析物气体之后输出信号达到其饱和值 90%所需要的时间。有时也使用稳态电阻的 63%的极限来计算响应-恢复时间。响应动力学主要由气体扩散过程、固体表面和气体分子之间的化学反应决定。通常,具有较高孔隙率的传感材料可提供更多的气体扩散通道,从而确保比固体材料更快地响应和恢复[60-68]。通过提高反应速率和工作温度还可以促进响应和恢复行为。

2.6.6 检测限

气体传感器的检测限通常定义为可检测的最低气体浓度,这是评估气体传感装置整体性能的不可或缺的指标。通常需要使用高性能气体传感器来精确监测低含量有害气体的浓度,如 CO、NO_2、SO_2,以有效避免安全事故的发生,降低这些有

毒气体对人体健康的风险,并实现对人身安全的最大保护。例如,有研究表明[69],当人体短时间暴露于浓度超过 200 μg/m³(约 100 ppb)的 NO_2 时,可能导致严重的呼吸道疾病,尤其是有冠状病毒病(如新冠肺炎)风险的人。此外,在重大疾病的早期筛查过程中,人体呼出的大多数标志气体(如丙酮)通常处于 ppm 或 ppb 水平[70]。在环境保护方面,微量 NO 可导致酸雨的形成,酸雨可能危害动物、植物和生态系统,并且可能导致光化学烟雾,危害人体健康[71]。对于食品安全检测,食品变质期间分泌的特征气体(如 H_2S)浓度也低至 ppb 水平[72]。因此,气体传感器具有低检测限的能力具有重要的现实意义。

参考文献

[1] Yamazoe N. Toward innovations of gas sensor technology[J]. Sens. Actuators B, 2005, 108: 2-14.

[2] Kim H, Lee J. Highly sensitive and selective gas sensors using p-type oxide semiconductors: overview[J]. Sens. Actuators B, 2014, 192: 607-627.

[3] Zhou X, Lee S, Xu Z, et al. Recent progress on the development of chemosensors for gases [J]. Chem. Rev., 2015, 115: 7944-8000.

[4] Zhou X, Cheng X, Zhu Y, et al. Ordered porous metal oxide semiconductors for gas sensing[J]. Chin. Chem. Lett., 2018, 29: 405-416.

[5] Miller D, Akbar S, Morris P. Nanoscale metal oxide-based heterojunctions for gas sensing: a review[J]. Sens. Actuators B, 2014, 204: 250-272.

[6] Zhu Y, Zhao Y, Ma J, et al. Mesoporous tungsten oxides with crystalline framework for highly sensitive and selective detection of foodborne pathogens[J]. J. Am. Chem. Soc., 2017, 139: 10365-10373.

[7] Xiao X, Liu L, Ma J, et al. Ordered mesoporous tin oxide semiconductors with large pores and crystallized walls for high-performance gas sensing[J]. ACS Appl. Mater. Interfaces, 2018, 10: 1871-1880.

[8] Wang Z, Zhu Y, Luo W, et al. Controlled synthesis of ordered mesoporous carbon-cobalt oxide nanocomposites with large mesopores and graphitic walls[J]. Chem. Mater., 2016, 28: 7773-7780.

[9] Han J, Wang T, Li T, et al. Enhanced NO_x gas sensing properties of ordered mesoporous WO_3/ZnO prepared by electroless plating[J]. Adv. Mater. Interfaces, 2018, 5: 1701167.

[10] Gao J, Wang L, Kan K, et al. One-step synthesis of mesoporous Al_2O_3-In_2O_3 nanofibres with remarkable gas-sensing performance to NO_x at room temperature[J]. J. Mater. Chem. A, 2014, 2: 949-956.

[11] Sun Y, Chen L, Wang Y, et al. Synthesis of MoO_3/WO_3 composite nanostructures for highly sensitive ethanol and acetone detection[J]. J. Mater. Sci., 2017, 52: 1561-1572.

[12] Alali K T, Lu Z, Zhang H, et al. P-p heterojunction CuO/$CuCo_2O_4$ nanotubes synthesized via electrospinning technology for detecting n-propanol gas at room temperature[J]. Chem. Front., 2017, 4: 1219-1230.

[13] Wang Y, Zhang H, Sun X. Electrospun nanowebs of NiO/SnO$_2$ p-n heterojunctions for enhanced gas sensing[J]. Appl. Surf. Sci., 2016, 389: 514-520.

[14] Kim H, Choi K, Kim K, et al. Ultra-fast responding and recovering C$_2$H$_5$OH sensors using SnO$_2$ hollow spheres prepared and activated by Ni templates[J]. Chem. Commun., 2010, 46: 5061-5063.

[15] Wang L L, Deng J N, Fei T, et al. Template-free synthesized hollow NiO-SnO$_2$ nanospheres with high gas-sensing performance[J]. Sens. Actuators B, 2012, 16: 90-95.

[16] Gai L, Ma L, Jiang H, et al. Nitrogen-doped In$_2$O$_3$ nanocrystals constituting hierarchical structures with enhanced gas-sensing properties[J]. Cryst. Eng. Comm., 2012, 14: 7479-7486.

[17] Kim H, Choi K, Kim K, et al. Highly sensitive C$_2$H$_5$OH sensors using Fe-doped NiO hollow spheres[J]. Sens. Actuators B, 2012, 171: 1029-1037.

[18] Kim J S, Na C W, Kwak C H, et al. Humidity-independent gas sensors using Pr-doped In$_2$O$_3$ macroporous spheres: role of cyclic Pr^{3+}/Pr^{4+} redox reactions in suppression of water-poisoning effect[J]. ACS Appl. Mater. Interfaces, 2019, 11: 25322-25329.

[19] Kim K, Park J K, Lee J, et al. Synergistic approach to simultaneously improve response and humidity-independence of metal-oxide gas sensors[J]. J. Haz. Mater., 2022, 424: 127524-127534.

[20] Zhang W, Ding S, Zhang Q, et al. Rare earth element-doped porous In$_2$O$_3$ nanosheets for enhanced gas-sensing performance[J]. Rare Metals, 2021, 40: 1662-1668.

[21] Bai J, Luo Y, Chen C, et al. Functionalization of 1D In$_2$O$_3$ nanotubes with abundant oxygen vacancies by rare earth dopant for ultra-high sensitive ethanol detection[J]. Sens. Actuators B Chem., 2020, 324: 128755.

[22] Wu K, Debliquy M, Zhang C. Room temperature gas sensors based on Ce doped TiO$_2$ nanocrystals for highly sensitive NH$_3$ detection[J]. Chem. Eng. J., 2022, 444: 136449.

[23] Liu Y, Guo R, Yuan K, et al. Engineering pore walls of mesoporous tungsten oxides via Ce doping for the development of high-performance smart gas sensors[J]. Chem. Mater., 2022, 34: 2321-2332.

[24] Yoon J W, Kim J S, Kim T H, et al. A new strategy for humidity independent oxide chemiresistors: dynamic self-refreshing of In$_2$O$_3$ sensing surface assisted by layer-by-layer coated CeO$_2$ nanoclusters[J]. Small, 2016, 12: 4229-4240.

[25] Ma J, Ren Y, Zhou X, et al. Pt nanoparticles sensitized ordered mesoporous WO$_3$ semiconductor: gas sensing performance and mechanism study[J]. Adv. Funct. Mater., 2018, 28: 1705268.

[26] Zhang J, Song P, Li Z, et al. Enhanced trimethylamine sensing performance of single-crystal MoO$_3$ nanobelts decorated with Au nanoparticles[J]. J. Alloys Compd., 2016, 685: 1024-1033.

[27] Liu X, Chang Z, Luo L, et al. Sea urchin-like Ag-alpha-Fe$_2$O$_3$ nanocomposite microspheres: synthesis and gas sensing applications[J]. J. Mater. Chem., 2012, 22: 7232-7238.

[28] Cui S, Qin J, Liu W. Ultrafine Pt-doped SnO$_2$ mesopore nanofibers-based gas sensor for

enhanced acetone sensing[J]. Chinese J. Analyt. Chem., 2022, 51: 100188.

[29] Fan F, Zhang J, Li J, et al. Hydrogen sensing properties of Pt – Au bimetallic nanoparticles loaded on ZnO nanorods[J]. Sens. Actuators B Chem., 2017, 241: 895 – 903.

[30] Li G, Cheng Z, Xiang Q, et al. Bimetal PdAu decorated SnO_2 nanosheets based gas sensor with temperature-dependent dual selectivity for detecting formaldehyde and acetone[J]. Sens. Actuators B Chem., 2019, 283: 590 – 601.

[31] Wang D, Deng L, Cai H, et al. Bimetallic PtCu nanocrystal sensitization WO_3 hollow spheres for highly efficient 3-hydroxy-2-butanone biomarker detection[J]. ACS Appl. Mater. Interfaces, 2020, 12: 18904 – 18912.

[32] Li G, Wang X, Yan L, et al. PdPt bimetal-functionalized SnO_2 nanosheets: controllable synthesis and its dual selectivity for detection of carbon monoxide and methane[J]. ACS Appl. Mater. Interfaces, 2019, 11: 26116 – 26126.

[33] Zong B, Xu Q, Mao S. Single-atom Pt-functionalized $Ti_3C_2T_x$ field-effect transistor for volatile organic compound gas detection[J]. ACS Sens., 2022, 7: 1874 – 1882.

[34] Liu B, Zhang L, Luo Y, et al. The dehydrogenation of H-S bond into sulfur species on supported Pd single atoms allows highly selective and sensitive hydrogen sulfide detection [J]. Small, 2021, 17: 2105643 – 2105656.

[35] Ye X, Lin S, Zhang J, et al. Boosting room temperature sensing performances by atomically dispersed Pd stabilized via surface coordination[J]. ACS Sens., 2021, 6: 1103 – 1110.

[36] Moon Y K, Jeong S Y, Jo Y M, et al. Highly selective detection of benzene and discrimination of volatile aromatic compounds using oxide chemiresistors with tunable Rh – TiO_2 catalytic overlayers[J]. Adv. Sci., 2021, 8: 2004078 – 2004088.

[37] Tian R, Wang S, Hu X, et al. Novel approaches for highly selective, room-temperature gas sensors based on atomically dispersed non-precious metals[J]. J. Mater. Chem. A, 2020, 8: 23784 – 23794.

[38] Liu B, Zhu Q, Pan Y, et al. Single-atom tailoring of two-dimensional atomic crystals enables highly efficient detection and pattern recognition of chemical vapors[J]. ACS Sens., 2022, 7: 1533 – 1543.

[39] Shin H, Jung W G, Kim D H, et al. Single-atom Pt stabilized on one-dimensional nanostructure support via carbon nitride/SnO_2 heterojunction trapping[J]. ACS Nano, 2020, 14: 11394 – 11405.

[40] Rothschild A, Komem Y. The effect of grain size on the sensitivity of nanocrystalline metal-oxide gas sensors[J]. J. Appl. Phys., 2004, 95: 6374 – 6380.

[41] Xu C, Tamaki J, Miura N, et al. Grain size effects on gas sensitivity of porous SnO_2-based elements[J]. Sens. Actuators B, 1991, 3: 147 – 155.

[42] Righettoni M, Tricoli A, Pratsinis S E. Thermally stable, silica-doped ε – WO_3 for sensing of acetone in the human breath[J]. Chem. Mater., 2010, 22: 3152.

[43] Barsan N, Schweizer-Berberich M, Göpel W. Fundamental and practical aspects in the design of nanoscaled SnO_2 gas sensors: a status report[J]. J. Anal. Chem., 1999, 365: 284 – 304.

[44] Korotcenkov G, Cho B K. The role of the grain size on thermal stability of nanostructured

SnO$_2$ and In$_2$O$_3$ metal oxides films aimed for gas sensor application[J]. Prog. Cryst. Growth Charact. Mater., 2012, 58: 167-208.

[45] Zhu Y, Zhao Y, Ma J, et al. Mesoporous tungsten oxides with crystalline framework for highly sensitive and selective detection of foodborne pathogens[J]. J. Am. Chem. Soc., 2017, 139: 10365-10373.

[46] Pulkkinen U, Rantala T T, Rantala T S, et al. Kinetic monte carlo simulation of oxygen exchange of SnO$_2$ surface[J]. Mol. Catal. A: Chem., 2001, 166: 15-21.

[47] Cui H, Zheng K, Xie Z, et al. Tellurene nanoflake-based NO$_2$ sensors with superior sensitivity and a sub-parts-per-billion detection limit[J]. ACS Appl. Mater. Interfaces, 2020, 12: 47704.

[48] Miller D R, Akbar S A, Morris P A. Nanoscale metal oxide-based heterojunctions for gas sensing: A review[J]. Sens. Actuators B, 2014, 204: 250-272.

[49] Wang Y, Cui X, Yang Q, et al. Preparation of Ag-loaded mesoporous WO$_3$ and its enhanced NO$_2$ sensing performance[J]. Sens. Actuators B, 2016, 225: 544-552.

[50] Arunkumar S, Hou T, Kim Y B, et al. Au decorated ZnO hierarchical architectures: Facile synthesis, tunable morphology and enhanced CO detection at room temperature[J]. Sens. Actuators B, 2017, 243: 990-1001.

[51] Ma J, Li Y, Zhou X, et al. Au nanoparticles decorated mesoporous SiO$_2$-WO$_3$ hybrid materials with improved pore connectivity for ultratrace ethanol detection at low operating temperature[J]. Small, 2020, 16: 2004772.

[52] Jing Z, Zhan J. Fabrication and gas-sensing properties of porous ZnO nanoplates[J]. Adv. Mater., 2008, 20, 4547-4551.

[53] Jo Y, Jeong S, Moon Y, et al. Exclusive and ultrasensitive detection of formaldehyde at room temperature using a flexible and monolithic chemiresistive sensor[J]. Nat. Commun., 2021, 12: 4955.

[54] Broek J, Abegg S, Pratsinis S E, et al. Highly selective detection of methanol over ethanol by a handheld gas sensor[J]. Nat. Commun., 2019, 10: 4220.

[55] Hoa N D, Duy N V, El-Safty S A, et al. Meso-/nanoporous semiconducting metal oxides for gas sensor applications[J]. Nanomater., 2015, 16: 1-14.

[56] Korotcenkov G, Brinzari V, Ivanov M, et al. Structural stability of indium oxide films deposited by spray pyrolysis during thermal annealing[J]. Thin Solid Films, 2005, 479: 38-51.

[57] Gong J, Chen Q, Lian M, et al. Temperature feedback control for improving the stability of a semiconductor-metal-oxide (SMO) gas sensor[J]. IEEE Sens. J., 2006, 6: 139-145.

[58] Tiemann M. Porous metal oxides as gas sensors[J]. Chemistry, 2007, 13: 8376-8388.

[59] Zhang G, Hosomi T, Mizukami W, et al. A thermally robust and strongly oxidizing surface of WO$_3$ hydrate nanowires for electrical aldehyde sensing with long-term stability [J]. J. Mater. Chem. A, 2021, 9: 5815-5824.

[60] Liu J, Huang H, Zhao H, et al. Enhanced gas sensitivity and selectivity on aperture-controllable 3D interconnected macro-mesoporous ZnO nanostructures[J]. ACS Appl. Mater. Interfaces, 2016, 8: 8583-8590.

[61] Wagner T, Kohl C D, Froba M, et al. Gas sensing properties of ordered mesoporous SnO_2 [J]. Sensors, 2006, 6: 318-323.

[62] Wagner T, Sauerwald T, Kohl C D, et al. Gas sensor based on ordered mesoporous In_2O_3 [J]. Thin Solid Films, 2009, 517: 6170-6175.

[63] Li Y, Luo W, Qin N, et al. Highly ordered mesoporous tungsten oxides with a large pore size and crystalline framework for H_2S sensing[J]. Angew. Chem. Int. Ed., 2014, 53: 9035-9040.

[64] Wagner T, Haffer S, Weinberger C, et al. Mesoporous materials as gas sensors[J]. Chem. Soc. Rev., 2013, 42: 4036-4053.

[65] Qin Y, Wang F, Shen W, et al. Mesoporous three-dimensional network of crystalline WO_3 nanowires for gas sensing application[J]. J. Alloys. Comp., 2012, 540: 21-26.

[66] Sun X, Hao H, Ji H, et al. Nanocasting synthesis of In_2O_3 with appropriate mesostructured ordering and enhanced gas-sensing property [J]. ACS Appl. Mater. Interfaces, 2014, 6: 401-409.

[67] Waitz T, Wagner T, Sauerwald T, et al. Ordered Mesoporous In_2O_3: Synthesis by structure replication and application as a methane gas sensor[J]. Adv. Funct. Mater., 2009, 19: 653-661.

[68] Wagner T, Kohl C D, Morandi S, et al. Photoreduction of mesoporous In_2O_3: mechanistic model and utility in gas sensing[J]. Chemistry, 2012, 18: 8216-8223.

[69] Song Z, Tang W, Chen Z, et al. Temperature-modulated selective detection of part-per-trillion NO_2 using platinum nanocluster sensitized 3D metal oxide nanotube arrays[J]. Small, 2022, 18: 2203212.

[70] Motooka M, Uno S. Improvement in limit of detection of enzymatic biogas sensor utilizing chromatography paper for breath analysis[J]. Sensors, 2018, 18: 440-450.

[71] Chen Z, Ye W, Wang J, et al. Sensitive NO detection by lead-free halide Cs_2TeI_6 perovskite with Te-N bonding[J]. Sens. Actuators B Chem., 2022, 357: 131397-131404.

[72] Meng L, Li Y, Yang M, et al. Temperature-controlled resistive sensing of gaseous H_2S or NO_2 by using flower-like palladium-doped SnO_2 nanomaterials[J]. Microchim. Acta., 2020, 187: 297-306.

第3章
半导体金属氧化物：形态和传感性能研究

近年来，研究者们广泛研究了包括单组分和多组分金属氧化物在内的半导体电阻型气体传感器的设计与性能，报道了非晶态、玻璃态、纳米晶态、多晶和单晶等金属氧化物的多种结构状态，并研究了电阻式气体传感器的性能与这些金属氧化物材料的不同结构状态之间的关系。每种结构的特殊理化性质和特征都可以对传感性能产生很大的影响[1-3]。其中，纳米颗粒和多晶材料具有较小的微晶尺寸、低廉的设计成本、较高的结构稳定性和电物理性质，非常适用于固态气体传感器的应用。

纳米颗粒和多晶材料由于具有众多的物理化学参数，因此在研究其对气体传感器性能的影响时，是非常复杂的研究对象[4]。因此，为了清楚地了解纳米颗粒和多晶氧化物对气体传感器性能的影响，有必要研究形态和晶体结构如何影响气体传感器的性能。据报道，金属氧化物气体传感器的性能可受晶粒尺寸、晶粒团聚行为、晶粒间接触面积、孔隙率、主要晶面取向以及微晶的刻面和形成气敏表面等参数的影响[5]。

3.1 形貌和结构对气体传感的影响

3.1.1 颗粒大小的影响

据报道，"尺寸效应"可用以下公式表述，例如，晶粒尺寸(d)或颗粒连接宽度(X)与德拜长度(L_D)的比较。

$$L_D = \sqrt{\frac{\varepsilon k_B T}{q^2 N_d}} \tag{3.1}$$

其中，ε 是介电常数，k_B 是玻尔兹曼常数，T 是绝对温度，在开尔文温度中，q 是基本电荷，N_d 是掺杂剂（供体或受体）的密度。

颗粒的连接宽度对多晶金属氧化物导电性的影响以及颗粒间连接电位分布如图3.1所示。很明显，连接宽度确定了载流子势垒的高度，连接的长度证实了耗尽-势垒的层宽。还需要提到的是，连接宽度的增加会降低金属氧化物的导电性，进而降低其气体传感性能[6]。

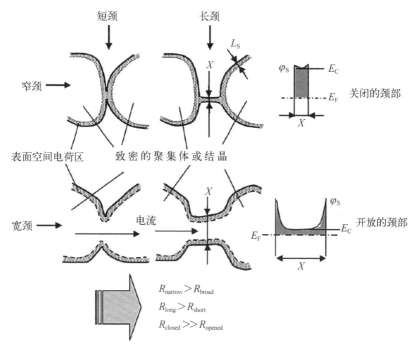

图 3.1 颗粒的连接宽度在多晶金属氧化物基体导电性中的
作用及颗粒间连接电位分布[6]

所得的金属氧化物的晶粒尺寸决定了其电位在多晶金属氧化物颗粒间的分布。一维结构的相应电位图如图 3.2 所示。在这之前，需要提供以下参数解释气体传感效应的"尺寸效应"[7]。对大晶粒而言，晶粒尺寸(d)往往远远大于 $2L_S$，其中 L_S 代表表面空间电荷的宽度；对于较小的颗粒连接宽度($d<L_S$)，薄膜和陶瓷的电导通常受到晶界处的肖特基势垒(V_S)的限制。在这种情况下，气体灵敏度实际上与晶粒尺寸无关。

在 $d\sim 2L_S$ 的情况下，晶粒之间连接处的每个导电沟道都是重叠的。如果颗粒间连接处的宽度大于内接触的宽度，则它们控制着气敏材料的导电性并确定了气体尺寸依赖性的灵敏度。

如果 $d<2L_S$，则每个晶粒都完全参与空间电荷层，并且转变电子可受到吸附物质电荷的影响。据报道，当晶粒尺寸与德拜长度的两倍相当时，空间电荷区域可以存在于整个结晶的金属氧化物中。由于这个模型可以将传感器的响应最大化，因此后一种情况对气体传感性能最有利[8]。

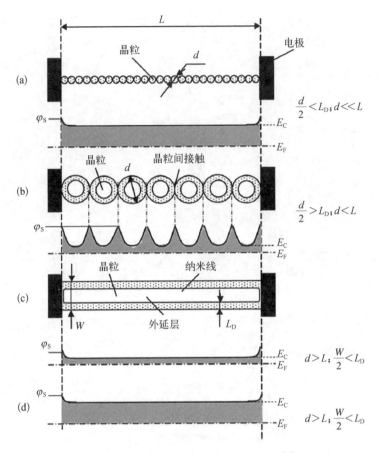

图 3.2 颗粒大小对颗粒间电位的影响[6]

通过对典型传感材料 SnO_2 进行分析,研究者总结了 5~80 nm 的不同晶粒尺寸的 SnO_2 纳米晶金属氧化物气体传感器的响应灵敏度。有效载流子浓度被认为是设计模型的重要表面状态[9],并且研究发现,随着表面态密度达到临界值,载流子浓度急剧下降,这是由于对应的所有电子都在颗粒表面上。

3.1.2 颗粒晶相的影响

在气体传感过程中,气固相互作用发生在金属氧化物纳米晶体的外部平面上,据报道,这些相互作用决定了纳米结构材料的气体传感性质。此外,如晶面、晶粒间接触、晶粒间接触面积、气体渗透性等不同的参数,都可能在气体传感中起着至关重要的作用[5]。每种金属氧化物晶体都具有独特的晶体结构,每个晶面都具有独特的表面电子参数,包括表面态密度、能级位置、相互作用气体分子间的吸附/解吸能、表面吸附态浓度、表面费米能级的能量位置、原生点缺陷的活化能等。由于不同的晶体表面取向导致化学吸附特性的不同行为，因此在原子水平上对用于吸附物质的化学键合存在较强的表面依赖性。

取向和晶粒尺寸是影响其吸附/解吸过程中特定表面能的重要参数。当晶体尺寸减小到纳米级别时，它的表面积增大、晶面增多，可以显著影响吸附性能。如前所述，金属氧化物晶体的形貌和尺寸对表面吸附物质与表面的键合类型有很大的影响。据报道，一些化学物质通常更倾向于通过不同类型的键合模型吸附在纳米晶体的边缘、角落和平面上。综上，纳米晶体独特的结构、电子特性以及吸脱附性能，因此，不同晶相组成会对气敏性能产生显著影响。

例如，原子和电子特性等表面能参数的变化可以引起气体传感特性的相应变化。又如，对于 SnO_2 晶体[10]，其(110)晶面和(101)晶面是 F (平)面，(111)晶面是 K (扭曲)面，而且与(110)晶面和(101)晶面相比，(111)晶面更粗糙。表面非饱和的阳离子和弱结合桥氧的表面浓度对催化活性也有很重要的影响，因此，不同的晶面表现出不同的催化活性(CA)，相对催化活性大小为 $CA_{(110)} < CA_{(001)} < CA_{(100)} < CA_{(101)}$。由于锡原子在化学吸附氧的中心，因此锡原子和氧原子之间的距离变化会影响解离氧化学吸附的速率，进而在很多情况下影响气敏特性[11]。

3.1.3 表面几何构型的影响

活性位点是多相催化过程中的基本参数与概念，意味着并非所有催化表面都是活性的，活性位点仅存在于表面原子(包括缺陷)的特殊排列中，或者只有特定的化学成分实际上是具有反应性的。例如，广泛的结果证明单原子是许多表面反应的高活性位点。据报道，扩散中的表面阶梯也发生在钙钛矿中，研究发现，表面氧空位的扩散主要发生在台阶边缘[12]。通过蒙特卡罗模拟建立 $SrTiO_3$ 表面结构理论模型发现，在激光分子束外延沉积过程中，晶体台阶位附近的氧空位浓度大于晶体表面的氧空位浓度[13]。由于它们的扩散速率较慢，因此可以推断出在晶体台阶位附近的氧空位倾向于积聚。其他结果也证明了氧空位的内扩散是速率决定步骤[13-14]。研究者们也证实了无论表面扩散的速率多大，氧化过程主要由靠近台阶边缘氧空位的扩散控制。研究结果表明，许多气体的吸附性能与不同金属氧化物特定表面的几何构型有关。不同表面几何构型对气体选择性不同，因此可以开发各种各样对不同气体响应的固体传感器。

贵金属，如铂(Pt)、金(Au)、钯(Pd)等因其优异的催化性能，已被广泛应用于气敏材料，并且贵金属的表面几何构型对气体传感性能也有很大的影响。研究表明，催化剂表面颗粒的分散状态和颗粒尺寸能影响气体传感性能[15]，且载体上具有较小粒径、高分散性的纳米颗粒可以促进催化过程，从而提高气体传感的灵敏度[16]。

簇和纳米粒子的尺寸取决于其金属本身的性质。例如，研究者详细研究了 Pt

和 Pd 原子在 ZrO_2 表面上的分散行为,结果表明,Pd 原子表现出比 Pt 原子更高的表面迁移率,因此 Pd 容易形成较大的金属簇;其结果与其他研究结果一致[17]。在 700 K 高温处理后,Pd 原子在 TiO_2(111)晶面上更容易聚集形成大粒径的金属簇。然而,经过相同条件的处理,Pt 团簇的尺寸不变,表明较小的 Pt 团簇在较高温度下可以稳定存在。此外,还有研究报道 Pd 原子在沉积期间可以吸附在 TiO_2 晶体的平台和台阶位置上,而 Pt 原子则仅在台阶上被观察到。

3.1.4 晶体孔隙度和晶粒间接触面积

通过分析可能影响传感器性能的众多参数,我们得出以下结论:通过增强表面电导率的作用或者减小颗粒间的接触面积和宽度来增加晶体间能垒的贡献,可以提高金属氧化物半导体的气体传感性能[7]。

此外,随着材料孔隙率的增加,金属氧化物半导体的气体敏感性显著地提高[18]。多孔材料相较于块体固体材料具有较高的催化活性表面和较突出的气体传感响应[11]。孔隙丰富的材料,其纳米颗粒相互接触的部分较少,因此与周围的气体相互作用更充分,有关 SnO_2 薄膜的研究结果能够非常好地和该结论吻合[19],尽管 SnO_2 纳米颗粒高度聚集,但 SnO_2 薄膜依然显示出较高的灵敏度。此外,研究发现,金属氧化物半导体的晶粒间接触面积较小的材料表现出较高的灵敏度[19]。随着孔隙率的增加,形成所谓封闭区域的气敏层体积也相应地减小,因此更有利于气体传感器材料达到最高的灵敏度。这里的封闭区域是指与气氛接触隔离的体相颗粒,它们的电阻几乎不受外部影响[20-22]。

需要提到的是,提高气体传感材料的比表面积,可以降低金属氧化物半导体气体传感器的工作温度以实现其最高灵敏度,因为材料的高比表面积有助于气体分子在气敏基质内的扩散,从而降低工作温度[21]。相应的金属氧化物的粒径大小可以通过分析 X 射线衍射(XRD)结果来计算,孔隙率可以通过经典的 Brunauer-Emmett-Teller(BET)吸附模型来计算。这些表征方法为认识金属氧化物气敏材料的精细结构提供了更加完整和可靠的信息[23]。

综上,可以得出结论,即金属氧化物气体传感器的灵敏度随着气敏材料的孔隙率和比表面积的增加而增加,该结论与参考文献中所得出的结论相一致[16]。孔隙率和比表面积是影响金属氧化物固体与反应气体相互作用并最终影响材料气敏性能的重要因素。同时这些结果还表明,具有更高孔隙率和更高比表面积的金属氧化物气体传感器材料可以实现对待检测气体更准确、更灵敏的响应。然而,需要说明的是,在某些情况下,无孔或低比表面积的金属氧化物半导体可以通过更高的热稳定性来补偿其较低的灵敏度和其他缺点,并且可以在更加苛刻的条件下进行测试和反应[24]。

需要注意的是,随着孔隙率的增加,膜厚度对气体传感器响应和响应速度的影响有明显的减弱。此外,Korotcenkov 等[25]报道了薄膜厚度对致密材料和多孔材料的影响,结果表明,致密块体金属氧化物材料的气体响应灵敏度随着薄膜厚度在 30~200 nm 范围内的增加而降低。然而,对于多孔的金属氧化物膜,膜厚度的变化并不影响气体传感器的灵敏度。

3.1.5 颗粒的聚集

众所周知,金属氧化物材料颗粒间的聚集是一种普遍的现象,几乎在自然界中无处不在。这种现象发生在金属多晶中会形成胶体聚集体和颗粒,以及脂质-蛋白质等黏弹性基质。此外,金属氧化物薄膜更易在高温处理下发生聚集作用[11]。

材料的聚集阻力是整体阻力,也就是说,整体材料包括三维的晶粒连接网络。因此,可以推断出,如果金属氧化物颗粒聚集体的孔隙度或气体传感器基质发生变化,上述提到的影响气体传感器灵敏度的参数,如:晶粒大小、聚集程度、颗粒间和颗粒聚集的接触面积、孔隙率、材料取向性和晶体的晶面都会对传感器灵敏度的变化产生影响。例如,减小晶粒尺寸可以提高纳米晶粒在气体传感效应中的影响。金属氧化物固体聚集体由于对气体的渗透率较低,有利于聚集体接触面对传感器响应的影响。这些结论与文献中所报道的观察结果相一致[26],从而证实了小晶体聚集成大晶体是导致表观响应特性发生巨大变化的关键因素。研究表明,传感器响应性能和动力学过程都取决于金属氧化物聚集体的本征物化性质。也就是说,更大、更密集的金属氧化物颗粒聚集体总是表现出更差的气体响应灵敏度和更长的恢复时间。重要的是,这些影响都与被检测气体的本征性质有关,特别与气体和金属氧化物的反应性以及该气体在金属氧化物基质中的扩散系数有关。还应该提到的是,高度聚集的金属氧化物颗粒聚集体也加速了封闭区域的形成,因此表现出较低的孔隙率。因此,通常来说密集和具有较高聚集程度的陶瓷具有较低的比表面积[27],如前所述,这种效应可能对气体传感性能产生较大的负面影响。

3.2 金属氧化物传感材料的合成方法

3.2.1 溶胶-凝胶法

通俗地说,溶胶-凝胶法大致可总结为通过化学方式将无机物前驱体溶液转化成固体氧化物的方法,如图 3.3 所示。其中,无机物前驱体可以是无机金属盐,也可以是金属有机盐,例如金属醇盐或乙酰丙酮化合物。在含水体系合成中,金属醇盐是最广泛使用的前驱体,因为由前驱体向金属氧化物的转化过程中涉及可控的

图 3.3 溶胶-凝胶法示意图[32]

金属醇盐的水解和缩合反应[28]。对于水性溶胶-凝胶过程,氧由水分子提供以形成氧化物。但是,对于非水体系的溶胶-凝胶过程,没有水供应氧原子形成金属氧化物。据报道,类似于非水解制备方法形成块体金属氧化物[29],体系中的氧由溶剂(醚类、醇类、酮类或醛类)或有机成分组成前驱体(醇盐或乙酰丙酮化合物)提供,从而形成金属氧化物纳米颗粒。例如,研究人员使用了钛异丙醇盐和氯化钛作为形成锐钛矿型纳米氧化钛晶体的前驱体[30]。据报道,醚消除反应是形成 M—O—M 键的重要反应,该机理通过两种金属醇盐在消除时发生缩合反应而发生,并且该反应机理也可用于解释氧化铪纳米颗粒的形成。消除酯类涉及金属羧酸盐和金属醇盐之间的反应,氧化锌、二氧化钛和氧化铟等金属氧化物均可由该方法合成。金属油酸盐与胺之间的反应类似于酯消除反应,例如控制合成二氧化钛纳米棒[31]。但是,使用酮类作为溶剂时,氧气的释放通常涉及醛醇缩合,两种羰基化合物发生(正式)消除反应并释放水分子,因此提供了溶胶-凝胶法合成过程中金属氧化物形成的供氧剂。有文献报道,ZnO 和 TiO_2 纳米颗粒可以用丙酮作为溶剂来合成。

1. 表面活性剂控制金属氧化物的合成

1993 年,Murray 等[33]以熔融的三辛基氧化膦(TOPO)为溶剂合成了单分散的 CdX(X 为 S、Se 或 Te)纳米晶。这项工作为热注射法合成提供了基础,即在表面活性剂存在时将室温下的前驱体溶液注入热溶剂中。这里所用的表面活性剂由配位头基和长链烷基链组成,具有以下几个优点:在纳米粒子上长链烷基链的保护下,纳米粒子在合成过程中可以避免聚集并提高胶体最终产物在有机溶剂中的稳定性;表面活性剂通常在颗粒生长过程中选择性吸附在特定晶面上,进而能够控制粒径、粒度分布、暴露面和颗粒形态学[34];此外,表面活性剂可在后修饰过程中与其他分子相交换,确保得到合适的纳米粒子的表面性质。

2. 溶剂控制金属氧化物的合成

与通过使用表面活性剂控制合成金属氧化物相比,通过溶剂控制合成金属氧化物的方法相对来说更为简单。初始反应体系仅包括两部分,即金属氧化物前驱体和常见的有机溶剂,并且相较于热注射法,溶剂控制合成方法的反应温度更低,通常在50~200℃的范围内。该简便的合成体系有利于化学反应的机理解释并且所得到的最终产物更容易被表征。同时,由于该反应体系采用无表面活性剂合成方法,因此所得到的产品更容易提纯。此外,通过使用表面活性剂的方法,纳米颗粒表面吸附的表面活性剂会钝化纳米粒子[35]并降低纳米颗粒表面的可接触性进而影响催化和传感性能,而这些缺点在不使用表面活性剂的合成方法中并不存在。在过去的几年里,研究者们已经系统研究了利用多种金属氧化物前驱体(如金属卤化物、乙酸盐、乙酰丙酮化物、醇盐和不同金属前体的混合物)来合成金属氧化物的无需表面活性剂的合成方法。通常使用含氧有机溶剂,如醇类、酮类或醛类以及具有短链烷基链的无胺溶剂(如胺或腈),甲苯或均三甲苯等"惰性"溶剂也可以使用。溶剂的选择主要取决于其在纳米粒子生长过程中起到的作用,例如晶面取向选择性以及所得到的最终产物组成等特征。正如刚才提到的,含氧溶剂通常提供氧原子以形成氧化物,而在使用不含氧的溶剂制备金属氧化物时,氧原子则由含氧的金属前驱体来提供。在此形成过程中,有机物种在反应开始时充当封端剂,与纳米表面结合,从而限制晶体生长和影响颗粒形态以及颗粒间的组装行为。吸附在特定晶面上的高度稳定的有机物质不仅抑制晶体生长,而且可以选择性地结合晶面并导致晶体各向异性生长。

3.2.2 水热和溶剂热法

水热和溶剂热法是纳米材料实验室和工业合成中最重要也是最完善的方法[36]。使用水热和溶剂热法合成的纳米材料具有操作简便、可大规模合成和可调反应参数等许多优点,它们提供了通过其他方法难以研究的亚稳相到纳米级形态体系中的状态。但是,该方法的合成原理仍然没有被研究者完全理解并形成较为普遍的理论,因此通过水热和溶剂热法来设计合成纳米材料仍然具有相当大的挑战。虽然水热法非常实用,但是其难以精确控制所合成金属氧化物及其相关材料的晶相和形貌。最近几年,研究者们系统研究了工业上利用高压水热釜合成金属氧化物的工程参数,目前,先进的进料装置和传感技术以及连续流动反应器已经在工业生产中得以实现[37]。除了传统的水热法,金属氧化物纳米材料的非水相合成方法也引起了研究者们极大的研究兴趣[38]。

近年来,研究者们对与金属氧化物相关的纳米材料水热合成法进行了广泛的研究,如最常见的二元材料氧化物(如SnO_2、ZnO或TiO_2)和多组分金属氧化物(如$SrWO_4$、$Ce_{0.6}Zr_{0.3}Y_{0.1}O_2$或$LiVMoO_6$)以及金属有机骨架(MOF)化合物等[39]。在此,我们列出了通过水热法可控合成金属氧化物的最新进展。表3.1包含常见的水热和溶剂热法合成金属氧化物的合成条件及形貌差异。

表 3.1 水热和溶剂热法合成金属氧化物的合成条件及形貌差异

氧 化 物	溶剂/添加剂	形 貌	参考文献
CeO_2、CuO、Co_3O_4、Fe_2O_3、MgO、NiO	水/葡萄糖	纳米颗粒嵌入空心球	[40]
CuO	水/柠檬酸钠	纳米片/棒/星状	[41]
Fe_2O_3、Fe_3O_4	水/Na_2SO_4/Na_2HPO_4	纳米管	[42]
Fe_3O_4	水/乙醇/油酸	纳米管	[43]
MoO_3	水/酸/离子添加剂	纳米棒	[44]
MoO_x	乙醇/水/十六胺	纳米带	[45]
SnO、SnO_2	水/HCl 或 NaOH	纳米颗粒/纳米片	[46]
SnO_2	水/乙醇	纳米片	[47]
TiO_2	水/NaOH	纳米棒/纳米管	[48]
TiO_2	水/HCl/醋酸	纳米颗粒	[49]
VO_x	丙酮/十六胺	纳米管	[50]
ZnO:Co/Mn	苯甲醇	纳米颗粒/棒/纤维	[51]
Bi_2WO_6	水/P123	纳米颗粒	[52]
Bi_2WO_6	水/P123	纳米花	[53]
$SrWO_4$	CTAB	纳米颗粒	[54]

由于水热合成中难以准确控制所合成的各个阶段,因此通过水热法控制纳米级金属氧化物的形貌仍然是一个棘手的问题。然而,基于各向异性结构的外延生长理论是一种获得各向异性形态的极其简单的方法。因此,水热合成中通常使用该方法来获得特定形貌的样品。由于天然产物和聚合物具有丰富、多功能的优点,其已经从众多有机化合物中被选出并广泛应用于纳米结构金属氧化物的水热合成中。例如,多糖通常用作水热合成中的形貌控制添加剂来控制双金属空心微球的合成[40]。在水热过程中,碳水化合物添加剂通常用于形成具有丰富亲水表面基团的金属氧化物微球,而且金属离子在球形形貌表面上的行为非常复杂。例如,各种中空金属氧化物球(Fe_2O_3、NiO、Co_3O_4、CeO_2、MgO 和 CuO)均可用有机模板的热分解来合成,这种策略还可以扩展到其他金属氧化物的合成体系中[55-69]。

通常,控制多组分氧化物的合成是开发新催化剂的重要先决条件。例如,Sanfiz 等[70]合成了 MoVTeNb 氧化物纳米颗粒(M1 相),对合成过程的反应时间、温度、温度、添加剂和搅拌方式等多个参数进行了优化,并通过精确调节水热过程使所获得的催化剂显示出高度的催化选择性。XANES/EXAFS 结果证明该水热过程可以使五种阳离子相均匀地混合。

近年来,水热法合成金属氧化物纳米材料由于具有易操作、可调变的设置和低成本等优点,引起了研究者们越来越多的关注。虽然仍有许多关于水热法机理方面的问题尚未达成共识,但越来越多且复杂的产品已通过水热技术合成出来。

Liu 等[71]采用改进的溶剂热法,以柠檬酸钠(Na_3Cit)为稳定剂合成了尺寸均匀的高水分散性磁铁矿颗粒(见图 3.4)。通过改变 $FeCl_3$ 或 Na_3Cit 的浓度,可对其粒径在 80~410 nm 的范围内进行灵活的调控。由于其表面附着有柠檬酸盐基团,磁铁矿颗粒具有优异的水分散性和分散稳定性。此外,反应温度是柠檬酸盐改性溶剂热还原法制备均匀磁铁矿球的关键因素,温度越高,颗粒尺寸越大。这些颗粒由许多大小为 5~10 nm 的原生磁铁矿纳米晶组成,表现出超顺磁性和 56~82 emu/g 的高磁化强度,增强了它们对外部磁场的响应,有利于其在实际应用过程中进行灵活的调节。

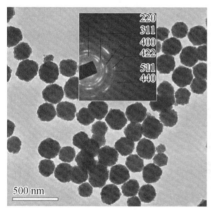 

图 3.4 溶剂热法合成的磁性 Fe_3O_4 微球[71]

3.2.3 自组装方法

自组装指结构单元达到最小的能量状态,自组装形成有序的纳米结构。值得注意的是,在该方法中,结构单元通常通过较弱的作用力(如范德瓦耳斯力、氢键)或硬质粒子来连接,而不是通过较强的共价键,如金属氧化物结晶过程中原子通过共价键相连接[72]。这种自组装的形式包括 DNA[73]、蛋白质[74]、脂质囊泡[75]、嵌段共聚物[76]、蛋白石[77]和纳米超晶格[78],可以通过施加诸如电/磁场或流体流动等外力来驱使自组装过程进行。在本节中,我们将重点阐述纳米晶体和嵌段共聚物的自组装过程,并详细介绍以上两种过程的合成机理。

1. 纳米粒子组装成超晶格

在本节中,根据之前的报道[79-80],我们总结了纳米晶体形貌和组装相关技术在过去十年的发展。纳米晶体可以被理解为金属氧化物、金属或介电晶体的碎片,可以被分散在溶液中并由配体(表面结合的分子)保护以形成稳定的相。经过几十年的发展,有报道说在前驱体的分解和合成过程中有必要使用有机表面活性剂来制备尺寸和形状均匀的纳米晶[33],纳米晶体表面有机表面活性剂封端类似于自组装形成的有机-无机界面,并且表面活性剂可以单层吸附在晶体表面[81],如已被广泛研究的金-硫醇单层体系[82]。各种形状和大小的单分散纳米晶体颗粒都可以通过调整合成参数(如前驱体、表面活性剂、反应温度和时间等)来合成。

通过挥发溶剂,这些表面活性剂封端的胶体纳米晶体可以自组装形成有序超晶格结构。通过使用球形纳米颗粒作为结构单元可以获得复杂的相行为,如图3.5所示,即使没有纳米级别的颗粒间相互作用,一种高度有序的排列方式仍然可以获得[83-84]。另外,可变超晶格组分和独特的结构确保了纳米晶自组装的可控合成,故被广泛应用于材料设计领域。

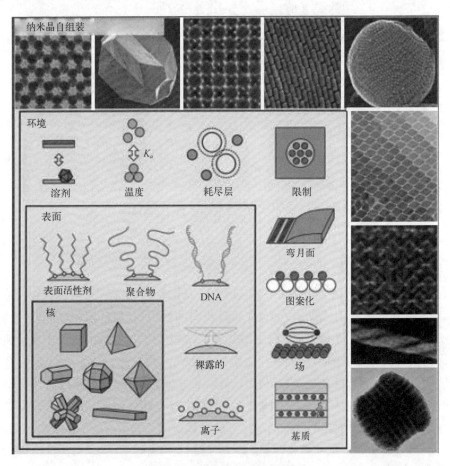

图 3.5　纳米晶体自组装过程[81]

纳米晶体自组装是涉及多个尺度控制的过程,纳米晶核(通常 1~100 nm)被一层表面配体包围(长度通常在 1 nm 到几十纳米之间),自组装环境可由粒子间相互作用和纳米晶体尺寸所控制,因此得到的超晶格结构通常介于 1 μm 和几毫米

挥发或者使溶液去稳定化是制备有序的胶体纳米晶体阵列最常见的策略。基于溶液挥发的自组装方法通常可形成超晶格薄膜,且超晶格组装通常发生在溶剂挥发过程的最后阶段。一般来说,表面活性剂可以覆盖裸露的纳米晶体表面,从而防止亚相溶剂的挥发以及较弱的颗粒间相互作用力,因此体系中过量的表面活性剂有利于形成长程有序超晶格[85]。图 3.6 显示了通过溶剂挥发制备纳米晶超晶格的方法。一种方法是在固体载体上挥发一定体积(约 10 μL)的稀释纳米晶体溶液。为了使碳氢化合物封端的纳米晶体形成长程有序超晶格结构,己烷和辛烷的混合物(体积比为 9∶1)通常被用作载体溶剂[86]。此外,在固体表面简单地沉积一滴纳米晶体溶液,通过气-液界面处的粒子组装可以制备二维超晶格薄膜[87]。自组装方法可以在一个小的容量瓶中实施,基质材料可以倾斜地放入体系中,进一步控制溶液以及基质材料的运动方向[88]。也可通过在更大的载体基质平面上扩散纳米晶溶液来形成更大规模的薄膜。极性液体如二甘醇由于与非极性纳米晶体不混溶,可以作为纳米晶体组装的界面,并且以相同的方式通过在固体表面沉积纳米晶体溶液来获得固体超晶格薄膜,进而进行下一步的表征[89]。这个自组装方法可以与 Langmuir-Blodgett 方法结合,通过施加横向表面压力可控地使致密的纳米晶单层膜形成较大的有序区域[90]。

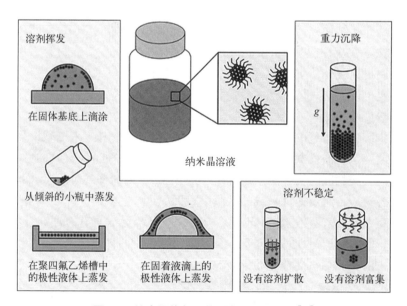

图 3.6　纳米晶体超晶格制备的实验方法[81]

包括通过溶剂挥发技术(左)合成超晶格薄膜和不稳定或沉积技术(右)形成三维超晶格结构

图 3.6 简述了相邻纳米晶表面配体趋向于重叠时的组装方式,该方法有助于纳米晶体在溶液中的逐渐聚集。另外,通过缓慢增加溶液极性来诱导溶液絮凝的

方法可以有效控制有机配体配位的纳米晶体在非溶剂体系中的受控扩散。一般来说,可以通过在上述纳米晶体溶液中引入少量非溶剂来实施(例如,在甲苯中引入乙醇);同时应当注意,这两种溶剂要避免相互溶解[91]。将两种可混溶的溶剂缓慢混合数天后,可以从小瓶底部收集到三维超晶格。与蛋白质结晶行为相似[92],如溶剂、温度和底物等几个参数,都可以有效地影响纳米晶体的自组装过程。但是,如果在挥发溶剂之前体系就发生了颗粒间的絮凝行为,那么此时只能够得到无序的超晶格结构。因此,选择适合脂肪族封端配体的溶剂(烃类液体如己烷、辛烷和甲苯,氯化烃如氯仿、四氯乙烯和氯苯)有利于促进纳米晶体的分散。另外,将组装溶液加热到适当的温度也可以促进超晶格中纳米晶体的有序排列。但是,在加热过程中纳米晶体会在热力学驱动下减小比表面积,进而颗粒会逐渐长大,因此组装的温度有一个上限[88]。还应该提到的是,溶液挥发自组装过程中,溶剂的蒸气压也是至关重要的因素。由于自组装过程中纳米晶体要在溶液中发生扩散,使用易挥发的溶剂可能会加速纳米晶体扩散从而排列成无序的结构。此外,通过使用适当的载体也可以影响超晶格的取向和维度。

2. 介孔材料的合成

有序介孔材料由于具有丰富的多孔结构、均匀的孔径分布、大比表面积和多组分等优点,是最重要的功能材料之一,如二氧化硅、碳和金属氧化物等[93]。高结晶度和独特的介孔结构以及大比表面积的介孔金属氧化物(OMMO)可以提供丰富的表面活性位点,而由于气体测试和吸附的氧物种的反应发生在材料表层,因此介孔接口可以提高气体传感器的灵敏度。此外,介孔结构较大的孔隙率和相互连通的孔结构也可以促进气体分子在介孔载体上的扩散,进而提高传感性能,以实现高灵敏度和选择性[94]。

与用传统纳米铸造的方法合成有序介孔金属氧化物相比,采用软模板法(即表面活性剂或两亲性嵌段共聚物)合成有序介孔金属氧化物是最有效的方法。因为两亲性嵌段共聚物同时具有亲水链和疏水链,因此在溶剂挥发阶段,表面活性剂分子可以很容易地自组装成球形或柱状胶束,且以疏水端为核,亲水端为壳。在无机物前驱体存在下,表面活性剂可以与无机物共同组装形成具有有序介孔结构的复合材料,进而通过选择性去除模板分子即可获得有序介孔氧化物。

在过去的20年中,商业化软模板包括表面活性剂(如十六烷基三甲基溴化铵,CTAB)和两亲性嵌段共聚物[如聚(环氧乙烷)-b-聚(环氧丙烷)-b-聚(环氧乙烷),PEO-b-PPO-b-PEO,如Pluronic P123和F127]已被广泛地应用于合成有序介孔金属氧化物中。然而,使用商业化软模板合成的有序介孔金属氧化物的孔径通常小于10 nm,这会限制它们在许多领域中的应用,如需要较快的物质扩散和高效的负载的大尺寸客体分子(如RNA、基因和染料)和纳米颗粒(如贵金属)等[95]。此外,有序介孔金属氧化物通常呈无定形或半结晶状态,热稳定性低,电性能差。因此,一系列具有大分子量的两亲性嵌段共聚物被开发并应用于具有高结晶度和大比表面积的有序介孔金属氧化物的合成中。

到目前为止,各种基于 PEO 的嵌段共聚物包括聚环氧乙烷-聚苯乙烯(PEO-b-PS)、聚(乙烯-共-丁烯)(KLE)和聚(乙烯基吡啶)基共聚物,如聚(2-乙烯基吡啶)-b-聚苯乙烯(P2VP-b-PS)和聚(4-乙烯基吡啶)-b-聚苯乙烯(P4VP-b-PS),均已被设计为合成有序介孔金属氧化物的软模板剂[95]。在这些两亲性嵌段共聚物中,PEO-b-PS 应用相对较为广泛。无机低聚物可与嵌段共聚物的亲水链段(PEO 链段)通过氢键连接在一起。由于两亲性嵌段共聚物具有丰富的相行为,因此这些复合胶束可以形成丰富的介观结构,如 WO_3、Nb_2O_5、TiO_2 和 Al_2O_3。

Zhang 等[96]以二嵌段共聚物(PEO-b-PS)为模板,异丙醇钛(TIPO)为前驱体,研究了一种新型配体辅助组装方法,用于合成具有高度结晶骨架的热稳定大孔有序介孔二氧化钛(见图 3.7)。X 射线小角散射(SAXS)、X 射线衍射(XRD)、透射电子显微镜(TEM)、高分辨率扫描电子显微镜(HRTEM)和 N_2 吸脱附实验表明,所获得的 TiO_2 材料具有有序的立方介观结构,具有大而均匀的孔径(约 16.0 nm)、高比表面积(约 112 m^2/g),以及高热稳定性(约 700℃)。HRTEM 和广角 XRD 测量清楚地说明了孔壁中具有锐钛矿结构的介孔二氧化钛的高结晶度。值得一提的是,在这一过程中,除了以四氢呋喃作为溶剂外,还使用乙酰丙酮作为配位剂,以避免钛前驱体的快速水解。此外,该研究采用了分级蒸发和加热工艺来控制聚合速度,促进了有序介观结构的组装,并确保形成全多晶锐钛矿相二氧化钛框架,同时不会破坏介观结构。

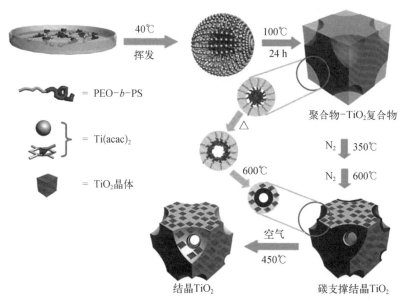

图 3.7 配体辅助组装法制备具有立方结构的有序高结晶大介孔二氧化钛材料的工艺[96]

3.2.4 微乳液介导的合成

微乳液是一种热力学稳定、各向同性、透明或半透明的稳定体系，由"水""油"、两亲性表面活性剂和助表面活性剂按适当比例自发形成，其中"水"通常指极性亲水性液体，"油"通常指非极性疏水性液体[97-98]。一般来说，微乳液可分为四种类型：水包油(O/W)乳液、油包水(W/O)乳液、双连续乳液和单相各向同性乳液[99]。其中，W/O乳液也称为反胶束微乳液，其中两亲分子的亲水基团朝向内水相，而疏水基团指向外油相，其结构实体如图 3.8 所示，形成由表面活性剂和助表面活性剂分子组成的单层界面包围的微小水核，它们彼此不相容，可以形象地称为"水池"。由于尺寸小、分散性好，这些微小的"水池"可以用作微型反应器，为合成金属氧化物纳米颗粒提供理想的化学微环境。微乳液技术在方便和可靠地构建纳米结构方面引起了广泛关注，它能够通过封装前驱体来避免纳米颗粒的聚集，从而将其直径限制在窄分布范围内。通过调整微乳液系统的参数，还可以控制粒径和形态以获得具有不同尺寸和形状的单分散纳米颗粒。

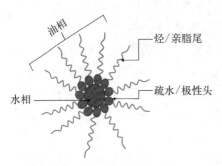

图 3.8 反胶束微乳液的结构模型[100]

金属氧化物纳米颗粒的反相微乳液介导合成机理如图 3.9 所示。首先将前驱体溶解在两个相同的微乳液体系中，分别获得乳液 A 和乳液 B。两种微乳液在剧烈搅拌下混合，两亲性束同时碰撞和聚合，诱导水池中的反应物发生化学反应，包括沉淀、水解、氧化还原反应等。在生成产物成核、聚结和凝聚后，获得固相纳米颗粒。应强调的是，这些纳米颗粒的生长过程取决于水核的大小，因为水池外的表面活性剂将吸附在其表面，以稳定并防止其积聚成大颗粒。微乳液体系中的化学反应完成后，通过离心或添加水和丙酮的混合物去除附着在纳米颗粒上的油相和表面活性剂，进一步干燥和煅烧后最终可获得金属氧化物纳米颗粒。使用反相微乳液方法制备金属氧化物纳米颗粒与表面活性剂的性质、水相与表面活性剂的比例以及反应条件（如温度、反应时间、反应物浓度、pH 等）密切相关。具体而言，表面活性剂的性质决定了水池的界面性质，在调节生成的纳米颗粒的形态和尺寸方面起着关键作用。水相与表面活性剂的比例与水池的大小密切相关，而水池的大小又决定了金属氧化物的最终尺寸。同时，成核和凝固过程以及产物的生长速度也受到反应条件的影响。Zhou 等[101]通过微乳液法构建了氧化铁纳米材料，并与传统的合成策略（包括水热、热分解和共沉淀）进行了比较。该研究的结论表明，通过微乳液法制备的氧化铁纳米颗粒具有最高的纯度（高达 99%）、最大的产率

(86%)和优异的稳定性。此外,微乳液法也可用于合成多组分金属氧化物。例如,Lim 等[102]通过微乳液法成功制备了 $Nd_{0.67}Sr_{0.33}CoO_{3-\delta}$(NSC)钙钛矿纳米颗粒,其具有集中的粒径分布(20~50 nm)和大的比表面积(12.759 m^2/g),并在析氧反应(OER)中表现出优异的电催化活性。此外,将微乳液法与其他合成策略(如超声诱导技术、水热法和溶胶-凝胶法)相结合已被证明是改善金属氧化物纳米颗粒物理化学性质的有效途径。例如,Zhang 等[103]通过微乳液静电纺丝(ME-ES)合成了分级二氧化钛(TiO_2)纳米纤维,如图 3.10 所示。有趣的是,通过调节微乳液系统中钛酸四丁酯(TBT)/石蜡油的进料比,实现了 TiO_2 的精确剪裁,其指导了具有多种结构的 TiO_2,如具有混合晶体的多通道、中空、不规则中空和介孔 TiO_2 纳米纤维的合成。

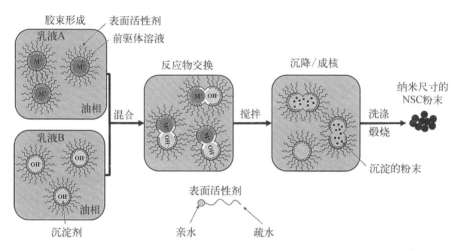

图 3.9 $Nd_{0.67}Sr_{0.33}CoO_{3-\delta}$(NSC)钙钛矿纳米颗粒的反相微乳液介导合成机理[102]

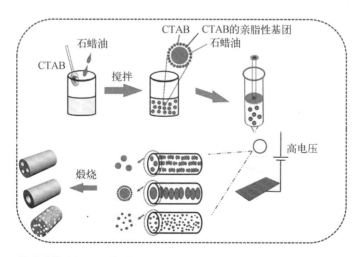

图 3.10 微乳液静电纺丝制备的具有独特结构的多孔 TiO_2 纳米纤维的形成机理[103]

3.2.5 化学气相沉积法

化学气相沉积(CVD)法是合成金属氧化物薄膜最常用的方法。该方法通过将前驱体沉积在玻璃或钢等基质的表面,并微调操作参数来获得均匀组分的薄膜[104],因此前驱体的性质和纯度是影响材料成分和结构的最重要因素。通常,CVD 过程一般采用具有挥发性的前驱体,但是,在气溶胶辅助(AA)CVD 过程中,如图 3.11 所示,AACVD 工艺结合了均匀的气相成核阶段和在基质上的非均匀生长阶段,因此前驱体的溶解度比挥发性更重要。前驱体通常被溶于一种溶剂中,然后通过超声波加湿器/雾化器形成气溶胶借由载气被输送到 CVD 反应器中[105]。该方法的优点是将气溶胶液滴输送到反应室中比依靠挥发具有挥发性的前驱体可行性更高。此外,通过优化诸如沉积温度、溶剂和超声波频率等参数可以精确调控沉积膜颗粒的大小和形貌。因此,CVD 工艺通常可以用于生产具有特定形貌的纳米材料[106-107],包括 In_2O_3、SnO_2、ZnO、Ga_2O_3 和三元透明导电氧化物,如氧化铟锡(ITO)、掺杂镓氧化铟、氟掺杂氧化锡(FTO)等[108]。

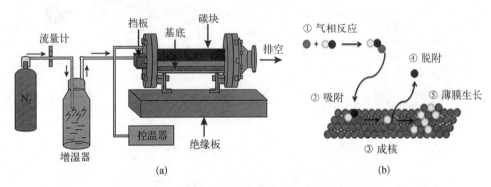

图 3.11 气溶胶辅助化学气相沉积(AACVD)[109]

(a) AACVD 操作说明;(b) AACVD 工艺原理图

参考文献

[1] Korotcenkov G. Practical aspects in design of one-electrode semiconductor gas sensors: Status report[J]. Sens. Actuator B-Chem., 2007, 121: 664 - 678.

[2] Korotcenkov G. Gas response control through structural and chemical modification of metal oxide films: state of the art and approaches[J]. Sens. Actuator B-Chem., 2005, 107: 209 - 232.

[3] Tsiulyanu D, Marian S, Liess H, et al. Effect of annealing and temperature on the NO_2 sensing properties of tellurium based films [J]. Sens. Actuator B-Chem., 2004, 100: 380 - 386.

[4] Korotcenkov G, Brinzari V, Schwank J, et al. Peculiarities of SnO_2 thin film deposition by

spray pyrolysis for gas sensor application[J]. Sens. Actuator B-Chem., 2001, 77: 244-252.

[5] Brinzari V, Korotcenkov G, Golovanov V. Factors influencing the gas sensing characteristics of tin dioxide films deposited by spray pyrolysis: understanding and possibilities of control[J]. Thin Solid Films, 2001, 391: 167-175.

[6] Korotcenkov G. The role of morphology and crystallograpsshic structure of metal oxides in response of conductometric-type gas sensors[J]. Wood Mater. Sci. Eng., 2008, 61: 1-39.

[7] Barsan N, Weimar U. Conduction model of metal oxide gas sensors[J]. J. Electroceram., 2001, 7: 143-167.

[8] Ogawa H, Nishikawa M, Abe A. Hall measurement studies and an electrical conduction model of tin oxide ultrafine particle films[J]. J. Appl. Phys., 1982, 53: 4448-4455.

[9] Rothschild A, Komem Y. The effect of grain size on the sensitivity of nanocrystalline metal-oxide gas sensors[J]. J. Appl. Phys., 2004, 95: 6374-6380.

[10] Kawamura F, Takahashi T, Yasui I, et al. Impurity effect on ⟨1 1 1⟩ and ⟨1 1 0⟩ directions of growing SnO_2 single crystals in SnO_2 - Cu_2O flux system[J]. J. Cryst. Growth., 2001, 233: 259-268.

[11] Nicolae B, Schweizer-Berberich M, Göpel W. Fundamental and practical aspects in the design of nanoscaled SnO_2 gas sensors: a status report[J]. J. Anal. Chem., 1999, 365: 287-304.

[12] Chen F, Zhao T, Fei Y, et al. Surface segregation of bulk oxygen on oxidation of epitaxially grown Nb-doped $SrTiO_3$ on $SrTiO_3$ (001)[J]. Appl. Phys. Lett., 2002, 80: 2889-2891.

[13] Wang X, Sui Y, Yang Q, et al. Effect of doping Zn on the magnetoresistance of polycrystalline Sr_2FeMoO_6[J]. J. Alloys Compd., 2007, 431: 6-9.

[14] Meier J, Schiøtz J, Liu P, et al. Nano-scale effects in electrochemistry[J]. Chem. Phys. Lett., 2004, 390: 440-444.

[15] Castañeda L. Effects of palladium coatings on oxygen sensors of titanium dioxide thin films[J]. Mater. Sci. Eng., 2007, 139: 149-154.

[16] Matko I, Gaidi M, Chenevier B, et al. Pt doping of SnO_2 thin films: a transmission electron microscopy analysis of the porosity evolution[J]. J. Electrochem. Soc., 2002, 149: H153-H158.

[17] Alfredsson M, Richard C, Catlow A. Predicting the metal growth mode and wetting of noble metals supported on c-ZrO_2[J]. Surf. Sci., 2004, 561: 43-56.

[18] Hyodo T, Abe S, Shimizu Y, et al. Gas-sensing properties of ordered mesoporous SnO_2 and effects of coatings thereof[J]. Sens. Actuators B-Chem., 2003, 93: 590-600.

[19] Giorgio S, Silvio G, Paolo N, et al. A new technique for the preparation of highly sensitive hydrogen sensors based on SnO_2(Bi_2O_3) thin films[J]. Sens. Actuators B-Chem., 1991, 5: 253-255.

[20] McAleer J, Moseley P, Norris J, et al. Tin dioxide gas sensors. Part 1.—Aspects of the surface chemistry revealed by electrical conductance variations[J]. J. Chem. Soc., 1987, 1: 1323-1346.

[21] Korotcenkov G, Brinzari V, Stetter J R, et al. The nature of processes controlling the kinetics of indium oxide-based thin film gas sensor response[J]. Sens. Actuators B-Chem., 2007, 128: 51-63.

[22] Wang J, Gan M, Shi J. Detection and characterization of penetrating pores in porous materials[J]. Mater. Charact., 2007, 58: 8-12.

[23] Rumyantseva M, Gaskov A, Rosman N, et al. Raman surface vibration modes in nanocrystalline SnO_2: correlation with gas sensor performances[J]. Chem. Mater., 2005, 17: 893-901.

[24] Min B, Choi S. SnO_2 thin film gas sensor fabricated by ion beam deposition[J]. Sens. Actuators B-Chem., 2004, 98: 239-246.

[25] Korotcenkov G, Ivanov M, Blinov I, et al. Kinetics of indium oxide-based thin film gas sensor response: The role of "redox" and adsorption/desorption processes in gas sensing effects[J]. Thin Solid Films, 2007, 515: 3987-3996.

[26] Williams D E, Pratt K F E. Microstructure effects on the response of gas-sensitive resistors based on semiconducting oxides[J]. Sens. Actuators B-Chem., 2000, 70: 214-221.

[27] Tan O, Zhu W, Yan Q, et al. Size effect and gas sensing characteristics of nanocrystalline $xSnO_2-(1-x)\alpha-Fe_2O_3$ ethanol sensors[J]. Sens. Actuators B-Chem., 2000, 65: 361-365.

[28] Hench L, West J. The sol-gel process[J]. Chem. Rev., 1990, 90: 33-58.

[29] Vioux A. Nonhydrolytic sol-gel routes to oxides[J]. Chem. Mater., 1997, 9: 2292-2299.

[30] Trentler T, Denler T, Bertone J, et al. Synthesis of TiO_2 nanocrystals by nonhydrolytic solution-based reactions[J]. J. Am. Chem. Soc., 1999, 121: 1613-1614.

[31] Zhang Z, Zhong X, Liu S, et al. Aminolysis route to monodisperse titania nanorods with tunable aspect ratio[J]. Angew. Chem. Int. Ed., 2005, 44: 3466-3470.

[32] Esposito S. "Traditional" sol-gel chemistry as a powerful tool for the preparation of supported metal and metal oxide catalysts[J]. Materials, 2019, 12: 668-693.

[33] Murray C, Norris D, Bawendi M. Synthesis and characterization of nearly monodisperse CdE (E=sulfur, selenium, tellurium) semiconductor nanocrystallites[J]. J. Am. Chem. Soc., 1993, 115: 8706-8715.

[34] Kumar S, Nann T. Shape control of Ⅱ-Ⅵ semiconductor nanomaterials[J]. Small, 2006, 2: 316-329.

[35] Andre N, Xia T, Lutz M, et al. Toxic potential of materials at the nanolevel[J]. Science, 2006, 311: 622-627.

[36] Cushing B L, Kolesnichenko V L, O'Connor C J. Recent advances in the liquid-phase syntheses of inorganic nanoparticles[J]. J. Chem. Rev., 2004, 104: 3893-3946.

[37] Cabanas A, Darr J A, Poliakoff M, et al. A continuous and clean one-step synthesis of nano-particulate $Ce_{1-x}Zr_xO_2$ solid solutions in near-critical water[J]. Chem. Commun., 2000, 11: 901-902.

[38] Djerdj I, Arčon D, Jagličić Z, et al. Nonaqueous synthesis of metal oxide nanoparticles: Short review and doped titanium dioxide as case study for the preparation of transition metal-doped oxide nanoparticles[J]. J. Solid State Chem., 2008, 181: 1571-1581.

[39] Rao C, Cheetham A, Thirumurugan A. Hybrid inorganic‐organic materials: a new family in condensed matter physics[J]. J. Phys. Condens. Matter., 2008, 20: 083202-083211.

[40] Titirici M M, Antonietti M, Thomas A. A generalized synthesis of metal oxide hollow spheres using a hydrothermal approach[J]. Chem. Mater., 2006, 18: 3808-3812.

[41] Xiao H, Fu S, Zhu L, et al. Controlled synthesis and characterization of CuO nanostructures through a facile hydrothermal route in the presence of sodium citrate[J]. Eur. J. Inorg. Chem., 2007, 14: 1966-1971.

[42] Jia L, Luo F, Han X, et al. Large-scale synthesis of single-crystalline iron oxide magnetic nanorings[J]. J. Am. Chem. Soc., 2008, 130: 16968-16977.

[43] Wang S, Min Y, Yu S. Synthesis and magnetic properties of uniform hematite nanocubes [J]. J. Phys. Chem. C, 2007, 111: 3551-3554.

[44] Greta R, Alexej M, Frank K, et al. One-step synthesis of submicrometer fibers of MoO_3 [J]. Chem. Mater., 2004, 16: 1126-1134.

[45] Nagappa B, Chandrappa G, Livage J. Synthesis, characterization and applications of nanostructural/nanodimensional metal oxides[J]. J. Phys., 2005, 65: 917-213.

[46] Uchiyama H, Ohgi H, Imai H. Selective preparation of SnO_2 and SnO crystals with controlled morphologies in an aqueous solution system[J]. Growth Des., 2006, 6: 2186-2190.

[47] Cheng B, Shi W, Zhang L, et al. Large-scale, solution-phase growth of single-crystalline SnO_2 nanorods[J]. J. Am. Chem. Soc., 2004, 126: 5972-5973.

[48] Robert M, James R, Milo S. Impact of hydrothermal processing conditions on high aspect ratio titanate nanostructures[J]. Chem. Mater., 2006, 18: 6059-6068.

[49] Reyes-Coronado D, Rodríguez-Gattorno G, Espinosa-Pesqueira M, et al. Phase-pure TiO_2 nanoparticles: anatase, brookite and rutile[J]. Nanotechnology, 2008, 19: 145605-145613.

[50] Chandrappa G T, Steunou N, Cassaignon S, et al. Hydrothermal synthesis of vanadium oxide nanotubes from V_2O_5 gels[J]. J. Catal. Today, 2003, 78: 85-89.

[51] Clavel G, Willinger M G, Zitoun D, et al. Solvent dependent shape and magnetic properties of doped ZnO nanostructures[J]. Adv. Funct. Mater., 2007, 17: 3159-3169.

[52] Zhang L, Wang W, Zhou L, et al. Bi_2WO_6 nano- and microstructures: shape control and associated visible-light-driven photocatalytic activities[J]. Small, 2007, 3: 1618-1625.

[53] Zhou Y, Vuille K, Heel A, et al. Studies on nanostructured Bi_2WO_6: convenient hydrothermal and TiO_2-coating pathways[J]. ChemInform, 2009, 635: 1848-1855.

[54] Sun L, Wu X, Luo S, et al. Synthesis and photoluminescent properties of strontium tungstate nanostructures[J]. J. Phys. Chem. C, 2007, 111: 532-537.

[55] Kiebach R, Pienack N, Bensch W, et al. Hydrothermal formation of W/Mo-oxides: a multidisciplinary study of growth and shape[J]. Chem. Mater., 2008, 20: 3022-3033.

[56] Michailovski A, Wörle M, Sheptyakov D, et al. Hydrothermal synthesis of anisotropic alkali and alkaline earth vanadates[J]. J. Mater. Res., 2011, 22: 5-18.

[57] Lee S H, Kim T W, Park D H, et al. Single-step synthesis, characterization, and application of nanostructured $K_xMn_{1-y}Co_yO_{2-\delta}$ with controllable chemical compositions

and crystal structures[J]. Chem. Mater., 200719: 5010-5017.

[58] Hu Y, Gu H, Hu Z, et al. Controllable hydrothermal synthesis of $KTa_{1-x}Nb_xO_3$ nanostructures with various morphologies and their growth mechanisms[J]. Cryst. Growth Des., 2008, 8: 832-837.

[59] Wei X, Xu G, Ren Z, et al. Composition and shape control of single-crystalline $Ba_{1-x}Sr_xTiO_3$ ($x=0-1$) nanocrystals via a solvothermal route[J]. J. Cryst. Growth., 2008, 310: 4132-4137.

[60] Prades M, Beltrán H, Masó N, et al. Phase transition hysteresis and anomalous Curie-Weiss behavior of ferroelectric tetragonal tungsten bronzes $Ba_2RETi_2Nb_3O_{15}$: RE=Nd, Sm[J]. J. Appl. Phys., 2008, 104: 104118-104125.

[61] Su Y, Li L, Li G. Synthesis and optimum luminescence of $CaWO_4$-based red phosphors with codoping of Eu^{3+} and Na^+[J]. Chem. Mater., 2008, 20: 6060-6067.

[62] Zhang L, Fu H, Zhang C, et al. Effects of Ta^{5+} substitution on the structure and photocatalytic behavior of the $Ca_2Nb_2O_7$ photocatalyst[J]. J. Phys. Chem. C, 2008, 112: 3126-3133.

[63] Hu Y. Hydrothermal synthesis of nano Ce-Zr-Y oxide solid solution for automotive three-way catalyst[J]. J. Am. Ceram. Soc., 2006, 89: 2949-2951.

[64] Gözüak F, Köseoğlu Y, Baykal A, et al. Synthesis and characterization of $Co_xZn_{1-x}Fe_2O_4$ magnetic nanoparticles via a PEG-assisted route[J]. J. Magn. Magn. Mater., 2009, 321: 2170-2177.

[65] Zhang T, Jin C G, Qian T, et al. Hydrothermal synthesis of single-crystalline $La_{0.5}Ca_{0.5}MnO_3$ nanowires at low temperature[J]. J. Mater. Chem., 2004, 14: 2787-2796.

[66] Niu J, Deng J, Liu W, et al. Nanosized perovskite-type oxides $La_{1-x}Sr_xMO_{3-\delta}$ (M = Co, Mn; $x = 0, 0.4$) for the catalytic removal of ethylacetate[J]. Catal. Today, 2007, 126: 420-429.

[67] Chen T, Fung K. Synthesis of and densification of oxygen-conducting $La_{0.8}Sr_{0.2}Ga_{0.8}Mg_{0.2}O_{2.8}$ nano powder prepared from a low temperature hydrothermal urea precipitation process[J]. J. Eur. Ceram. Soc., 2008, 28: 803-810.

[68] Zhou L, Liang Y, Hu L, et al. Much improved capacity and cycling performance of $LiVMoO_6$ cathode for lithium ion batteries[J]. J. Alloys Compd., 2008, 457: 389-393.

[69] Zhang Q, Zhu M, Zhang Q, et al. Synthesis and characterization of carbon nanotubes decorated with manganese-zinc ferrite nanospheres[J]. Mater. Chem. Phys., 2009, 116: 658-662.

[70] Sanfiz A C, Hansen T, Girgsdies F, et al. Preparation of phase-pure M_1 MoVTeNb oxide catalysts by hydrothermal synthesis—influence of reaction parameters on structure and morphology[J]. Top Catal., 2008, 50: 19-32.

[71] Liu J, Sun Z, Deng Y, et al. Highly water-dispersible biocompatible magnetite particles with low cytotoxicity stabilized by citrate groups[J]. Angew. Chem. Int. Ed., 2009, 121: 5975-5879.

[72] Min Y, Akbulut M, Kristiansen K, et al. The role of interparticle and external forces in

nanoparticle assembly[J]. Nat. Mater., 2008, 7: 527-538.

[73] Pinheiro A V, Han D, Shih W M, et al. Challenges and opportunities for structural DNA nanotechnology[J]. Nat. Nanotechnol., 2011, 6: 763-772.

[74] Dill K A, MacCallum J L. The protein-folding problem, 50 years on[J]. Science, 2012, 338: 1042-1046.

[75] Chen I A, Walde P. From self-assembled vesicles to protocells[J]. Cold Spring Harb. Perspect. Biol., 2010, 2: 1-13.

[76] Bates F S, HiUmyer M A, Lodge T P, et al. Multiblock polymers: panacea or pandora's box? [J]. Science, 2012, 336: 434-440.

[77] Kim S, Lee S, Yang S, et al. Self-assembled colloidal structures for photonics[J]. NPG Asia Mater., 2011, 3: 25-33.

[78] Sun S, Murray C B, Weller D, et al. Monodisperse FePt nanoparticles and ferromagnetic FePt nanocrystal superlattices[J]. Science, 2000, 287: 1989-1992.

[79] Murray C, Kagan C. Synthesis and characterization of monodisperse nanocrystals and close-packed nanocrystal assemblies[J]. Annu. Rev. Mater. Sci., 2000, 30: 545-610.

[80] Claridge S, Castleman A, Khanna S, et al. Cluster-assembled materials[J]. ACS Nano, 20093: 244-245.

[81] Ulman A. Formation and structure of self-assembled monolayers[J]. Chem. Rev., 1996, 96: 1533-1554.

[82] Vericat C, Vela M, Benitez G, et al. Self-assembled monolayers of thiols and dithiols on gold: new challenges for a well-known system[J]. Chem. Soc. Rev., 2010, 39: 1805-1813.

[83] Silvera Batista C A, Larson R G, Kotov N A. Nonadditivity of nanoparticle interactions[J]. Science, 2015, 350: 1242471-1242477.

[84] Piner R D, Zhu J, Xu F, et al. "Dip-pen" nanolithography[J]. Science, 1999, 283: 661-663.

[85] Lin X M, Jaeger H M, Sorensen C M, et al. Formation of long-range-ordered nanocrystal superlattices on silicon nitride substrates[J]. J. Phys. Chem. B, 2001, 105: 3353-3357.

[86] Talapin D V, Murray C B. PbSe nanocrystal solids for n- and p-channel thin film field-effect transistors[J]. Science, 2005, 310: 86-89.

[87] Bigioni T P, Lin X M, Nguyen T T, et al. Kinetically driven self assembly of highly ordered nanoparticle monolayers[J]. Nat. Mater., 2006, 5: 265-270.

[88] Bodnarchuk M I, Kovalenko M V, Heiss W, et al. Energetic and entropic contributions to self-assembly of binary nanocrystal superlattices: temperature as the structure-directing factor[J]. J. Am. Chem. Soc., 2010, 132: 11967-11977.

[89] Dong A, Chen J, Vora P, et al. Binary nanocrystal superlattice membranes self-assembled at the liquid-air interface[J]. Nature, 2010, 466: 474-477.

[90] Aleksandrovic V, Greshnykh D, Randjelovic I, et al. Preparation and electrical properties of cobalt-platinum nanoparticle monolayers deposited by the Langmuir-Blodgett technique[J]. ACS Nano, 2008, 2: 1123-1130.

[91] Rupich S M, Shevchenko E V, Bodnarchuk M I, et al. Size-dependent multiple twinning in

nanocrystal superlattices[J]. J. Am. Chem. Soc., 2010, 132: 289-296.

[92] Chayen N. Turning protein crystallisation from an art into a science[J]. Curr. Opin. Struct. Biol., 2004, 14: 577-583.

[93] Yang P, Zhao D, Margolese D I, et al. Generalized syntheses of large-pore mesoporous metal oxides with semicrystalline frameworks[J]. Nature, 1998, 396: 152-155.

[94] Zhu Y, Zhao Y, Ma J, et al. Mesoporous tungsten oxides with crystalline framework for highly sensitive and selective detection of foodborne pathogens[J]. J. Am. Chem. Soc., 2017, 139: 10365-10373.

[95] Deng Y, Wei J, Sun Z, et al. Large-pore ordered mesoporous materials templated from non-Pluronic amphiphilic block copolymers[J]. Chem. Soc. Rev., 2013, 42: 4054-4070.

[96] Zhang J, Deng Y, Gu D, et al. Ligand-assisted assembly approach to synthesize large-pore ordered mesoporous titania with thermally stable and crystalline framework[J]. Adv. Energy Mater., 2011, 1: 241-248.

[97] Bera A, Mandal A. Microemulsions: a novel approach to enhanced oil recovery: a review [J]. J. Pet. Explor. Prod. Te., 2015, 5: 255-268.

[98] Li X, Wang B, Dai S, et al. Ionic liquid-based microemulsions with reversible microstructures regulated by CO_2[J]. Langmuir, 2020, 36: 264-272.

[99] Hejazifar M, Lanaridi O, Bica-Schröder K. Ionic liquid based microemulsions: A review [J]. J. Mol. Liq., 2020, 303: 112264-112286.

[100] Kumar H, Sarma A, Kumar P. A comprehensive review on preparation, characterization, and combustion characteristics of microemulsion based hybrid biofuels[J]. Renew. Sust. Energ. Rev., 2020, 117: 109498-109515.

[101] Zhou L, Zhu Q, Feng Y, et al. Preparation and characterization of iron oxide low-dimensional nanomaterials[J]. Integr. Ferroelectr., 2022, 225: 240-254.

[102] Lim C, Kim C, Gwon O, et al. Nano-perovskite oxide prepared via inverse microemulsion mediated synthesis for catalyst of lithium-air batteries[J]. Electrochim. Acta, 2018, 275: 248-255.

[103] Zhang J, Hou X, Pang Z, et al. Fabrication of hierarchical TiO_2 nanofibers by microemulsion electrospinning for photocatalysis applications[J]. Ceram. Int., 2017, 43: 15911-15917.

[104] Knapp C, Carmalt C. Solution based CVD of main group materials[J]. J. Chem. Soc. Rev., 2016, 45: 1036-1064.

[105] Marchand P, Hassan I A, Parkin I P, et al. Aerosol-assisted delivery of precursors for chemical vapour deposition: expanding the scope of CVD for materials fabrication[J]. Dalton. Trans., 2013, 42: 9406-9422.

[106] Condorelli G G, Malandrino G, Fragalà I L. Engineering of molecular architectures of β-diketonate precursors toward new advanced materials[J]. Coord. Chem. Rev., 2007, 251: 1931-1950.

[107] Bekermann D, Barreca D, Gasparotto A, et al. Multi-component oxide nanosystems by chemical vapor deposition and related routes: challenges and perspectives [J]. CrystEngComm, 2012, 14: 6347-6355.

[108] Guo X, Xue C, Sathasivam S, et al. Fabrication of robust superhydrophobic surfaces via aerosol-assisted CVD and thermo-triggered healing of superhydrophobicity by recovery of roughness structures[J]. J. Mater. Chem. A, 2019, 7: 17604–17612.

[109] Powell M, Potter D, Wilson R, et al. Scaling aerosol assisted chemical vapour deposition: Exploring the relationship between growth rate and film properties[J]. Mater. Des., 2017, 129: 116–124.

第 4 章
半导体金属氧化物：组分及其气敏性能

随着社会的快速发展，人们对气体传感器的要求也在不断提高，希望气体传感器具有高的灵敏度、高的选择性、快速的响应/恢复，以及低功耗和低成本。然而，对于电阻型气体传感器，单组分半导体金属氧化物难以满足人们对理想传感器的所有需求。最近的许多研究表明，通过多组分的整合可以显著提高传感器的气体传感性能[1-2]。研究较多的用于气体传感的多组分复合材料主要有三类：① 二元金属氧化物异质结；② 贵金属修饰；③ 特定元素的掺杂，如非金属元素、金属元素、稀土元素等。当然，还有其他组分与金属氧化物进行复合的材料，如无机碳材料、有机材料等。

将不同类型（n 型半导体或 p 型半导体）的半导体金属氧化物进行复合以形成异质结是改善传感器性能最常用的方式之一。异质结指两种不同材料之间互相接触的物理界面。一方面，由于两种不同禁带宽度或者不同类型的半导体金属氧化物材料在界面处紧密接触，费米能级可以平衡到相同的能量，因此通常会导致电荷转移和电荷耗尽层的形成。这种独特的效果是改善气体传感性能的基础。另一方面，对于某些特殊气体，异质结材料的表面也可能发生协同反应。当两种半导体材料的表面同时暴露时，目标分子可首先与一种材料反应，反应的产物再与另一种材料反应以获得最终产物[2]。贵金属修饰的金属氧化物材料是另一类重要的复合材料。与用氧化物进行复合不同，修饰的贵金属含量通常是很少的。研究表明，贵金属起着化学敏化剂和电子敏化剂的作用[3-5]。元素掺杂则与前两种复合材料都不同，它们没有两个特定的不同的相。掺杂元素可以进入金属氧化物晶格，改变主体材料的晶体结构或电子结构，甚至有时掺杂剂也可以改变材料的形态。

根据研究报道，气敏材料的组分对气体传感性能的影响可归因于三个主要因素：① 电子效应，包括由费米能级平衡引起的能带弯曲、载流子分离、耗尽层变化和界面势垒能量增加等；② 化学效应，包括表面反应活化能的降低、催化剂催化活性和协同表面反应等；③ 几何效应，包括晶粒尺寸细化、比表面积增加和气体可及性增强。通过了解这些影响性能的因素，对材料的优化设计非常有帮助。

4.1 双金属氧化物异质结

半导体金属氧化物可以根据载流子的不同区分为 n 型半导体和 p 型半导体两种类型。n 型半导体的载流子是自由电子,而 p 型半导体的载流子则是空穴。因此,当两种金属氧化物结合形成异质结时,就会存在三种类型的异质结:p-n 异质结、n-n 异质结和 p-p 异质结。

4.1.1 p-n 异质结

构筑 p-n 异质结是一种非常常见的半导体金属氧化物的界面调制方式,可以利用其来改善材料的气体传感特性。在 p-n 异质结中,由于两种材料的费米能级不同,"能带弯曲"可以在 n 型和 p 型金属氧化物的界面处产生。在高能态下占据的电子可以通过氧化物界面转移到未占据的低能态,以平衡费米能级,这被称为"费米能级介导的电荷转移",其相当于电子和空穴复合。"能带弯曲"可以调节 p 侧和 n 侧的能量结构,从而使传感材料具备更好的气体传感性能。

以 p 型半导体 NiO/n 型半导体 SnO_2 异质结为例来分析 p-n 异质结中的电子效应。在这个例子中,由于 NiO 和 SnO_2 的能带高度和带隙不同,p-n 异质结在 n 型 SnO_2 和 p 型 NiO 的晶粒界面形成[6]。图 4.1 是能带结构的示意图,当 NiO 和 SnO_2 的纳米颗粒相互接触时,由于载流子间存在着巨大的浓度梯度,SnO_2 侧的电子会扩散到 NiO 侧,而 NiO 中的空穴则向相反的方向扩散,直到载流子扩散达到平衡,最终在界面处形成弯曲的能带。随着电子和空穴的转移,最终体系的费米能级达到平衡,并且在 p-n 异质结界面处形成较宽的耗尽区。因此,NiO/SnO_2 复合材料的电阻比纯 SnO_2 或 NiO 都要高。在该实例中,选择还原气体(三乙胺,TEA)作为目标测试气体。一旦复合材料暴露于 TEA 气体,TEA 气体就与表面上吸附的氧物种反应并将电子传输回块体材料中,从而导致整个材料的电阻降低。此外,TEA 的氧化反应还将电子释放到 p 型 NiO 中,电子与空穴的复合降低了空穴浓度。NiO 中空穴的减少意味着电子的增加,则 p-n 异质结两侧相同载流子的浓度梯度大大减小,载流子的扩散也大幅减弱,导致界面处的耗尽层变薄。因此,可以进一步降低 p-NiO/n-SnO_2 复合物在还原性气氛中的电阻。考虑到传感器响应的定义($S=R_a/R_g$),由于 SnO_2 和 NiO 传感器之间形成了 p-n 异质结,在空气中的电阻增加而在目标气体氛围中的电阻减少,因此其灵敏度大大提升。

p-n 异质结的形成不仅会影响气体灵敏度,还会影响选择性。以 p 型 Co_3O_4 纳米颗粒修饰的 n 型 ZnO 纳米线复合材料为例(见图 4.2)[7],在该实例中,通过使用 p 型 Co_3O_4 修饰的 n 型 ZnO 纳米线(NW)气体传感器,实现了气体 NO_2 和

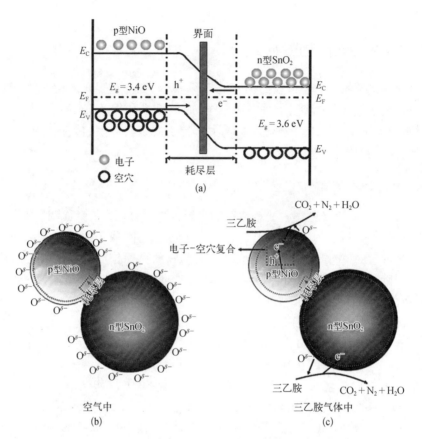

图 4.1 p 型 NiO/n 型 SnO₂ 异质结的三乙胺传感机理示意图[6]
(a) p 型 NiO/n 型 SnO₂ 异质结的能带结构图;(b)、(c) 暴露于空气中与暴露于 TEA 气体中时 p 型 NiO/n 型 SnO₂ 异质结传感器的模型示意图

C_2H_5OH 的选择性检测。结果表明,p 型 Co_3O_4 纳米颗粒修饰的 n 型 ZnO 纳米线在 400℃左右对 C_2H_5OH 具有较高的响应,对 NO_2 的响应可忽略不计,而纯 ZnO 在相同条件下对两者的响应均较低。由于 p-n 异质结的存在,p 型 Co_3O_4 纳米颗粒修饰的 n 型 ZnO 纳米线复合材料的电阻远高于纯 ZnO 纳米线的电阻。当暴露于诸如 NO_2 的氧化气体时,任何吸附导致的 Co_3O_4-ZnO 电阻的额外增加都被最小化。然而,当引入 C_2H_5OH(还原性气体)时,Co_3O_4-ZnO 的电阻显著降低。R_g/R_a 和 R_a/R_g 的值通常分别用作对氧化性和还原性气体的响应灵敏度,因此,由于具有 p-n 异质结,Co_3O_4-ZnO 复合材料较纯 ZnO 纳米线对还原性气体更敏感。该团队还将乙醇的增强响应归因于 360℃ 的工作温度下反应完全转化为 CO_2 和 H_2O。据报道,在 Co_3O_4 催化剂的帮助下,乙醇气体完全氧化为 CO_2 和 H_2O 的温度低至 380℃。这种材料可以减少路边车辆经常排放的二氧化氮的干扰,提高对乙醇的检测能力,有助于在酒驾检测中检测出呼出气体中的酒精含量。

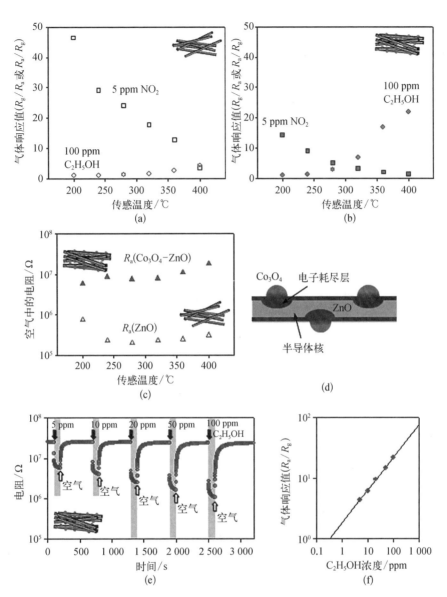

图 4.2 ZnO 纳米线(NWs)和 Co_3O_4 - ZnO NWs 的 NO_2 和乙醇传感特性研究[7]

(a)、(b) ZnO NWs 和 Co_3O_4 - ZnO NWs 对 5 ppm NO 和 100 ppm C_2H_5OH 在 200~400℃的气体响应(R_a/R_g 或 R_g/R_a,其中 R_a:空气中的电阻,R_g:气体中的电阻);(c) ZnO NWs 传感器和 Co_3O_4 - ZnO NWs 传感器在空气中的电阻;(d) Co_3O_4 - ZnO NWs 的电子结构示意图;(e) Co_3O_4 - ZnO NWs 传感器对 C_2H_5OH 传感的动态电阻变化;(f) 400℃下 C_2H_5OH 的灵敏度-浓度曲线

该理论模型也可用于解释其他具有 p-n 异质结的复合材料,如 CuO/ZnO 和 NiO/WO_3[8-9]。Xiao 等[9]采用了一种简便的溶剂蒸发诱导多组分共组装方法,结合碳支撑结晶策略,以 PEO-b-PS 为结构导向剂、钨(Ⅵ)氯化物作为 WO_3 前驱

体、乙酰丙酮镍[Ni(AcAc)$_2$]作为 NiO 前驱体,在酸性 THF/H$_2$O 溶液中可控地合成了 NiO 掺杂的结晶介孔 WO$_3$(见图 4.3)。所得材料具有面心立方介孔结构、大孔径(约 30 nm)、大比表面积(30~50 m^2/g)、大孔体积(0.15~0.19 cm^3/g)和连接相邻介孔的超大孔窗口(12~16 nm),并且介孔 WO$_3$ 骨架由超细 NiO 纳米晶体修饰。由于其连接良好的多孔结构且具有丰富 WO$_3$/NiO 界面的高比表面积,该复合材料具有超快响应(约 4 s)的优异气敏性能、高灵敏度($R_a/R_g=58\pm5.1$),以及在相对较低的工作温度(250 ℃)下对 50 ppm H$_2$S 的高选择性。化学机理研究揭示了 WO$_3$/NiO 基气敏元件的复杂表面反应,发现在气敏过程中会产生 SO$_2$、WS$_2$ 和 NiS 中间体。

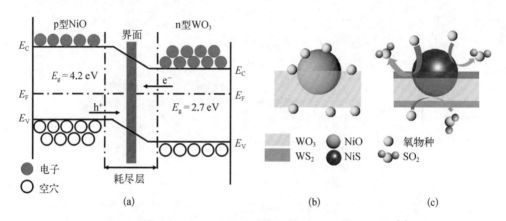

图 4.3 基于介孔 WO$_3$/NiO 复合材料的 H$_2$S 传感机理研究[9]

(a) WO$_3$/NiO 在 H$_2$S 传感过程中的能带结构变化和电子转移过程;(b)、(c) 暴露在空气中和 H$_2$S-空气混合物中时的传感机理示意图,其中 E_C:导带边缘,E_F:费米能级,E_V:价带

除了上述 p-n 异质结提升气体传感器性能的一般性机理,在某些特定气体存在时,金属氧化物(n 型或 p 型)中的一种可与气体反应,导致 p-n 异质结消失,也可以进一步增强响应。Shao 等[10]报道了 p 型 CuO(纳米粒子)修饰的 n 型 SnO$_2$(纳米线)器件,用于 H$_2$S 的选择性检测。p-CuO 颗粒和 n-SnO$_2$ 在界面处形成 p-n 异质结,产生的耗尽层大大减少了 SnO$_2$ 纳米线中电子的有效传导通道,从而导致材料具有更高的电阻。在暴露于 H$_2$S 气氛之后,p 型 CuO 颗粒与 H$_2$S 反应并变为具有金属导电性的 CuS,导致 p-n 异质结消失,纳米线中的导电通道变宽,因此电阻值变低、响应变高。当引入空气时,CuS 可以被氧化成 CuO,从而又可使 p-n 异质结恢复(见图 4.4)。

贵金属与金属氧化物之间也可能存在 p-n 异质结。例如,Koo 等[11]将制备的 Pd@ZnO-WO$_3$ 纳米线用于甲苯传感。在空气中,Pd 以 PdO 形式存在,而 PdO 为 p 型半导体,可以与 ZnO 形成 p-n 异质结,电阻进一步增大;当暴露于还

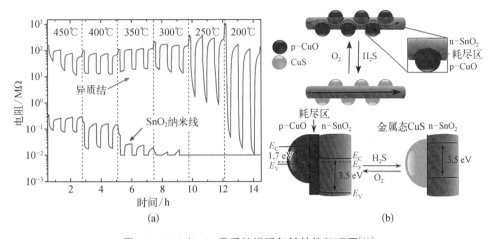

图 4.4　CuO/SnO₂ 异质结增强气敏性能机理图[10]

(a) CuO/SnO₂ 的 H_2S 传感特性研究；(b) CuO/SnO₂ 在 H_2S 传感过程中的能带结构变化和电子转移过程

原性气体甲苯中时，PdO 被还原，p-n 异质结消失，此时有额外电子注入材料中，电阻变小，因此显著影响了材料的电阻变化，从而提高了材料的灵敏度。

p-n 异质结的微观结构和暴露方式对材料的性质具有巨大影响。Woo 等[12]报道了一维 $ZnO-Cr_2O_3$ 异质结纳米线作为高灵敏度和选择性三甲胺（TMA）传感器的应用。结果表明，与原始 ZnO 纳米线相比，Cr_2O_3 纳米颗粒修饰的复合材料的响应显著增强，而外部为均匀 Cr_2O_3 涂层的复合材料对气体的响应则显著降低（见图 4.5）。当 Cr_2O_3 纳米颗粒在 ZnO 纳米线表面离散分布时，与 p 型 Co_3O_4 修饰的 n 型 ZnO 改善气体传感性能的实例相似，响应增强可归因于 p-n 异质结的形成和 Cr_2O_3 的催化作用。然而，当在 ZnO 纳米线上均匀涂覆厚度为 30～40 nm 的连续 Cr_2O_3 壳层时，p 型 Cr_2O_3 层成为主要电子传导通路，因此在暴露于 TMA 时表现为 p 型半导体的响应行为并降低了对 TMA 的响应（与纯 ZnO 相比）。

Mashock 等[13]进行的一项用 SnO_2 修饰 CuO 纳米线的气敏性能研究也得到了类似的结果。该纳米结构具有相反的性质，利用 n 型 SMO（SnO_2）来修饰 p 型纳米线（CuO）。与上面讨论的 p-n 异质结一样，其在两种不同类型的金属氧化物界面处形成耗尽区，并且 p-n 异质结的效果可能被放大，因为 p 型 SMO 电阻由具有氧吸附形成的累积层的表面传导支配。在沉积离散的 SnO_2 纳米颗粒之后，由于 p-n 异质结的形成，电阻可以增大。沉积时间加倍后，更多的 SnO_2 纳米颗粒可以彼此接触并在表面上变得连续。较长的沉积时间使得材料具有更高的电阻，但与具有较短沉积时间的纳米线相比，其对 NH_3 的响应却只有较小程度的增加。该项研究提出了两种有助于解释这一点的机制：① NH_3 通过提供电子，增加 p 型半

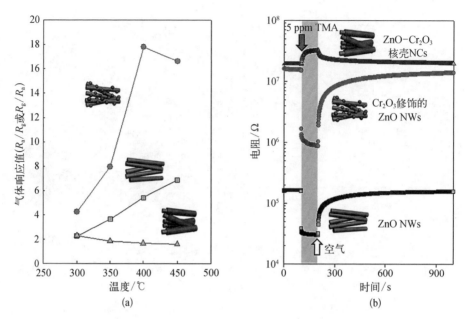

图 4.5 ZnO-Cr₂O₃ 复合材料的 TMA 传感特性研究[12]

(a)、(b) 未改性的 ZnO 纳米纤维、Cr₂O₃ 修饰的 ZnO 纳米纤维和 ZnO-Cr₂O₃ 核壳结构纳米电缆 (NC)对 TMA 气体的响应和电阻变化

导体的电阻;② NH_3 通过降低 SnO_2 纳米颗粒的吸附氧来降低 CuO 纳米线表面上的空穴浓度,从而产生更强的 p-n 异质结,进而增加了电阻,强化了第一种机制的影响。当通过较长时间的沉积产生连续的 SnO_2 覆盖时,n 型 SnO_2 成为主要的导电通路,而 p 型 CuO 不参与反应。但离散粒子的修饰允许这两种机制发生,因此显示出最佳的传感性能。

上述例子仅考虑沿纳米线轴向的电子传导,其由可用的传导横截面积决定,即不受耗尽区的影响,不考虑颗粒之间界面处存在的势垒的变化,因为它随氧吸附程度而变化。Jain 等[14]报道,Ni 掺杂的 SnO_2 在液化石油气(LPG)中具有更高的灵敏度。由于 NiO 存在于 SnO_2 颗粒表面,因此 Ni 掺杂 SnO_2 的纳米结构可以被建模为 SnO_2@NiO 核-壳球形纳米颗粒。Jain 等将改进的性能归因于 p-n 异质结的形成所增加的势垒高度。与 1D 结构不同,SnO_2 纳米颗粒上存在的薄 NiO 层可作为 SnO_2 导带中电子重组的屏障,这种 p-n 异质结将促进 NiO 和 SnO_2 之间的电子转移,从而提高在 NiO 表面上的还原氧化反应速率。

4.1.2 n-n 异质结

由于两种不同的 n 型或 p 型金属氧化物之间的能级差异,能带弯曲也可以在 n-n 异质结和 p-p 异质结中发生。电荷可以从较高费米能级的半导体金属氧化物转移到较低费米能级的半导体金属氧化物中,导致能量势垒增加和较高的复合物电阻,从而促进氧气在表面的吸附以提高气体传感性能。

Zeng 等[15]报道了通过简单地将 TiO_2 与 SnO_2 胶体溶液混合制备的 TiO_2-SnO_2 纳米复合材料,提高了对挥发性有机化合物的响应性能。纯 SnO_2 的气体传感机制与其他 n 型半导体金属氧化物材料类似,如图 4.6(a)～图 4.6(c)所示:预吸附的氧气可以在 n 型半导体金属氧化物表面附近产生耗尽层,导致表面周围的能带弯曲。能带弯曲可以增加能量势垒并因此增加复合材料的电阻。在负载 TiO_2 纳米颗粒之后,n-n 异质结可以在 SnO_2 和 TiO_2 界面形成。TiO_2 和 SnO_2 的功函数分别为 5.58 eV 和 4.9 eV,由于带隙和功函数存在较大的梯度,电子可以从 TiO_2 侧转移到 SnO_2 侧,导致表面周围的能带弯曲。在 n-n 异质结的界面处,由于电子转移到较低能量的导带,在 SnO_2 侧产生累积层而不是耗尽层,这与 p-n 异质结完全不同。由于电子和空穴复合,p-n 异质结处的界面具有较少的自由电子,通常会在两侧形成耗尽层(见图 4.7)。SnO_2 表面上的氧吸附可以消耗累积层,促进氧吸附,这可以进一步增强其气敏性能。

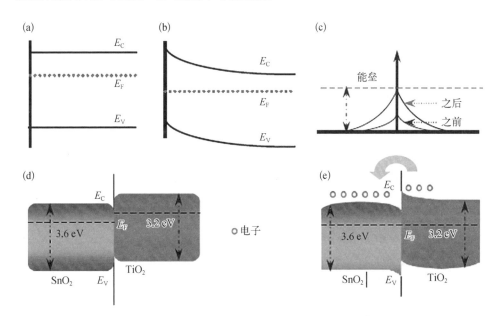

图 4.6 TiO_2-SnO_2 复合体系的能带结构变化和电子转移过程[15]

(a)、(b) 在氧吸附之前和氧吸附之后的 TiO_2-SnO_2 复合体系的能带结构;(c) 氧吸附引起的能垒变化;
(d)、(e) 在 SnO_2 与 TiO_2 接触之前和之后的能带

Zhao 等[16]选用 SnO_2 纳米晶(NCs,3～5 nm)为锡源、钛酸正丁酯为钛源、实验室自合成两亲性嵌段共聚物 PEO-b-PS 为模板剂,开发了一种溶剂挥发诱导取向共组装策略,合成了具有 n-n 异质结的树枝状多级介孔 TiO_2-SnO_2 复合材

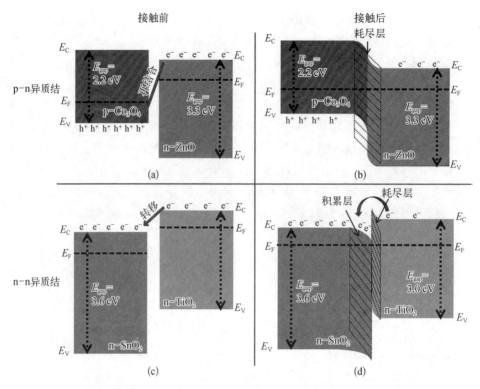

图 4.7 没有表面吸附物质的异质结界面处能带弯曲示意图[2]

(a) p 型 Co_3O_4 和 n 型 ZnO 之间即将形成的 p-n 异质结,可用的较低能量价带状态(空穴)刺激跨界面的电子转移和费米能量(E_F)的平衡;(b) 由于复合而在 p-n 异质结的两侧形成的耗尽层,产生了电子流动的势垒;(c) 在 n 型 SnO_2 和 n 型 TiO_2 之间即将形成的 n-n 异质结,可用的较低能量导带状态刺激电子转移至 n 型 SnO_2;(d) 由于电子的损失在 n 型 TiO_2 表面形成的耗尽层,以及由于电子的增加在 SnO_2 表面形成的累积层,增强了氧吸附,在界面处形成了势垒

料(SHMT)(见图 4.8)[16]。SnO_2 纳米晶表面丰富的羟基使其可以通过氢键和模板剂的 PEO 段相互作用,并通过共组装过程嵌入多级介孔 TiO_2 材料的孔壁中。所获得的多级介孔复合材料具有树枝状形貌、均匀的粒径(500 nm)和发散型的介孔孔道(9 nm),其比表面积以及孔容分别可达 76 m^2/g 和 0.11 cm^3/g。凭借着独特的形貌和结构,基于 SHMT 的气体传感器展现出优异的乙醇气体传感性能,响应/恢复速度仅为 7 s/5 s,检测限低至 200 ppb。

Zhu 等[17]结合硬模板法、原子层沉积法和水热法,在 MEMS 基底上原位合成了单层有序 SnO_2 纳米碗支化 ZnO 纳米线的多级异质纳米材料,并以此作为气体传感器,对浓度低至 1 ppm 的硫化氢实现了高灵敏度和高选择性的探测(见图 4.9)。具体而言,此多级异质纳米材料在 250℃ 的工作温度下,对 1 ppm 硫化氢的响应(R_a/R_g)高达 6.24,其响应变化率(5.24)约为单层有序 SnO_2 纳米碗材料的

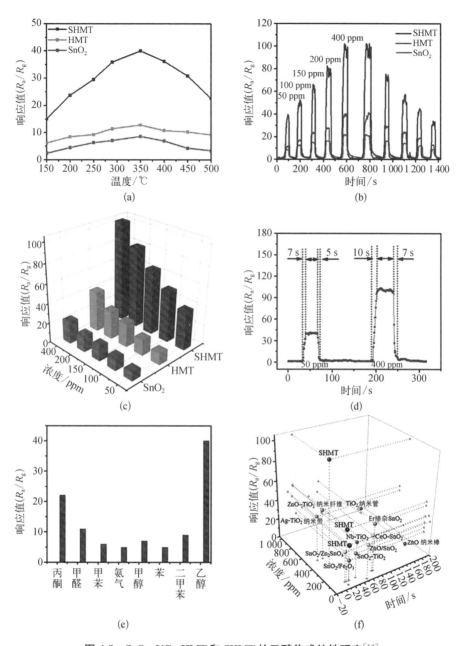

图 4.8 SnO_2 NCs、HMT 和 SHMT 的乙醇传感特性研究[16]

(a) HMT、SnO_2 NCs 和基于 SHMT 的传感器在不同工作温度(150~500℃)下对 50 ppm 乙醇的响应;(b) 基于 SnO_2 NCs、HMT 和 SHMT 的气体传感器在 350℃下对不同浓度乙醇(50~400 ppm)的响应曲线;(c) 基于 SnO_2 NC、HMT 及 SHMT 的传感器在不同浓度(50~400 ppm)下对乙醇蒸气的响应;(d) SHMT 对 50 ppm 和 400 ppm 乙醇的动态响应恢复曲线;(e) 基于 SHMT 的传感器对 50 ppm 下各种气体的响应;(f) 基于 SMO 传感器的响应和响应/恢复时间与此工作的比较

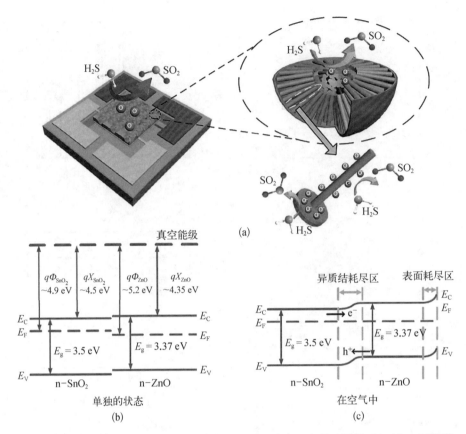

图 4.9 单层有序的分级 SnO_2@ZnO 纳米碗(NWs)在还原气体中的传感机制示意图[17]

(a) 单层有序的分级 SnO_2@ZnO NWs 暴露在 H_2S 气体中的三维示意图;(b)、(c) SnO_2@ZnO NWs 处于单独状态和处于空气中的能带示意图

2.6 倍,同时具有较快的响应/恢复速率。此外,对该多级异质纳米材料的气敏性能重复测量一个月,结果证实其具有较好的长期稳定性和可重复性。多级异质结构不仅有效增加了材料的比表面积,提升了材料的气体吸附能力,同时异质结还提高了材料的气敏响应能力。

研究表明,在空气氧吸附条件下调节异质结界面处的势垒高度是非常重要的。以上例子中,研究仍集中在还原性气体的检测上。空气环境中具有 n-n 异质结的复合材料的电阻大于纯电阻,这有利于在还原性气体存在下得到更大的电阻下降。然而,为了检测氧化性气体,在空气条件下产生较低的电阻是必要且有益的,这样才能获得更大的电阻增加。Sen 等[18]通过在 SnO_2 纳米线上生长 $W_{18}O_{49}$ 纳米线,成功地合成了具有刷状结构的复合材料,其表现出优异的氯气检测选择性。n 型 $W_{18}O_{49}$ 向 SnO_2 提供了电子,这增加了界面处产生的势垒。然而,$W_{18}O_{49}$ 对氧化性气体的灵敏度不高,SnO_2 仍然直接暴露在气体中。在这个体系中,异质结界面不是电荷传导路径，SnO_2 才是主要的电荷传导路径。因此，从 $W_{18}O_{49}$ 注入电子导致 SnO_2 在空气氛围下具有较低的电阻和对氯气的高响应。当引入还原性气体 H_2S 时，H_2S 与 $W_{18}O_{49}$ 的相互作用很弱，因为 $W_{18}O_{49}$ 纳米线表面已经处于缺乏吸附氧的状态，从而导致响应较低（见图 4.10）。

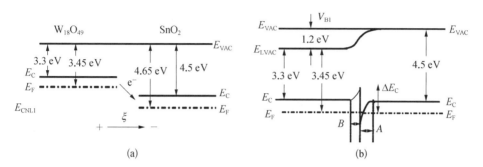

图 4.10　$W_{18}O_{49}$ 和 SnO_2 异质结传感机理[18]

(a) 分离的 $W_{18}O_{49}$ 和 SnO_2 材料的能带结构示意图；(b) 相应具有理想界面的异质结结构

4.1.3　p-p 异质结

除了 n-n 异质结之外，p-p 异质结是同种类型半导体的另一种复合方式。具有适当能带的 p 型半导体之间进行复合也有助于改善材料的气体传感性能。Suh 等[19]报道了通过多步掠射角沉积法合成的氧化镍修饰的氧化钴纳米棒（NRs）。在 NiO 和 Co_3O_4 的界面可以形成许多 p-p 异质结，这有助于提高对苯的灵敏度和选择性。图 4.11 是相应的能级结构图，显示了两种材料接触前后的能带弯曲。应当注意，Co_3O_4 和 NiO 的主要载流子是空穴，因为它们都是 p 型半导体，这是与上述 p-n 异质结和 n-n 异质结不同的关键点。Co_3O_4 的价带为 -6.30 eV，低于

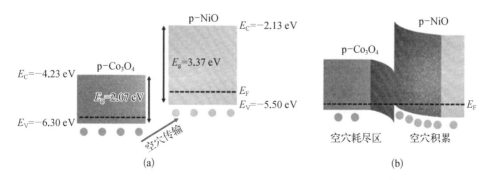

图 4.11　Co_3O_4/NiO 异质结的能带结构变化和电子转移过程[19]

(a)、(b) 在接触之前和之后的能带结构示意图

NiO 的价带（−5.50 eV）。因此，载流子从 Co_3O_4 转移到 NiO，并在 Co_3O_4 侧形成空穴耗尽层，在 NiO 处形成空穴积累层。与我们上面讨论的一维模型一样，该复合材料的主要有效载流子传导通道是 Co_3O_4 纳米棒。这项研究提出，如果电荷载体的量相同，则较低载流子浓度的材料在吸附气体分子时产生较大的电阻水平变化。p-p 异质结的形成可导致 Co_3O_4 中的空穴浓度较低，因此与纯 Co_3O_4 纳米棒相比，其具有更高的电阻水平和增强的气体传感行为。

4.2 贵金属修饰

在大多数气体传感过程中，电导率的强烈变化依赖于目标气体与传感层上吸附的氧物种之间的表面反应效率。因此，改善表面的催化反应是提高气体传感性能最有效的方法之一。由于贵金属纳米粒子是最常见和有效的催化剂，它们被广泛应用于提高基于金属氧化物的半导体传感器的气体传感性能。大量研究表明，Au[20-29]、Pt[30-36]、Ag[37-40]、Pd[41-51]、Rh[52]等贵金属可以显著提高气体传感特性，包括提高灵敏度和降低响应/恢复速度等。因为贵金属可以充当催化剂，从而促进表面反应进行。此外，负载贵金属还可以改善选择性，这是由于不同贵金属的特殊物理化学特性。例如，Au(111)晶面上吸附 CO 的结合能低，因此 Au 即使在低温下也是 CO 氧化的优良催化剂[4,53-54]；Pt 和 Pd 由于溢出效应而可以增强对 H_2 的选择性[55,5,42]。事实上，贵金属修饰增强材料对气体敏感性的机理非常复杂。贵金属的敏化作用大致可分为电子敏化和化学敏化两种类型。电子敏化是指通过调节载体电荷浓度来增强气体响应，而化学敏化是指通过溢出效应和催化效应来改善敏感材料对气体的响应。这两种敏化机制通常具有协同作用。

通常，贵金属的费米能级低于应用于气体传感器的金属氧化物半导体的费米能级。当贵金属与金属氧化物半导体接触时，电子从半导体转移到贵金属中，直到两种材料的费米能级平衡。结果，在贵金属和金属氧化物的界面处，贵金属侧有多余的电子，而半导体一侧则有多余的空穴，从而导致能带弯曲和肖特基结的形成。肖特基势垒可以有效地防止分离的电子-空穴对重新组合并增强气体传感测试过程中的响应。Li 等[56]报道了 Au 负载的 $WO_3 \cdot H_2O$ 纳米方块用于二甲苯气体传感器。由于不同的带隙和功函数，Au 和 $WO_3 \cdot H_2O$ 将在界面处结合在一起形成肖特基结。如图 4.12 所示，在负载了 Au 纳米颗粒之后，与纯的 $WO_3 \cdot H_2O$ 相比，Au 负载的 $WO_3 \cdot H_2O$ 表面区域形成了更厚的电子耗尽层，目标气体可以与更多吸附的氧物种反应。因此，更多被捕获的电子可以返回到 $WO_3 \cdot H_2O$ 的导带，导致更大的电阻变化即更高的灵敏度响应。此外，该研究提出量子阱效应也是气体传感过程中的重要因素。如图 4.13 所示，由 Au 和 $WO_3 \cdot H_2O$ 组成的纳米级复合结构传感材料产生的众多能带弯曲可以看作量子阱。当暴露在空气环境中时,电子被限制在量子阱中,导致电阻增加。当暴露于二甲苯时,由于从目标气体的反应和吸附的氧物种中释放出大量的自由电子,能带弯曲度和势垒减小。结果,受约束的电子可以轻松地越过势垒,这导致电阻的大幅下降。同样的效应也表现在其他贵金属/半导体金属氧化物体系中,如 Pt/WO$_3$[33]、Au/ZnO[20]等。

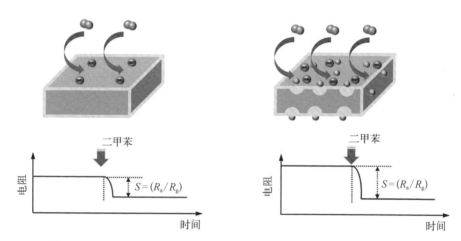

图 4.12 WO$_3$·H$_2$O 和 Au 负载的 WO$_3$·H$_2$O 的气体传感机理示意图[56]

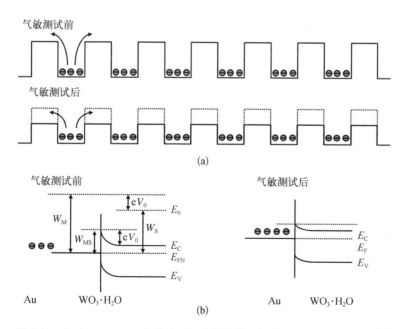

图 4.13 Au/WO$_3$·H$_2$O 在传感过程中的能带结构变化和电子转移过程[56]

(a) 气敏测试前后 Au 负载的 WO$_3$·H$_2$O 量子阱变化示意图;(b) 气敏测试前后功函数、能带能量和势垒变化

在上述情况下贵金属都是还原状态，还有一种情况是贵金属表面被部分氧化。虽然贵金属通常是化学稳定的，但贵金属纳米颗粒的表面在高温烧结过程中可以被部分氧化。贵金属氧化物通常是 p 型半导体，当部分氧化的贵金属纳米颗粒与 n 型金属氧化物接触时，p-n 异质结可以在界面形成，这与纯贵金属/金属氧化物半导体界面不同。Kim 团队[11,57-58]在贵金属功能化半导体金属氧化物一维纳米材料方面做了大量工作。在这些例子中，部分氧化的贵金属用于负载一维纳米材料并显著改善气体传感性能。以 Pd/PdO 修饰的 ZnO-WO$_3$ 纳米纤维为例，在这种情况下，与其他传感器相比，Pd@ZnO-WO$_3$ NF 具有最大的基线电阻（见图 4.14）。增加的基线电阻不仅是因为在 Pd 和 ZnO 之间形成肖特基势垒，还可因为 PdO 和 ZnO 之间形成 p-n 异质结。当引入还原性气体如甲苯时，部分 PdO 被还原成 Pd 并向 ZnO 传输电子，导致表面耗尽层的有效调制和电阻的降低，从而造成了最大的电阻变化，获得了最佳的灵敏度[11]。

溢出效应在贵金属修饰的化学敏化机制中起重要作用，可以描述如下：气体吸附在一相（这里指贵金属颗粒）上并离解；然后，离解的活性物种迁移到第二相（这里指的是作为载体的金属氧化物半导体）并被活化。在气体传感过程中，氧气和目标气体都可以吸附在界面上，这意味着它们都有可能在贵金属作用下发生溢出效应。

研究指出，由于溢出效应，贵金属改性可以显著增加吸附的氧物种的含量。氧分子首先吸附在传感层中的贵金属表面上并离解成氧原子，然后这些物质迁移到半导体氧化物的表面。已知氧分子通过从金属氧化物表面捕获电子而吸附在表面，导致形成耗尽层。因此，增加的吸附氧含量可以改善气体传感性能。Li 等[20]报道了基于负载 Au 纳米颗粒的 ZnO 纳米棒的快速和超高灵敏度的乙醇传感。图 4.15(a)显示了分子氧的催化离解过程。为了进一步证明负载 Au 的 ZnO 纳米棒的吸附氧含量高于纯 ZnO 纳米棒的吸附氧含量，该团队在氩气氛围中进行了解吸氧物种的实验，发现当传感器暴露于没有氧气的氩气时，吸附的氧气从传感器表面解吸，由氧分子捕获的电子被返回到 ZnO 的导带中，导致传感器的电阻大大降低。负载 Au 纳米颗粒的 ZnO 纳米棒传感器在空气和氩气中的电阻比高达 40，然而，纯 ZnO 的这一比例仅为 18 左右。这清楚地说明了负载 Au 纳米颗粒的 ZnO 的氧吸附大大增强。

一些目标气体可以通过溢出效应激活。目标气体分子可以吸附在贵金属表面生成活性物质，然后这些物质被吸附到金属氧化物的表面并与吸附的氧物种反应，促进了表面化学反应的进行，从而增加了传感器的灵敏度。以 Pd/WO$_3$ 为例[59]，氢气分子可以在贵金属表面上离解成活性氢物种，活性氢物种溢出并迁移到金属氧化物传感层的表面与吸附氧物种进行反应。H$_2$ 氧化反应速度可以显著提高，这意味着 Pd 可以降低反应活化能。同时，这也可以减少响应/恢复时间并降低传感器的工作温度。该过程的示意图如图 4.16 所示。得益于贵金属的化学敏化和电子敏化效应，研究者们在研究贵金属功能化修饰提升金属氧化物半导体材料的气敏性能上做出了相当多的努力。

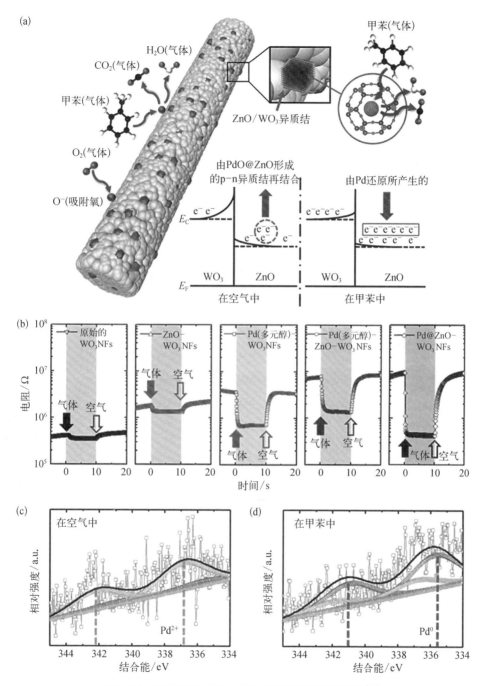

图 4.14 Pd@ZnO-WO₃ NFs 的甲苯传感特性研究[57]

(a) Pd@ZnO-WO₃ NFs的甲苯传感机理示意图；(b) 350℃下对 1 ppm 甲苯的动态电阻转变特性；(c)、(d) 在空气中和在 350℃下对 1 ppm 甲苯经过 5 次循环测试后 Pd@ZnO-WO₃ NFs 的非原位 XPS 分析 Pd 3d 谱图

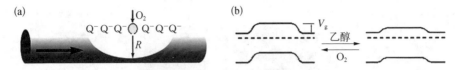

图 4.15　Au/ZnO 的乙醇传感机理[20]

(a) 负载在 ZnO 纳米棒上的 Au 纳米颗粒的化学敏化(溢出效应)示意图；(b) 在空气和乙醇蒸气中负载 Au 纳米颗粒的 ZnO 棒的能带图

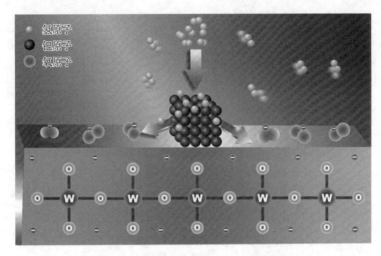

图 4.16　Pd/WO$_3$ 中溢出效应的示意图[59]

Deng 等[35]利用嵌段共聚物协同共组装技术，以自行设计的两亲性嵌段共聚物为结构导向剂，利用亲水端与亲水性钨源前驱体结合、疏水端与疏水性有机铂源结合，通过一步法直接共组装合成出一系列孔道高度连通、骨架高度晶化、铂纳米颗粒均匀负载的介孔 WO$_3$/Pt 复合材料，并首次将该材料用于高性能气体传感器的构建，实现了快速灵敏地监测一氧化碳气体（见图 4.17）。所合成的材料具有大的比表面积（112～128 m^2/g）、较大的孔径（13 nm）以及均匀分散的铂纳米颗粒（4 nm）。其独特的结构和组分使其在较低的工作温度下（125℃）对 100 ppm 的 CO 具有高灵敏度响应（R_a/R_g=10），超快的响应/恢复时间（16 s/1 s）以及高度的选择性，明显优于无孔材料以及商业化的可燃气体传感器。在对其机理的研究中，研究团队发现，该合成方法制备的 WO$_3$/Pt 材料中，一方面，Pt 与 WO$_3$ 载体存在的金属-载体强相互作用可以增强 Pt 的催化性能，同时 Pt 的引入可以明显增加载体 WO$_3$ 的缺陷，有利于表面吸附氧的增加。另一方面，利用 XPS 表征可以发现，在工作温度下，空气氛围中的 Pt 纳米颗粒存在少量氧化态，能在材料中形成 p-n 异质结，而其在一氧化碳气氛中却被还原成还原态的 Pt，从而显著地降低电阻，极大地提高灵敏度和选择性。这为优化设计贵金属负载半导体气体传感器提供了新思路。

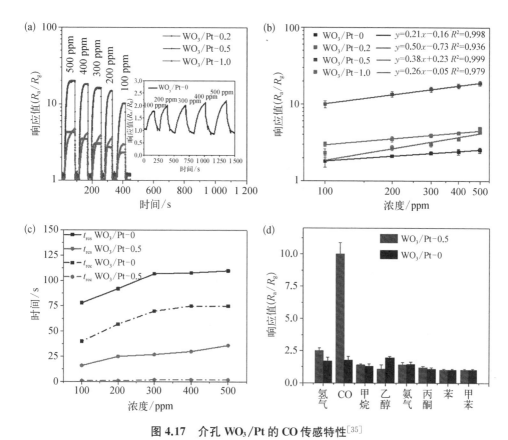

图 4.17 介孔 WO_3/Pt 的 CO 传感特性[35]

(a) 介孔 $WO_3/Pt-n$(n=0、0.2、0.5、1.0)在125℃和57%相对湿度下对不同浓度(100 ppm、200 ppm、300 ppm、400 ppm、500 ppm)CO的动态响应-恢复特性曲线;(b) CO浓度与所获得的介孔 $WO_3/Pt-n$ 的响应的关系;
(c) 在125℃、57%相对湿度下不同浓度(100 ppm、200 ppm、300 ppm、400 ppm、500 ppm)的 $WO_3/Pt-0.5$ 和 $WO_3/Pt-0$ 对CO的响应和恢复时间;(d) $WO_3/Pt-0.5$ 和 $WO_3/Pt-0$ 对100 ppm不同气体的响应

Ma 等[60]基于溶剂挥发诱导共组装策略,通过加入巯基硅烷偶联剂增强氯铂酸、氯金酸等贵金属前驱体与过渡金属盐、两亲性嵌段共聚物的相互作用,成功地以一锅法直接合成了一系列贵金属纳米颗粒修饰的过渡金属氧化物材料,包括 Au/WO_3、Au/TiO_2、Au/NbO_x 以及 Pt/WO_3 等体系(见图4.18)。这一系列复合材料均具有高度分散的小尺寸贵金属纳米颗粒修饰、大的比表面积,以及有序且互相贯通的孔道。以 Au/WO_3 作为典型研究体系,研究结果揭示了巯基硅烷偶联剂在组装过程中增强了前驱体间的相互作用,同时原位生成的 SiO_2 不仅显著提高了材料的热稳定性,也诱导生成了氧化钨的异相掺杂,极大地提高了材料的气体传感性能。得益于高度有序的介孔结构以及独特的组分,所合成的 Au/WO_3 材料在低的工作温度下(200℃)对50 ppm乙醇蒸气的响应灵敏度(R_a/R_g)达36,并且能对低至50 ppb的乙醇气体进行响应。

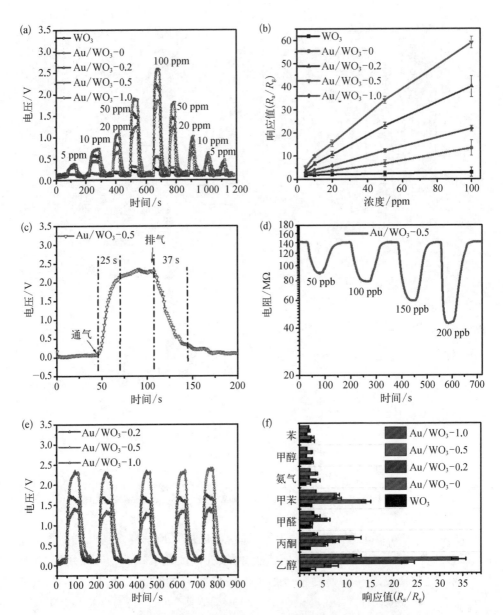

图 4.18 Au/WO₃ 的乙醇传感特性研究[60]

(a) WO₃ 和 Au/WO₃-n(n=0、0.2、0.5、1.0)对不同浓度(5 ppm、10 ppm、20 ppm、50 ppm 和 100 ppm)乙醇气体的动态响应-恢复特性曲线;(b) 乙醇气体浓度与 WO₃ 和 Au/WO₃-n 响应之间的关系;(c) Au/WO₃-0.5 向 50 ppm 乙醇气体的动态电压转变特性;(d) Au/WO₃-0.5 向低浓度(50~200 ppb)乙醇气体的动态电阻转变特性曲线;(e) Au/WO₃-n 对 50 ppm 乙醇气体的重复响应和回收曲线;(f) 五个样品对浓度为 50 ppm 的不同目标气体的气体响应

Ma 等[61]提出了一种结合孔道工程策略的多组分共组装方法,使用双功能桥联分子(3-巯基丙基)三甲氧基硅烷(MPTMS),直接合成出负载有高度分散的 Au 纳米颗粒

(5 nm)的、高热稳定性的有序介孔 Au/SiO_2-WO_3 复合材料(见图 4.19)。在此基础上，通过调控体系中的 SiO_2 含量并进行可控刻蚀，获得了孔道联通性显著提升的复合材料，证明了改善孔道联通性可以提升气体传感性能。该复合材料在较低的工作温度(150℃)下对痕量乙醇蒸气具有很高的灵敏度(对 50～250 ppb 的 R_a/R_g=2～14)。

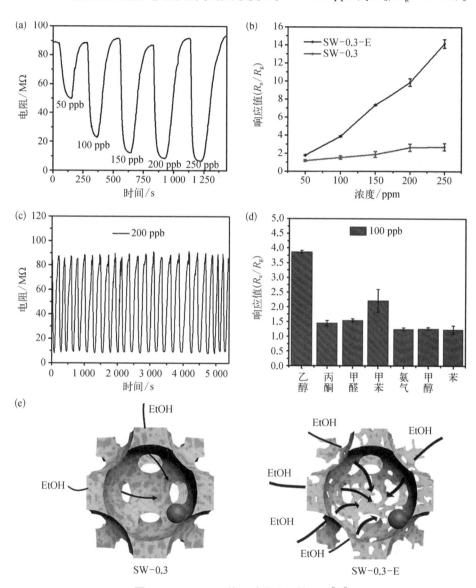

图 4.19　SW-0.3 的乙醇传感特性研究[61]

(a) SW-0.3-E 对不同浓度(50 ppb、100 ppb、150 ppb、200 ppb、250 ppb)乙醇气体的动态响应；(b) 乙醇气体浓度与 SW-0.3 和 SW-0.3-E 响应之间的关系；(c) 基于 SW-0.3-E 的传感器对 200 ppb 乙醇气体的动态电阻响应的 24 个循环；(d) SW-0.3-E 对浓度为 100 ppb 的不同目标气体的响应；(e) SW-0.3 和 SW-0.3-E 的气敏过程示意图

Ma 等[62]合成了有序双介孔 WO_3/Au 复合材料,其对单核细胞增生李斯特菌的代谢气体 3-羟基-2-丁酮展现出优异的传感性能,在较低的工作温度(175℃)下对 2.5 ppm 的 3-羟基-2-丁酮的灵敏度(R_a/R_g)高达 18.8,可检测低至 50 ppb 的气体,并且具有优异的长期稳定性、选择性和抗干扰能力(见图 4.20)。DFT 计算以及原位表征实验表明 3-羟基-2-丁酮分子容易吸附在 WO_3/Au 材料表面,并被逐步氧化。由于其出色的传感性能,将该气体传感器应用于可通过蓝牙连接

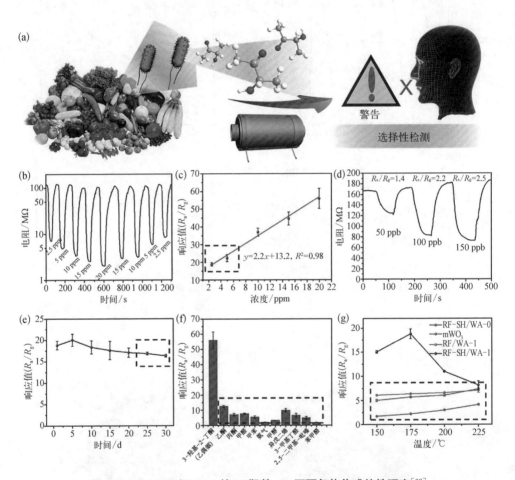

图 4.20 RF-SH/WA-1 的 3-羟基-2-丁酮气体传感特性研究[62]

(a) 用于选择性检测单核细胞增生李斯特菌生物标志物 3-羟基-2-丁酮(乙偶姻)的基于半导体金属氧化物的气体传感器的示意图;(b) 暴露于不同浓度(2.5 ppm、5 ppm、10 ppm、15 ppm、20 ppm)的 3-羟基-2-丁酮气体中的基于 RF-SH/WA-1 的传感器的动态响应-恢复特性曲线;(c) 3-羟基-2-丁酮气体浓度与基于 RF-SH/WA-1 的传感器响应之间的关系;(d) 暴露于低浓度(50~150 ppb)3-羟基-2-丁酮气体的基于 RF-SH/WA-1 的传感器的动态电阻转换特性曲线;(e) 基于 RF-SH/WA-1 的传感器的长期稳定性(对 2.5 ppm 3-羟基-2-丁酮气体的响应);(f) 基于 RF-SH/WA-1 的传感器暴露于浓度为 20 ppm 的不同目标气体下的气体响应;(g) 各种基于 WO_3 的纳米多孔传感器暴露于 2.5 ppm 3-羟基-2-丁酮气体的温度响应曲线

至手机的无线传感设备中,可在 1 分钟内快速、准确、实时读取 3-羟基-2-丁酮的浓度,有望解决 3-羟基-2-丁酮检测需要昂贵复杂设备和专业技术人员的难题,在食品安全中具有潜在应用前景。

4.3 元素掺杂

4.3.1 非金属元素掺杂

掺杂,即添加特定的外在元素,已被证明可以改善半导体金属氧化物的气敏性能。掺杂不仅可以影响材料的电子特性,而且能影响晶粒的结构特性,如尺寸和形状等。因此,掺杂可以通过影响电子特性和结构特性进一步影响材料的化学传感性能。

非金属元素掺杂剂,如 Si[63-66]、C[67]、N[68],毒性较低,价格相对便宜并且分布广泛。已经有一些研究表明,非金属元素掺杂可以提高气敏性能。非金属元素掺杂会影响晶粒的尺寸、晶相等,进一步影响传感性能。Pratsinis 团队[66]成功合成了 Si 掺杂的 WO_3,其具有超低的丙酮浓度检测限(低至 20 ppb)、良好的热稳定性,在理想环境(干燥空气)和实际环境(相对湿度高达 90%)条件下都具有非常高的信噪比。这种便携式丙酮分析仪可以通过人类呼出的气体方便快捷地诊断糖尿病,有希望作为常规血液分析的替代,以监测血糖,从而用于预防和减轻肥胖症和糖尿病。在 500℃下,Si 掺杂使 ε-WO_3 含量稳定为 87%。ε-WO_3 的自发电偶极矩导致其与具有高偶极矩的分析物(如丙酮)有强烈的相互作用,因此增强了其对丙酮的响应。此外,Si 掺杂可以显著抑制晶粒和晶体生长,如图 4.21(b)所示。适当的晶粒尺寸可以导致高的表面积和提高客体气体的扩散速度,有助于改善气体传感性能。

Xiao 等[67]以棉纤维为模板和碳源,通过分解过氧多钨酸,合成了 C 掺杂的 WO_3 多晶材料。基于 C 掺杂 WO_3 材料的传感器对丙酮气体具有高响应性、高选择性、快速的响应和恢复时间、低检测限(低至 0.2 ppm),以及在 300℃的工作温度下浓度和灵敏度优异的线性关系。结果表明,合成的 C 掺杂 WO_3 材料可以为糖尿病的无创诊断提供有效的替代方案。该项研究指出粒度发挥着重要作用。众所周知,当金属氧化物的晶粒尺寸(D)接近或小于空间电荷层的两倍厚度(德拜长度 2δ)时,传感器的灵敏度将增加。C 掺杂可以减小晶粒尺寸,使 C 掺杂 WO_3 的晶粒尺寸小于 2δ(WO_3 在 300℃下的德拜长度为 60 nm),从而获得较好的灵敏度。

最近,Zhang 等[68]使用两亲性二嵌段共聚物聚苯乙烯-b-聚(4-乙烯基吡啶)(PS-b-P4VP)作为结构导向剂,通过简单的溶剂蒸发诱导聚集组装(EIAA)方法合成了原位 N 掺杂有序介孔二氧化钛。来自 PS-b-P4VP 的氮原子可以原位掺入 TiO_2 的晶格中,导致更小的微晶尺寸和更多的材料缺陷,从而提高了气体的传感性能。N 掺杂 TiO_2 基传感器对 250 ppm 丙酮表现出高灵敏度($R_a/R_g=17.6$),

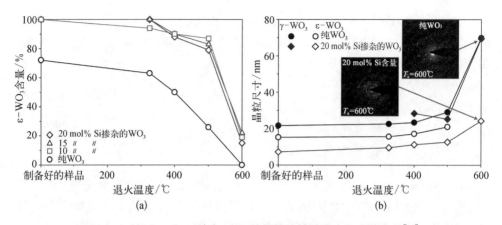

图 4.21　纯 WO_3 和 Si 掺杂 WO_3 的结构特性随退火温度的变化[66]

(a) 纯 WO_3 和 Si 掺杂的 WO_3 中 ε-WO_3 相含量在空气中不同温度下退火 5 h 的变化；(b) 纯 WO_3（圆）和 20 mol% Si 掺杂的 WO_3（菱形）的晶体尺寸随退火温度的变化，插图为在 600℃ 退火 5 h 后纯 WO_3 和 20 mol% Si 掺杂的 WO_3 的电子衍射谱图

是纯 TiO_2 基传感器的 11.4 倍（250 ppm 丙酮气体灵敏度为 1.54），且具有较短的响应和恢复时间（分别为 4 s 和 4 s）。

Wang 等[69]合成了三维有序的大孔/介孔（3DOM）C 掺杂 WO_3 材料，并对其进行了广泛的表征（见图 4.22），研究了 3DOM C 掺杂 WO_3 材料的气敏性能，优化了煅烧温度和工作温度，特别是确定了 3DOM 材料的孔径如何影响气敏性能。结果表明，孔径为 410 nm 的材料对丙酮的响应最高。当暴露于 NH_3、CO、甲醇、乙醇和甲醛时，仅观察到轻微的响应，这意味着该传感器具有极好的选择性。此外，该传感器具有良好的重复性和长期稳定性。更吸引人的是，他们发现该传感器对

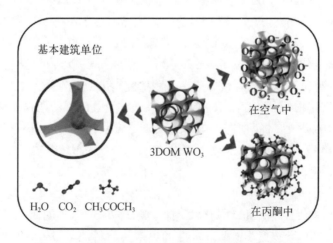

图 4.22　三维 C 掺杂 WO_3 传感器的丙酮传感机理[69]

丙酮的响应与 3DOM C 掺杂 WO₃ 材料的孔径有关。故认为，3DOM 结构中基本建筑单元的大小在丙酮响应中起着关键作用。

Zhou 等[70]设计了一系列具有不同二氧化硅含量的有序介孔氧化锌（SiO_2 - ZnO）复合材料，采用原位气相色谱仪-质谱（GC - MS）分析结合智能重量分析仪测量作为重要研究手段，系统揭示了二氧化硅的引入对材料选择性的影响及潜在的

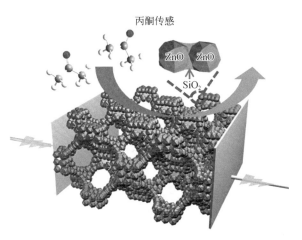

图 4.23 二氧化硅掺杂介孔氧化锌材料的丙酮传感示意图[70]

作用机制（见图 4.23）。研究发现，由于二氧化硅的掺入提高了材料表面的极性，特异性地增强了材料对丙酮气体的吸附能力，从而有效提高了 $mZnO - 2SiO_2$ 对丙酮的灵敏度和传感选择性。

4.3.2 金属元素掺杂

金属元素掺杂已被证明是提高半导体金属氧化物传感器气敏性能的有效方法，因为掺杂可以有效地调节晶胞参数和氧的吸附。由于一些特定金属元素的半径接近于半导体金属氧化物（W、Sn、In、Zn、Ni 等）的半径，从而很容易在半导体金属氧化物的晶格中进行取代，因此，金属元素的掺杂可以显著影响材料中的缺陷或载流子浓度，进一步影响传感性能。金属元素如 Fe[71-75]、Cr[76-79]、Cu[80-82]、Co[83-85]、Ti[86-87]、Mo[69,88-89]等已被报道用作增强气体传感性能的掺杂剂。

Wang 等[39]报道了具有分级结构的 Cr 掺杂 WO₃ 微球用于低温 H₂S 检测。在 80℃的工作温度下，基于 Cr 掺杂 WO₃ 微球的传感器对 0.1 vol% H₂S 的响应是基于纯 WO₃ 的传感器响应的 6 倍。由于 Cr^{3+} 的有效离子半径为 61.5 pm，与 W^{6+}（60 pm）相似，因此 Cr^{3+} 可以作为取代杂原子掺杂到 WO₃ 晶格中，从而增加氧空位的数量。该过程可以表示如下：

$$Cr_2O_3 \xrightarrow{WO_3} 2Cr'''_W + 3O^X_O + 2V^{\cdot\cdot}_O \tag{4.1}$$

众所周知，氧空位的增加可以增加氧的吸附，从而提高气敏性能。此外，氧化铬和 H₂S 之间存在强烈的键合作用，有助于增加材料对目标气体的吸附，进而提高气敏性能。

金属原子掺杂的电子敏化效应也可以在 p 型氧化物半导体中发生。对于 p 型氧化物半导体，几乎没有研究通过控制空穴密度来增强气体响应。经目标还原性气体与 p 型半导体金属氧化物表面吸附的带负电氧物种之间的电子-空穴复合反应，可以降低近表面空穴浓度，从而增加传感器的电阻。因此，可以预期在空气中检测到的较高空穴浓度将导致在气体传感过程期间载流子密度的较小变化，而较低空穴浓度将更有利于增强气体响应。Yoon 等[75]通过静电纺丝的方法制备了 Fe 掺杂的 NiO 纳米纤维。Fe 掺杂的 NiO 纳米纤维的响应和选择性显著提高，而纯 NiO 纳米纤维对所有分析物气体的响应都非常低。特别地，3.04 at% Fe 掺杂的 NiO 纳米纤维对 100 ppm C_2H_5OH 的响应增加至 217.86 倍。该研究的结果表明，气体传感响应取决于载流子密度的变化而不是粒径的变化，在 Ni^{2+} 位置取代 Fe^{3+} 可以通过产生电子或 Ni 空位来补偿，如式(4.2)~式(4.3)所示：

$$Fe_2O_3 \xrightarrow{NiO} Fe^{\cdot}_{Ni} + 2O^X_O + \frac{1}{2}O_2(g) + 2e^- \quad (4.2)$$

$$Fe_2O_3 \xrightarrow{3NiO} 2Fe^{\cdot}_{Ni} + 3O^X_O + V''_{Ni} \quad (4.3)$$

如图 4.24 所示，通过电子电荷补偿，Fe 掺杂的 NiO 传感器的电阻将增加，因为电子的产生将通过电子-空穴复合反应过程降低空穴浓度。当以形成的 Ni 空位为主要的电荷补偿机制时，空穴浓度不会发生很大变化。然而，在靠近 Ni 空位的相反电荷位置处的空穴也可以增加传感器电阻。因此，Fe 掺杂可以使传感器电阻显著增加并进一步改善气体传感响应。

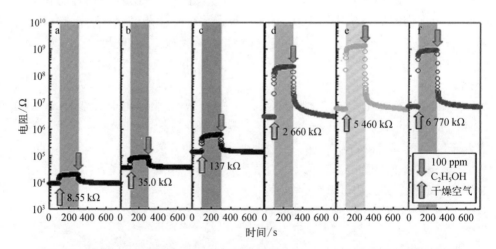

图 4.24 六种不同传感器在 475℃ 工作温度下对 100 ppm C_2H_5OH 的动态响应[75]

(a) 纯 NiO 纳米纤维；(b) 0.18 at% Fe 掺杂 NiO 纳米纤维；(c) 0.79 at% Fe 掺杂 NiO 纳米纤维；(d) 1.02 at% Fe 掺杂的 NiO 纳米纤维；(e) 3.04 at% Fe 掺杂的 NiO 纳米纤维；(f) 13.2 at% Fe 掺杂的 NiO 纳米纤维

4.3.3 稀土元素掺杂

稀土金属(rare earth element, RE, RE=La、Ce、Pr、Nd、Sm、Eu、Gd、Dy、Ho、Er 和 Yb,拥有 4f 壳层电子)是另一种可有效增强金属氧化物半导体气敏性能的掺杂元素。稀土元素是相对特殊的掺杂剂,其通过改变半导体金属氧化物的晶体结构或晶粒尺寸、形态、表面碱度、缺陷和氧空穴浓度,以及整个材料的电子结构来提高其他金属氧化物对气体传感的选择性和灵敏度。

大量研究表明,RE 元素掺杂可以有效地抑制晶粒的生长,减小晶粒尺寸。较小尺寸的晶粒或畴通常会导致较高的表面积和较大的有效尺寸暴露,从而增强气体传感性能。图 4.25 是 La 掺杂的 WO_3 纳米纤维的 FESEM 图,其中图 4.25(a)~图 4.25(d)分别代表 0 mol%、1 mol%、3 mol% 和 5 mol% La 掺杂量。随着掺杂量的增加,图 4.25(c)中构成骨架的晶粒尺寸明显大于图 4.25(b)中的晶粒尺寸,进一步提高掺杂量的图 4.25(d)的晶界已经很难观察到[90]。Mohanapriya 等[94]报道了通过静电纺丝合成的 Ce 掺杂的 SnO_2 中空纳米纤维。Ce 掺杂样品(3 mol% Ce 掺杂、6 mol% Ce 掺杂和 9 mol% Ce 掺杂)的晶粒尺寸均小于未掺杂样品的晶粒尺寸,如表 4.1 所示[91]。Wei 等[92]还观察到 Ce 掺杂 In_2O_3 中的类似现象,提出 Ce^{4+} 容

图 4.25 不同 La 掺杂量的 WO_3 的 FESEM 图[90]

(a) 0 mol% La 掺杂 WO_3;(b) 1 mol% La 掺杂 WO_3;(c) 3 mol% La 掺杂 WO_3;
(d) 5 mol% La 掺杂 WO_3

易位于晶界处,由于 Ce^{4+} 和 In^{3+} 之间的离子价态和离子半径不同,从而限制了晶粒的生长。

表 4.1 未掺杂和 Ce 掺杂的 SnO_2 中空纳米纤维的纤维直径、晶粒尺寸和比表面积[90]

样 品	直径/nm	晶粒尺寸/nm	比表面积/(m^2/g)
未掺杂	234 ± 34	41 ± 16	16.7
3 mol% Ce 掺杂	110 ± 15	22 ± 10	35.2
6 mol% Ce 掺杂	127 ± 15	27 ± 10	34.7
9 mol% Ce 掺杂	158 ± 22	27 ± 10	38.5

RE 元素掺杂会增加有利于吸附氧的材料的缺陷。据报道,当 Ce 原子取代 MoO_3 纳米带中的 Mo 原子时,可以促进氧空位的形成[93]。形成的具有双正电荷的氧空位可以用 Kröger - Vink 表示法表示:

$$CeO_2 \xrightarrow{MoO_3} Ce''_{Mo} + 2O_O + V_O^{\cdot\cdot} \tag{4.4}$$

研究提出氧空位的增加对氧气吸附很重要,从而提高了气体传感性能,如图 4.26 所示。

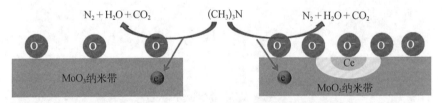

图 4.26 纯 MoO_3 和 Ce 掺杂 MoO_3 纳米带表面传感过程的吸附及反应模型[93]

有些时候,RE 元素掺杂还可以改变表面碱度。众所周知,还原性气体的氧化是通过脱氢或脱水过程进行的。具有高碱度的氧化物表面有利于脱氢过程进行,而酸性表面则有利于催化脱水进行。在应用于气体传感领域的半导体金属氧化物中,碱性氧化物占绝大多数。在半导体金属氧化物中的 RE 掺杂剂可以产生更多的碱性位点,这有利于通过脱氢途径实现还原性气体的氧化反应。

此外,RE 元素掺杂还可以通过改变载流子密度和原子取代来调节材料的电学性质。Mohanapriya 等[91]报道了 Ce 掺杂 SnO_2 中的载流子浓度可以通过用 Ce^{3+} 取代 Sn^{4+} 来降低,从而导致传感器的电阻增大。取代 Sn 位点的 Ce^{3+} 捕获 SnO_2 晶格中的自由电子,可由式(4.5)表示:

$$Ce_2O_3 + 2e' \rightarrow 2Ce'_{Mo} + 3O_O^X + V_O^X \tag{4.5}$$

实际上，Ce 离子具有 Ce^{4+} 和 Ce^{3+} 混合价态，因此 Ce 掺杂对电阻的影响是复杂的。Han 等[94]报道，3% Ce 掺杂可以降低 In_2O_3 多孔纳米球的电阻。研究提出，当 Ce^{4+} 掺入 In_2O_3 晶格时，它将取代 In^{3+} 位点，形成电子给体缺陷（$Ce_{In}\cdot$）。同时，为了保持电中性，Ce 离子将电子释放到导带中，这增加了自由电子的浓度。但是，Wei 等[95]则报道 Ce 掺杂可以增加 In_2O_3 花状微球分级结构的电阻，研究提出：① 一个 Ce^{4+} 可以捕获一个电子并转换成 Ce^{3+}，这会降低电子浓度；② 半径较大的 Ce^{4+} 掺杂原子可能会影响能量结构。这两种效应可能同时发生，其竞争结果将导致不同体系的电阻变化不同。

Liu 等[95]通过原位协同组装方法结合碳支撑结晶策略，合成了具有 $59\sim 72\ m^2/g$ 的高比表面积、稳定的晶体框架和精细剪裁的孔壁的新型介孔 Ce 掺杂 WO_3（见图 4.27）。Ce 原子在介孔 WO_3 孔壁中的掺杂可以有效地调节 W 原子的配位环境，使氧空位（O_v）显著增多，形成 $W^{\delta+}-O_v$ 位。因此，获得的介孔 Ce 掺杂 WO_3 在低工作温度（150℃）下表现出优异的 H_2S 传感性能，具有超高响应值（381 vs 50 ppm）、快速响应动力学（6 s）、优异的选择性和抗湿性能以及良好的长期稳定

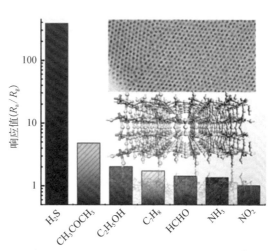

图 4.27 介孔 Ce 掺杂 WO_3 材料的纳米和化学结构以及 H_2S 传感性能[95]

性。优越的气敏性能归因于在敏感层的表面吸附催化反应期间，O_v 密度增加，表面吸附的 H_2S 转化为 SO_x 和 SO_4^{2-} 的能力增强。密度泛函理论（DFT）计算表明，Ce^{4+} 嵌入 WO_3 的晶格中以形成最稳定结构，而不是原子取代，并且 Ce 掺杂的 WO_3 显示出比纯 WO_3 更高的 H_2S 吸附能和更大的电荷转移量，说明 Ce^{4+} 掺杂的 WO_3 具有更好的 H_2S 传感响应。此外，Liu 等基于 Ce 掺杂的介孔 WO_3 开发了一种新型气体传感模块和智能便携式传感器装置，用于通过蓝牙通信在智能手机上高效、实时监测 H_2S 浓度。

4.4 与碳材料（石墨烯、碳纳米管）复合

上述讨论的都是最经典的通过组分的改变来影响金属氧化物半导体气体传感性能的方法。实际上，新材料层出不穷，半导体金属氧化物基的气体传感器组分也

在不断地与时俱进。近年来，碳纳米管、石墨烯等新型碳材料以其独特的化学稳定性、电学特性、机械强度引起了研究者们的极大兴趣。将这些材料具备的独特电学性质与半导体金属氧化物的气敏传感特性相结合，有可能开发出具有独特功能的新型气敏材料。本节通过文献报道讨论金属氧化物/石墨烯复合材料、金属氧化物/碳纳米管复合材料等的组分与气敏性能的关系。

自 2004 年英国曼彻斯特大学的科学家安德烈·盖姆和康斯坦丁·诺沃肖洛夫发现石墨烯以来，石墨烯就成为了许多研究领域的热点。石墨烯被认为是具有单原子层厚度的理想二维材料，其具有优异的导电性、高的比表面积，以及良好的机械强度，可以负载大量的金属氧化物纳米粒子，并为材料提供更多的电子传输通道，从而提高材料的导电性。同时，石墨烯巨大的表面可以抑制金属氧化物的迁移和团聚，保持纳米颗粒的高度分散。Liu 等[96]通过简单的溶剂热法制备了三维石墨烯气凝胶-ZnO 复合材料（ZnO/GAs）。以该材料制备的气体传感器在室温条件下即可对 10～200 ppm 的 NO_2 气体进行响应，具有良好的稳定性和选择性。同时，通过对比实验发现，简单地混合 ZnO 和石墨烯所得到的样品并不能拥有和溶剂热法合成的 ZnO/GAs 一样的气体传感性能（见图 4.28），说明了在 ZnO/GAs 材料中，ZnO 与石墨烯形成了异质结。石墨烯本身具有类似金属的导电性，当其与金属氧化物接触时，半导体的电子容易转移到石墨烯中，形成肖特基势垒，从而有利于增强传感性能。Zhang 等[97]利用晶种诱导生长的方法，在二维的石墨烯片上通过溶剂热法生长了一维的 SnO_2 纳米棒。该材料在 260℃的工作温度下对硫化氢气体的灵敏度比纯的 SnO_2 材料高出两倍，检测限低至 1 ppm。该研究认为性能提升的主要原因是二维石墨烯提供的基底大大提高了阵列的比表面积以及促进了石墨烯与金属氧化物之间异质结的形成。

碳纳米管（carbon nanotube，CNT）是另一种新型的碳材料。CNT 可以看作石墨烯卷曲一定角度形成的管状圆柱体，具有较高的长径比。根据其壁的层数，碳纳米管可以分为多壁碳纳米管（MWCNT）和单壁碳纳米管（SWCNT）。壁的层数不同，其物理特性也具有一定的差异。与石墨烯类似，CNT 具有较高的导电性、稳定性和机械强度。碳纳米管虽然和石墨烯同为碳的同素异形体，但性质却有所不同。不同于石墨烯的是，不同碳纳米管的电学性质不同，有的是金属型，有的则为半导体型。使用半导体型的 CNT 与金属氧化物进行复合，可以形成类似 p-n 结的异质结，从而提高气体传感性能。Dai 等[98]在多壁碳纳米管（MWCNT）外均匀生长了 α-Fe_2O_3，形成了 MWCNT@α-Fe_2O_3 复合材料。该材料在较低的工作温度（225℃）下展现出对丙酮气体较高的灵敏度（35 vs. 100 ppm 丙酮）、较快的响应/恢复时间（2 s/35 s）以及良好的长期稳定性，同时具有较低的检测限（0.5 ppm）。如图 4.29 所示，MWCNT 的能带结构类似于 p 型半导体，当与 n 型的金属氧化物复合时，界面处会形成 p-n 结势垒，从而增加了材料的电阻，引起在还原性气体丙

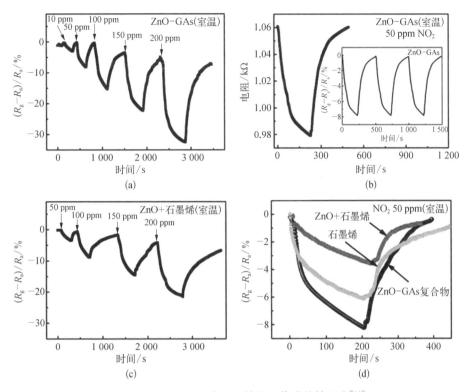

图 4.28 ZnO/GAs 对 NO₂ 的室温传感特性研究[96]

(a) 基于 ZnO/GAs 的传感器在室温下对不同浓度的 NO₂ 气体的响应;(b) ZnO/GAs 在室温下对 50 ppm NO₂ 气体的响应和周期性测试;(c) 基于 3D 石墨烯+ZnO 的简单混合物的传感器在室温下对不同浓度的 NO₂ 气体的响应;(d) ZnO/GAs 复合材料、3D 石墨烯+ZnO 的简单混合物和 3D 石墨烯在室温下对 50 ppm NO₂ 的响应的比较

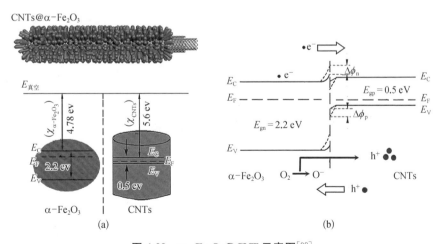

图 4.29 α-Fe₂O₃@CNT 示意图[98]

(a) α-Fe₂O₃@CNT 在空气中的能带图;(b) CNT@α-Fe₂O₃ 复合结构由空气到丙酮气体中的能带变化

酮中较大的电阻变化。Asad 等[99]合成的 CuO-SWCNT(单壁碳纳米管)复合材料在室温下即可对 100 ppb 的 H_2S 气体响应。

Zhao 等[100]针对金属氧化物材料的最佳气体传感工作温度通常较高(>250℃)的问题,将石墨烯与介孔 WO_3 材料进行复合,通过水热处理将片层的石墨烯材料组装成为具有气凝胶的宏观形貌,并以之为辅助模板,以 PEO-b-PS 为"软"模板,控制胶束的生成和在石墨烯表面的有序组装,实现了介孔氧化钨在石墨烯片层表面的均匀生长,介孔氧化钨@石墨烯气凝胶(mWO_3@GA)材料的比表面积高达 167 m^2/g,孔径尺寸为 20 nm。这种方法可以有效地抑制因石墨烯片层的团聚而导致的负载不均匀,具有一定的普适性,可以拓展至介孔 TiO_2/石墨烯和二氧化锡纳米晶/石墨烯复合材料的合成。所得到的 mWO_3@GA 材料由于具有独特的复合结构和组分间的协同效应,在较低的温度(150℃)下,展现出对丙酮气体的优异传感性能,对 50 ppm 的丙酮具有快速的响应(13 s)和恢复(65 s)时间、高的灵敏度以及良好的稳定性和气体选择性,远优于单一组分的材料(见图 4.30)。

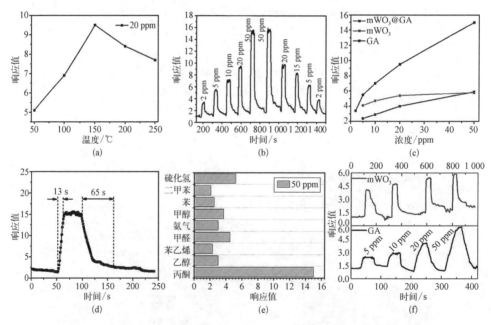

图 4.30 基于 mWO_3@GA 纳米复合材料的传感器对丙酮的传感特性研究[100]

(a) 在不同温度下,mWO_3@GA 对 20 ppm 丙酮的响应;(b) 150℃时,mWO_3@GA 对不同浓度(2~50 ppm)丙酮的响应恢复曲线;(c) 介孔 WO_3、GA 及 mWO_3@GA 传感器在 150℃下不同浓度(2~50 ppm)的丙酮的响应;(d) mWO_3@GA 传感器在 150℃下对 50 ppm 丙酮的响应恢复曲线;(e) 150℃时 mWO_3@GA 对 50 ppm 的不同气体的响应;(f) 介孔 WO_3 传感器和 GA 传感器在 150℃下对不同浓度(2~50 ppm)的丙酮的响应恢复曲线

参考文献

[1] Wang C, Yin L, Zhang L, et al. Metal oxide gas sensors: sensitivity and influencing factors[J]. Sensors, 2010, 10: 2088 - 2106.

[2] Miller D R, Akbar S A, Morris P A. Nanoscale metal oxide-based heterojunctions for gas sensing: A review[J]. Sens. Actuators B, 2014, 204: 250 - 272.

[3] Rai P, Majhi S M, Yu Y, et al. Noble metal@metal oxide semiconductor core@shell nano-architectures as a new platform for gas sensor applications[J]. RSC. Adv. 2015, 5: 76229 - 76248.

[4] Woo H S, Na C W, Lee J H. Design of highly selective gas sensors via physicochemical modification of oxide nanowires: overview[J]. Sensors, 2016, 16: 1531.

[5] Luo Y, Zhang C, Zheng B, et al. Hydrogen sensors based on noble metal doped metal-oxide semiconductor: A review[J]. Int. J. Hydrog. Energy, 2017, 42: 20386 - 20397.

[6] Ju D, Xu H, Xu Q, et al. High triethylamine-sensing properties of NiO/SnO_2 hollow sphere P - N heterojunction sensors[J]. Sens. Actuators B, 2015, 215: 39 - 44.

[7] Na C W, Woo H S, Kim I D, et al. Selective detection of NO_2 and C_2H_5OH using a Co_3O_4 - decorated ZnO nanowire network sensor[J]. Chem. Commun. 2011, 47: 5148 - 5150.

[8] Xie Y, Xing R, Li Q, et al. Three-dimensional ordered ZnO - CuO inverse opals toward low concentration acetone detection for exhaled breath sensing[J]. Sens. Actuators B, 2015, 211: 255 - 262.

[9] Xiao X, Zhou X, Ma J, et al. Rational synthesis and gas sensing performance of ordered mesoporous semiconducting WO_3/NiO composites[J]. ACS Appl. Mater. Interfaces, 2019, 11: 26268 - 26276.

[10] Shao F, Hoffmann M W G, Prades J D, et al. Heterostructured p - CuO (nanoparticle)/ n - SnO_2 (nanowire) devices for selective H_2S detection[J]. Sens. Actuators B, 2013, 181: 130 - 135.

[11] Koo W T, Choi S J, Kim S J, et al. Heterogeneous sensitization of metal-organic framework driven metal@metal oxide complex catalysts on an oxide nanofiber scaffold toward superior gas sensors[J]. J. Am. Chem. Soc., 2016, 138: 13431 - 13437.

[12] Woo H S, Na C W, Kim I D, et al. Highly sensitive and selective trimethylamine sensor using one-dimensional ZnO - Cr_2O_3 hetero-nanostructures[J]. Nanotechnology, 2012, 23: 245501.

[13] Mashock M, Yu K, Cui S, et al. Modulating gas sensing properties of CuO nanowires through creation of discrete nanosized p - n junctions on their surfaces[J]. ACS Appl. Mater. Interfaces., 2012, 4: 4192 - 4199.

[14] Jain K, Pant R P, Lakshmikumar S T. Effect of Ni doping on thick film SnO_2 gas sensor [J]. Sens. Actuators B, 2006, 113: 823 - 829.

[15] Zeng W, Liu T, Wang Z. Sensitivity improvement of TiO_2 - doped SnO_2 to volatile organic compounds[J]. Physica E, 2010, 43: 633 - 638.

[16] Zhao T, Qiu P, Fan Y, et al. Hierarchical branched mesoporous TiO_2 - SnO_2

nanocomposites with well-defined n-n heterojunctions for highly efficient ethanol sensing[J]. Adv. Sci., 2019, 6: 1902008.

[17] Zhu L Y, Yuan K P, Yang J H, et al. Hierarchical highly ordered SnO_2 nanobowl branched ZnO nanowires for ultrasensitive and selective hydrogen sulfide gas sensing[J]. Microsyst. Nanoeng., 2020, 6: 30.

[18] Sen S, Kanitkar P, Sharma A, et al. Growth of $SnO_2/W_{18}O_{49}$ nanowire hierarchical heterostructure and their application as chemical sensor[J]. Sens. Actuators B, 2010, 147: 453-460.

[19] Suh J M, Sohn W, Shim Y S, et al. p-p Heterojunction of nickel oxide-decorated cobalt oxide nanorods for enhanced sensitivity and selectivity toward volatile organic compounds[J]. ACS Appl. Mater. Interfaces., 2018, 10: 1050-1058.

[20] Li C, Li L, Du Z, et al. Rapid and ultrahigh ethanol sensing based on Au-coated ZnO nanorods[J]. Nanotechnology, 2007, 19: 035501.

[21] Yang X, Salles V, Kaneti Y V, et al. Fabrication of highly sensitive gas sensor based on Au functionalized WO_3 composite nanofibers by electrospinning[J]. Sens. Actuators B, 2015, 220: 1112-1119.

[22] Hosseini Z S, Mortezaali A, Fardindoost S. Sensitive and selective room temperature H_2S gas sensor based on Au sensitized vertical ZnO nanorods with flower-like structures[J]. J. Alloy Compd., 2015, 628: 222-229.

[23] Wang Y, Lin Y, Jiang D, et al. Special nanostructure control of ethanol sensing characteristics based on Au@In_2O_3 sensor with good selectivity and rapid response[J]. RSC Adv., 2015, 5: 9884-9990.

[24] Vallejos S, Stoycheva T, Umek P, et al. Au nanoparticle-functionalised WO_3 nanoneedles and their application in high sensitivity gas sensor device[J]. Chem. Commun., 2011, 47: 565-567.

[25] Ramgir N S, Kaur M, Sharma P K, et al. Ethanol sensing properties of pure and Au modified ZnO nanowires[J]. Sens. Actuators B, 2013, 187: 313-318.

[26] Kaneti Y V, Moriceau J, Liu M, et al. Hydrothermal synthesis of ternary α-Fe_2O_3-ZnO-Au nanocomposites with high gas-sensing performance[J]. Sens. Actuators B, 2015, 209: 889-897.

[27] Chung F C, Wu R J, Cheng F C. Fabrication of a Au@SnO_2 core-shell structure for gaseous formaldehyde sensing at room temperature[J]. Sens. Actuators B, 2014, 190: 1-7.

[28] Chung F C, Zhu Z, Luo P Y, et al. Au@ZnO core-shell structure for gaseous formaldehyde sensing at room temperature[J]. Sens. Actuators B, 2014, 199: 314-319.

[29] Ramgir N S, Sharma P K, Datta N, et al. Room temperature H_2S sensor based on Au modified ZnO nanowires[J]. Sens. Actuators B, 2013, 186: 718-726.

[30] D'Arienzo M, Armelao L, Cacciamani A, et al. One-step preparation of SnO_2 and Pt-doped SnO_2 as inverse opal thin films for gas sensing[J]. Chem. Mater., 2010, 22: 4083-4089.

[31] Shin J, Choi S J, Lee I, et al. Thin-wall assembled SnO_2 fibers functionalized by catalytic

Pt nanoparticles and their superior exhaled-breath-sensing properties for the diagnosis of diabetes[J]. Adv. Funct. Mater., 2013, 23: 2357 - 2367.

[32] Karmaoui M, Leonardi S G, Latino M, et al. Pt-decorated In_2O_3 nanoparticles and their ability as a highly sensitive (<10 ppb) acetone sensor for biomedical applications[J]. Sens. Actuators B, 2013, 230: 697 - 705.

[33] Wang Y, Liu J, Cui X, et al. NH_3 gas sensing performance enhanced by Pt-loaded on mesoporous WO_3[J]. Sens. Actuators B, 2017, 238: 473 - 481.

[34] Yu M R, Wu R J, Chavali M. Effect of 'Pt' loading in ZnO - CuO hetero-junction material sensing carbon monoxide at room temperature[J]. Sens. Actuators B, 2011, 153: 321 - 328.

[35] Ma J, Ren Y, Zhou X, et al. Pt nanoparticles sensitized ordered mesoporous WO_3 semiconductor: gas sensing performance and mechanism study[J]. Adv. Funct. Mater., 2018, 28: 1705268.

[36] Wang K, Zhao T, Lian G, et al. Room temperature CO sensor fabricated from Pt-loaded SnO_2 porous nanosolid[J]. Sens. Actuators B, 2013, 184: 33 - 39.

[37] Hwang I S, Choi J K, Woo H S, et al. Facile control of C_2H_5OH sensing characteristics by decorating discrete Ag nanoclusters on SnO_2 nanowire networks[J]. ACS Appl. Mater. Interfaces, 2011, 3: 3140 - 3145.

[38] Mirzaei A, Janghorban K, Hashemi B, et al. Synthesis, characterization and gas sensing properties of Ag@α - Fe_2O_3 core - shell nanocomposites[J]. Nanomaterials, 2015, 5: 737 - 749.

[39] Wang Y, Cui X, Yang Q, et al. Preparation of Ag-loaded mesoporous WO_3 and its enhanced NO_2 sensing performance[J]. Sens. Actuators B, 2016, 225: 544 - 552.

[40] Zhu G, Liu Y, Xu H, et al. Photochemical deposition of Ag nanocrystals on hierarchical ZnO microspheres and their enhanced gas-sensing properties[J]. CrystEngComm, 2012, 14: 719 - 725.

[41] Kolmakov A, Klenov D O, Lilach Y, et al. Enhanced gas sensing by individual SnO_2 nanowires and nanobelts functionalized with Pd catalyst particles[J]. Nano Lett., 2005, 5: 667 - 673.

[42] Adams B D, Ostrom C K, Chen S, et al. High-performance Pd-based hydrogen spillover catalysts for hydrogen storage[J]. J. Phys. Chem. C, 2010, 114: 19875 - 19882.

[43] Lou Z, Deng J, Wang L, et al. Toluene and ethanol sensing performances of pristine and PdO - decorated flower-like ZnO structures[J]. Sens. Actuators B, 2013, 176: 323 - 329.

[44] Kim S, Park S, Park S, et al. Acetone sensing of Au and Pd-decorated WO_3 nanorod sensors[J]. Sens. Actuators B, 2015, 209: 180 - 185.

[45] Tian S, Ding X, Zeng D, et al. A low temperature gas sensor based on Pd-functionalized mesoporous SnO_2 fibers for detecting trace formaldehyde[J]. RSC Adv., 2013, 3: 11823 - 11831.

[46] Hong Y J, Yoon J W, Lee J H, et al. One-pot synthesis of Pd-loaded SnO_2 yolk-shell nanostructures for ultraselective methyl benzene sensors[J]. Chem. Eur. J., 2014, 20: 2737 - 2741.

[47] Li W, Shen C, Wu G, et al. New model for a Pd-doped SnO_2 - based CO gas sensor and catalyst studied by online in-situ X-ray photoelectron spectroscopy[J]. J. Phys. Chem. C, 2011, 115: 21258–21263.

[48] Moon J, Park J A, Lee S J, et al. Pd-doped TiO_2 nanofiber networks for gas sensor applications[J]. Sens. Actuators B, 2010, 149: 301–305.

[49] Kim J C, Jun H K, Huh J S, et al. Tin oxide-based methane gas sensor promoted by alumina-supported Pd catalyst[J]. Sens. Actuators B, 1997, 45: 271–277.

[50] Wang Z, Li Z, Jiang T, et al. Ultrasensitive hydrogen sensor based on Pd^0 - loaded SnO_2 electrospun nanofibers at room temperature[J]. ACS Appl. Mater. Interfaces, 2013, 5: 2013–2021.

[51] Xue J, Liu J, Liu Y, et al. Recent advances in synthetic methods and applications of Ag_2S - based heterostructure photocatalysts[J]. Nanoscale, 2013, 5: 2505–2513.

[52] Kim S J, Hwang I S, Na C W, et al. Ultrasensitive and selective C_2H_5OH sensors using Rh-loaded In_2O_3 hollow spheres[J]. J. Mater. Chem, 2011, 21: 18560–18567.

[53] Kandoi S, Gokhale A A, Grabow L C, et al. Why Au and Cu are more selective than Pt for preferential oxidation of CO at low temperature[J]. Catal. Lett, 2004, 93: 93–100.

[54] Schubert M M, Hackenberg S, Van Veen A C, et al. CO oxidation over supported gold catalysts—"Inert" and "active" support materials and their role for the oxygen supply during reaction[J]. J. Catal, 2001, 197: 113–122.

[55] Tien L C, Sadik P W, Norton D P, et al. Hydrogen sensing at room temperature with Pt-coated ZnO thin films and nanorods[J]. Appl. Phys. Lett., 2005, 87: 222106.

[56] Li F, Guo S, Shen J, et al. Xylene gas sensor based on Au-loaded $WO_3 \cdot H_2O$ nanocubes with enhanced sensing performance[J]. Sens. Actuators B, 2017, 38: 364–373.

[57] Kim S J, Choi S J, Jang J S, et al. Mesoporous WO_3 nanofibers with protein-templated nanoscale catalysts for detection of trace biomarkers in exhaled breath[J]. ACS Nano, 2016, 10: 5891–5899.

[58] Choi S J, Kim S J, Cho H J, et al. WO_3 nanofiber-based biomarker detectors enabled by protein-encapsulated catalyst self-assembled on polystyrene colloid templates[J]. Small, 2016, 12: 911–920.

[59] Liu B, Cai D, Liu Y, et al. Improved room-temperature hydrogen sensing performance of directly formed Pd/WO_3 nanocomposite[J]. Sens. Actuators B, 2014, 193: 28–34.

[60] Ma J, Xiao X, Zou Y, et al. A general and straightforward route to noble metal-decorated mesoporous transition-metal oxides with enhanced gas sensing performance[J]. Small, 2019, 15: 904240.

[61] Ma J, Li Y, Zhou X, et al. Au nanoparticles decorated mesoporous SiO_2 - WO_3 hybrid materials with improved pore connectivity for ultratrace ethanol detection at low operating temperature[J]. Small, 2020, 16: 2004772.

[62] Ma J, Li Y, Li J, et al. Rationally designed dual-mesoporous transition metal oxides/noble metal nanocomposites for fabrication of gas sensors in real-time detection of 3-hydroxy-2-butanone biomarker[J]. Adv. Funct. Mater., 2022, 32: 2107439.

[63] Tricoli A, Graf M, Pratsinis S E. Optimal doping for enhanced SnO_2 sensitivity and

thermal stability[J]. Adv. Funct. Mater., 2008, 18: 1969 - 1976.

[64] Righettoni M, Tricoli A, Gass S, et al. Breath acetone monitoring by portable Si: WO_3 gas sensors[J]. Anal. Chim. Acta, 2012, 738: 69 - 75.

[65] Righettoni M, Tricoli A, Pratsinis S E, et al. Si: WO_3 sensors for highly selective detection of acetone for easy diagnosis of diabetes by breath analysis[J]. Anal. Chem., 2010, 82: 3581 - 3587.

[66] Righettoni M, Tricoli A, Pratsinis S E. Thermally stable, silica-doped ε - WO_3 for sensing of acetone in the human breath[J]. Chem. Mater, 2010, 22: 3152 - 3157.

[67] Xiao T, Wang X Y, Zhao Z H, et al. Highly sensitive and selective acetone sensor based on C-doped WO_3 for potential diagnosis of diabetes mellitus[J]. Sens. Actuators B, 2014, 99: 210 - 219.

[68] Zhang Y, Yang Q, Yang X, et al. One-step synthesis of in-situ N-doped ordered mesoporous titania for enhanced gas sensing performance[J]. Micropor. Mesopor. Mater, 2018, 270: 75 - 81.

[69] Wang M D, Li Y Y, Yao B H, et al. Synthesis of three-dimensionally ordered macro/mesoporous C-doped WO_3 materials: Effect of template sizes on gas sensing properties[J]. Sens. Actuators B-Chem, 2019, 288: 656 - 666.

[70] Zhou X, Zou Y, Ma J, et al. Cementing mesoporous ZnO with silica for controllable and switchable gas sensing selectivity[J]. Chem. Mater, 2019, 31: 8112 - 8120.

[71] Woo H S, Kwak C H, Kim I D, et al. Selective, sensitive, and reversible detection of H_2S using Mo-doped ZnO nanowire network sensors[J]. J. Mater. Chem. A, 2014, 2: 6412 - 6418.

[72] Zhao J, Yang T, Liu Y, et al. Enhancement of NO_2 gas sensing response based on ordered mesoporous Fe-doped In_2O_3[J]. Sens. Actuators B, 2014, 191: 806 - 812.

[73] Galatsis K, Cukrov L, Wlodarski W, et al. p-and n-type Fe-doped SnO_2 gas sensors fabricated by the mechanochemical processing technique[J]. Sens. Actuators B, 2003, 93: 562 - 565.

[74] Yu A, Qian J, Pan H, et al. Micro-lotus constructed by Fe-doped ZnO hierarchically porous nanosheets: preparation, characterization and gas sensing property[J]. Sens. Actuators B, 2011, 158: 9 - 16.

[75] Yoon J W, Kim H J, Kim I D, et al. Electronic sensitization of the response to C_2H_5OH of p-type NiO nanofibers by Fe doping[J]. Nanotechnology, 2013, 24: 444005.

[76] Al-Hardan N, Abdullah M J, Aziz A A. Impedance spectroscopy of undoped and Cr-doped ZnO gas sensors under different oxygen concentrations[J]. Appl. Surf. Sci, 2011, 257: 8993 - 8997.

[77] Kim H J, Yoon J W, Choi K I, et al. Ultraselective and sensitive detection of xylene and toluene for monitoring indoor air pollution using Cr-doped NiO hierarchical nanostructures [J]. Nanoscale, 2013, 5: 7066.

[78] Ruiz A M, Sakai G, Cornet A, et al. Cr-doped TiO_2 gas sensor for exhaust NO_2 monitoring[J]. Sens. Actuators B, 2003, 93: 509 - 518.

[79] Wang Y, Liu B, Xiao S, et al. Low-temperature H_2S detection with hierarchical Cr-doped

WO$_3$ microspheres[J]. ACS Appl. Mater. Interfaces. 2016, 8: 9674-9683.

[80] Kumar V, Sen S, Muthe K P, et al. Copper doped SnO$_2$ nanowires as highly sensitive H$_2$S gas sensor[J]. Sens. Actuators B, 2009, 138: 587-590.

[81] Teleki A, Bjelobrk N, Pratsinis S E. Flame-made Nb-and Cu-doped TiO$_2$ sensors for CO and ethanol[J]. Sens. Actuators B, 2008, 130: 449-457.

[82] Gong H, Hu J Q, Wang J H, et al. Nano-crystalline Cu-doped ZnO thin film gas sensor for CO. Sens[J]. Actuators B, 2006, 115: 247-251.

[83] Parthibavarman M, Renganathan B, Sastikumar D. Development of high sensitivity ethanol gas sensor based on Co-doped SnO$_2$ nanoparticles by microwave irradiation technique[J]. Curr. Appl. Phys. 2013, 13, 1537-1544.

[84] Jing Z, Wu S. Synthesis, characterization and gas sensing properties of undoped and Co-doped γ-Fe$_2$O$_3$-based gas sensors[J]. Mater. Lett. 2006, 60: 952-956.

[85] Liu L, Li S, Zhuang J, Wang L, et al. Improved selective acetone sensing properties of co-doped ZnO nanofibers by electrospinning. Sens. Actuators B, 2011, 55: 782-788.

[86] Zheng K, Gu L, Sun D, et al. The properties of ethanol gas sensor based on Ti doped ZnO nanotetrapods[J]. Mater. Sci. Eng. C, 2010, 166: 104-107.

[87] Guo P, Pan H. Selectivity of Ti-doped In$_2$O$_3$ ceramics as an ammonia sensor[J]. Sens. Actuators B, 2006, 114: 762-767.

[88] Song X C, Yang E, Liu G, et al. Preparation and photocatalytic activity of Mo-doped WO$_3$ nanowires[J]. Nanopart. Res. 2010, 12: 2813-2819.

[89] Mai L Q, Hu B, Hu T, et al. Electrical property of Mo-doped VO$_2$ nanowire array film by melting-quenching sol-gel method[J]. J. Phys. Chem. B, 2006, 110: 19083-19086.

[90] Feng C, Wang C, Cheng P, et al. Facile synthesis and gas sensing properties of La$_2$O$_3$-WO$_3$ nanofibers[J]. Sens. Actuators B, 2015, 221: 434-442.

[91] Mohanapriya P, Segawa H, Watanabe K, et al. Enhanced ethanol-gas sensing performance of Ce-doped SnO$_2$ hollow nanofibers prepared by electrospinning[J]. Sens. Actuators B, 2013, 188: 872-878.

[92] Wei D, Huang Z, Wang L, et al. Hydrothermal synthesis of Ce-doped hierarchical flower-like In$_2$O$_3$ microspheres and their excellent gas-sensing properties[J]. Sens. Actuators B, 2018, 255: 1211-1219.

[93] Li Z, Wang W, Zhao Z, et al. One-step hydrothermal preparation of Ce-doped MoO$_3$ nanobelts with enhanced gas sensing properties[J]. RSC Adv, 2017, 7: 28366-28372.

[94] Han D, Song P, Zhang S, et al. Enhanced methanol gas-sensing performance of Ce-doped In$_2$O$_3$ porous nanospheres prepared by hydrothermal method[J]. Sens. Actuators B, 2015, 216: 488-496.

[95] Liu Y, Guo R, Yuan K, et al. Engineering pore walls of mesoporous tungsten oxides via Ce doping for the development of high-performance smart gas sensors[J]. Chem. Mater. 2022, 34: 2321-2332.

[96] Liu X, Sun J, Zhang X. Novel 3D graphene aerogel-ZnO composites as efficient detection for NO$_2$ at room temperature[J]. Sens. Actuators B, 2015, 211: 220-226.

[97] Zhang Z, Zou R, Song G, et al. Highly aligned SnO$_2$ nanorods on graphene sheets for gas

sensors[J]. J. Mater. Chem. 2011, 21: 17360.

[98] Dai M, Zhao L, Gao H, et al. Hierarchical assembly of α-Fe_2O_3 nanorods on multiwall carbon nanotubes as a high-performance sensing material for gas sensors[J]. ACS Appl. Mater. Interfaces, 2017, 9: 8919-8928.

[99] Asad M, Sheikhi M H. Highly sensitive wireless H_2S gas sensors at room temperature based on CuO-SWCNT hybrid nanomaterials[J]. Sens. Actuators B, 2016, 231: 474-483.

[100] Zhao T, Ren Y, Jia G, et al. Facile synthesis of mesoporous WO_3@graphene aerogel nanocomposites for low-temperature acetone sensing[J]. Chin. Chem. Lett. 2019, 30: 2032-2038.

第5章
半导体金属氧化物：微结构与传感性能

5.1 半导体金属氧化物的基本特征

能源短缺和环境污染被认为是十分重要和迫切需要解决的问题，其解决依赖于有效和独特的材料或技术的发明。快速发展的纳米技术和材料科学已为各种应用开发了许多新型的多功能纳米材料，包括碳族材料（即石墨烯、碳纤维、碳纳米管、富勒烯、活性炭）、金属氧化物基材料（即 Fe_3O_4、Co_3O_4、MnO_2、TiO_2、SnO_2、WO_3 等）、金属硫化物基材料（即 MoS_2、Li_2S、CoS_2、VS_2、TiS_2）等。纳米材料存在各种微观结构，如球形、多孔、层状、核壳、蛋黄壳、纳米线等，通常具有优异的物理化学性质，如高磁性、高导电性、热稳定性，高的表面积，大的孔体积和丰富的活性位点，可以提供主要的反应位点，颇具应用潜力，涉及催化、传感、生物工程、环境修复和能量转换/存储等领域。

近几十年来，固态电阻型半导体金属氧化物气体传感器已广泛应用于多个跨学科领域，涉及健康（即医疗诊断、食品加工、有毒和爆炸性气体检测）和燃烧过程中的能源效率检测[1-5]。根据前述内容，半导体金属氧化物在气体传感器中具有优越的性能，包括高灵敏度、高响应性、低成本、设备简单且易于加工成装置等。然而，半导体的这些主要特征在很大程度上取决于原始传感材料的微观结构，它由许多典型因子组成，如微相组成、晶粒尺寸、孔隙率、结晶性和晶格取向[6-7]。某个因子或参数可以通过诱导电子结构或晶格排列的变化而对气体传感性能产生显著影响。在本章中，我们将系统总结和讨论半导体金属氧化物微结构对气体传感器的影响。

5.2 结构类型和典型架构

根据电子特性的不同，SMO 材料可分为两类：n 型半导体和 p 型半导体。相比之下，不同的气体传感器（如 n 型和 p 型）具有截然不同的电子受体、传导路径和相互作用机制，对于这些内容的理解有待深入研究。其中上述指标直接决定了材料在传感领域的应用范畴[8-11]。n 型半导体的主要载流子是电子（e^-），因其原理是氧空位产生电子并形成电子耗尽层而成为主要的商业传感器,常见的有 ZnO、SnO_2、In_2O_3、Fe_2O_3 和 WO_3[10,12-15],可以通过串行路径(即半导体粒子核和电阻粒子接触,其可以导出典型的化学电阻变化)来进行传导。而 p 型半导体的主要载流子是空穴(h^+),通常可借助负电荷吸附氧形成空穴累积层(HAL),常见的有 CuO、NiO、Cr_2O_3、Co_3O_4 和 Mn_3O_4[16-18],通过平行路径传导,包括电阻粒子核和半导体近表面区域,其也可以在近表面区域产生化学电阻变化。

针对气体传感的应用可追溯到 1931 年,Brauer 率先设计了 Cu_2O 半导体并将其应用于气体传感器检测,发现通过改变水蒸气或氧空位的吸附含量可以调节 Cu_2O 的电阻[7]。此后,半导体金属氧化物为气体传感器领域的快速发展打开了一扇门。对于 SMO 材料的传感机制,典型的认识是当 n 型半导体暴露在还原气体中时,内部电子可以通过吸附的氧物种和气体客体之间的表面反应返回到耗氧层[19]。然而,p 型半导体对氧化气体较敏感,并且对氧化气体和还原气体的耐受性的变化是截然不同的。因此,n 型和 p 型氧化物适用的应用领域差异较大。纯 SMO 材料可用作各种气体的传感材料,早在 2014 年,笔者课题组便设计合成了高度有序的介孔 n 型 WO_3 半导体,并首次将其应用于 VOC 的检测。如图 5.1 所示,借助自主设计的两亲性嵌段共聚物 PEO-b-PS 作为模板剂,通过典型的溶剂挥发诱导自组装技术合成的高度有序、骨架晶化的 WO_3 介孔材料具有较高的比表面积(121 m^2/g)和大的孔径(10.9 nm),这种的大比表面积、大孔结构和高度晶化的 WO_3 颗粒有利于在极低浓度(约 0.25 ppm)下感应 H_2S,具有快速响应时间(2.0 s)和恢复时间(38 s),并表现出优异的选择性[20]。在后续研究中,研究人员借助同样的溶剂挥发技术诱导 In^{6+} 与 PEO-b-PS 共组装,为避免合成的 In_2O_3 在高温煅烧下结晶坍塌,引入热稳定性高的 CaO 作为辅助烧结剂,成功制备了一种具有大孔径(14.5 nm)的高结晶 In_2O_3(n 型半导体),该多孔 In_2O_3 纳米片可以促进 NO_2 的扩散/传输,对 50 ppb NO_2 在较低的工作温度下(150℃)仍具有高的选择性、灵敏度和快速响应/恢复能力[21]。

除了 n 型半导体之外,研究人员还开发了 p 型半导体并应用于许多特殊的氧化性气体的检测。例如,Wen 等[22]设计了具有独特中孔和准单晶结构的介孔 Co_3O_4 纳米针阵列,气体传感分析系统如图 5.2(a)~图 5.2(b)所示,该材料对 100 ppm C_2H_5OH 表现出相对较高的灵敏度①(约 89.6)。此外,Nguyen 和 El-Safty[23]以 $CoCl_2$ 和尿素为前驱体,在不引入额外模板剂的前提下通过简单的水热合成技术制备出结晶性高的中孔/大孔 Co_3O_4 纳米棒,其中设计的气体传感测量系统如图 5.2(c)所示,该材料可作为丙酮、苯和 C_2H_5OH 的有效检测材料。此外,该研究还发现,中孔/大孔纳米棒的灵敏度分别是同类型氧化物的纳米颗粒和多孔块体材料的 7.1 倍和 4 倍。在整个传感检测过程中,表面吸附氧发挥了重要的作

---

① 即较高的响应值。

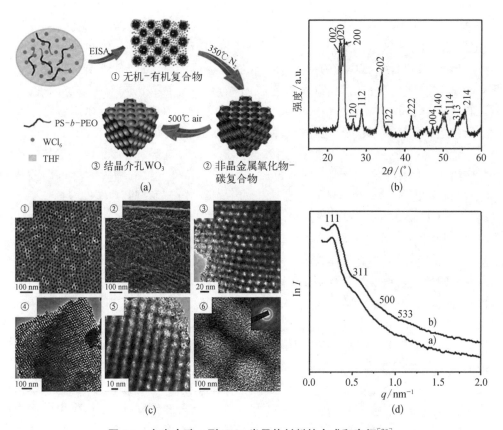

图 5.1 有序介孔 n 型 WO₃ 半导体材料的合成和表征[20]

(a) 模板碳化技术合成晶化有序介孔 WO₃ 半导体材料的技术路线；(b) 高温煅烧后的有序介孔 WO₃ 材料 XRD 图(350℃ N₂ 碳化和500℃空气脱除残碳)；(c) 不同晶面的 FESEM 图和 TEM 图，其中①为表面形貌，②为截面形貌，③为(110)晶面 TEM 图，④为(100)晶面 TEM 图，⑤为(211)晶面 TEM 图，⑥为(211)晶面 HETEM 图

用。一系列研究发现，一旦两种传感器的形态结构相同，p 型半导体对相同气体的响应就弱于传统的 n 型半导体[3,24-26]。

在实际应用中，各种结构的组合可以影响固-气界面的相互作用。为了解释结构和分散状态对材料性能的影响，可定义四种典型的结构-骨架类型：① 两种成分的简单物理混合物，其大多数随机分布在整个材料中；② 在某些方面添加了额外氧化物的基础材料；③ 两种基于氧化物的特殊合成策略之间明确定义的分区或界面；④ 两种或多种纳米颗粒的混合物沉积在基底上以形成双层[3,27-30]。然而，对于半导体金属氧化物，根据引入组分的差异性，上述结构可分为三种类型，即嵌入或掺杂的纳米颗粒、引入的额外金属氧化物和形成的合金[31-32]。上述高效的功能化或表面修饰将基于异构体的协同效应增强纯半导体的响应性能，且常见的功能化策略主要包括掺杂活性元素、表面涂层、与目标客体混合等[32-34]。各种结构在操纵主体分子的电子结构和微结构中起重要作用，特别是对于低活性金属氧化物。例

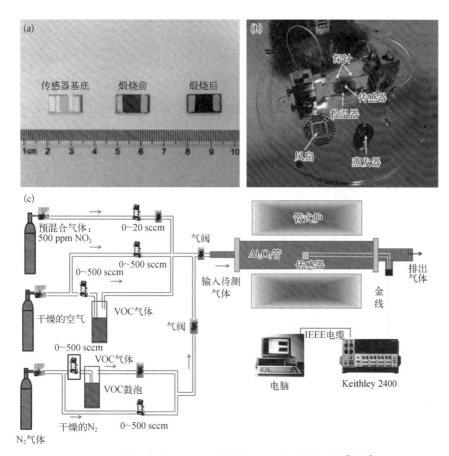

图 5.2 基于介孔 Co_3O_4 纳米针阵列的传感器及系统[22-23]

(a) 介孔 Co_3O_4 纳米针阵列的传感器基板和样品传感器的俯视图；(b) 气体传感分析系统的照片；
(c) 气体传感测量系统示意图（测试和参考气体使用一系列质量流量控制器混合，其中传感器的电流和电阻使用由 PC 计算机通过 IEEE 电缆和 Labview 控制的 Keithley 2400 型自动测量程序测量）

如，WO_3 作为一种有巨大应用前景的气敏材料可以掺杂各种客体分子，包括贵金属（即 Au、Pt、Ag、Pd 和 Ru）、过渡金属（即 Fe、Mn、Cu、Ni）、稀土金属（即 La、Pr、Nd、Sm）和游离金属（即 C、N、Si）[34-39]。研究发现，质量分数为 0.03% 的 Au 掺杂的 WO_3 粉末对 2.0 ppm H_2S 显示出优异的响应（响应值约 12.40），明显高于纯 WO_3（约 4.85）[40]。2017 年，笔者课题组[12] 报道了一系列不同浓度 Pt 纳米粒子敏化的有序介孔 WO_3 传感材料，制备的 WO_3/Pt 复合材料最高比表面积可达 128 m^2/g，同时具备大孔容（0.32 cm^3/g）和孔径结构（约 13 nm），其对低浓度的 CO 表现出优异的催化传感响应[良好的灵敏度、高选择性和超短响应/恢复时间（16 s/1 s）]，远远高于非掺杂的有序介孔 WO_3，这主要是由于高活性的超小 Pt 纳米颗粒（约 4 nm）的敏化效应和构筑的异质结效应。其 CO 传感机制如图 5.3 所示，表面耗尽层的有效调节和额外产生的电子可以显著降低传感反应的阻力并增强 WO_3 对 CO 的响应。

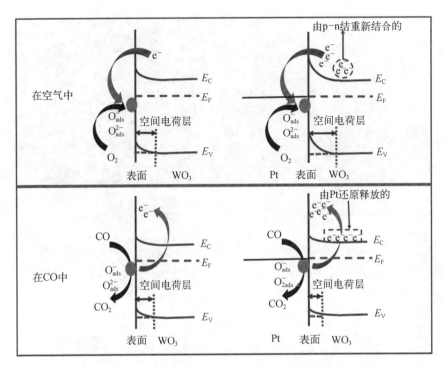

图 5.3 介孔 WO_3/Pt 和介孔 WO_3 的 CO 传感机制示意图[12]

近年来,Kruefu 等[41]通过典型的水热/浸渍途径合成了具有一维纳米结构的 0.50% Ru 功能化 WO_3 复合材料,该复合材料不仅在形貌上表现为均一的纳米棒结构,而且具有较高结晶度和取向排列。0.50% Ru - WO_3 对 10 ppm H_2S 表现出约 192 的超高传感器响应及非常短的响应时间(0.8 s)。此外,该材料对 H_2S 表现出优异的选择性,其选择性远远高于未引入 Ru 的 WO_3 纳米棒,表明贵金属纳米颗粒在气体传感过程中发挥了重要的促进作用,提供了大量的活性催化位点(见图 5.4)。类似地,Kim 等[42]通过一步水热法制备了直径约为 2.0 μm 的 1.67%(原子百分比)Rh 负载的 In_2O_3 空心球,Rh 负载的 In_2O_3 空心球对 100 ppm C_2H_5OH 表现出超高响应值(约 4 748),接近纯 In_2O_3 空心球的 180 倍,且 Rh 的引入降低了材料的最佳工作温度,提高了材料对 C_2H_5OH 的选择性(C_2H_5OH 气体的响应值是其他气体响应值的 15.1~24.7 倍)。针对 In_2O_3 半导体,Wang 等[43]以常温常压裂解及化学还原技术合成了纳米级贵金属 Ag 负载的向日葵型 In_2O_3 分级材料,该材料具有较高的孔隙率、分支状的多级结构和 Ag 纳米颗粒敏化效应,从而对 20 ppm HCHO 表现出优异的气敏性能:响应时间(约 0.9 s)短,灵敏度/选择性高,恢复时间(14 s)快。尽管上述材料中的掺杂元素不同,但这些贵金属的功能类似且对于材料敏化/提供活性位点而言贡献是一致的。

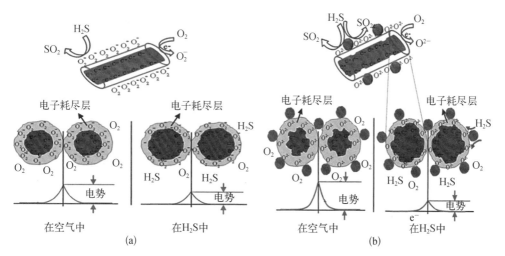

图 5.4 H_2S 的气体传感机理[41]

(a) WO_3 纳米棒；(b) $Ru-WO_3$ 纳米棒

除贵金属外，过渡金属或稀土金属也可以作为掺杂元素改善原始 SMO 材料的表面性质。2014 年，Zhao 等[44]通过纳米铸造工艺以三维立方介孔 KIT-6 型氧化硅为硬模板成功合成了 Fe 掺杂的有序介孔 In_2O_3，该介孔 In_2O_3 材料具有大的比表面积、有序的介观结构和较高的结晶度，且对 NO_2 的响应也比纯介孔 In_2O_3 高得多。同年，Vyas 等[45]采用 ZnO 纳米线作为基底材料，以气相转移技术合成了 Fe 原位掺杂的 ZnO 纳米线。该 ZnO 纳米线具有 20~50 nm 的均一直径和暴露的 (002) 晶面，该 (002) 晶面展现出优异的氧气传感性能 (见图 5.5)。这些新型纳米线阵列均显示出高灵敏度 (约 31)，在 140℃时具有快速响应/恢复时间 (11 s/11 s)。特别地，该研究认为初始的 ZnO 充当具有空 Zn $4s^2$ 轨道和 O $2p^6$ 轨道的本征半导体，并且过渡金属掺杂和氧缺陷增加了易电离的有效 Zn $4s^2$ 电子。此外，Fe 在晶格区域的掺杂能引发带隙间电子跃迁，从而降低了电荷俘获效率[46-47]。Rossinyol 等[48]则分别以二维六方 SBA-15 和三维立方 KIT-6 型氧化硅为模板设计了 Cr 掺杂的有序介孔 WO_3，并将其用作检测 NO_2 的传感材料，该传感材料在低 NO_2 浓度下显示出较短的响应时间和较高的传感器响应，这是由于电子受主型杂质的调制和掺入 Cr 增加的氧空位导致了性能提升。区别于传统的掺杂工艺，高浓度引入另一种组分构成的复合材料也可以改善传感性能。Tabassum 等[49]采用纯铜层作为载体并在铜层上生长薄氧化锌，形成高活性的金属-电介质界面，合成的 Cu-ZnO 薄膜对 0~100 ppm 的 H_2S 表现出优异的传感性能。

此外，金属氧化物合金在传感应用中具有多重功能的特征。Qin 等[50]借助 PMMA 模板策略设计了三维有序的大孔 $LaFeO_3$ 材料，该材料显示出对甲醇的高传感响应 (响应值为 96) 和快速响应/恢复时间，并且这种新型金属氧化物合金可

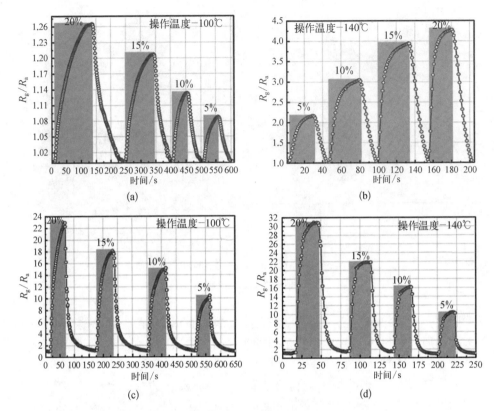

图 5.5　ZnO 和 Fe 掺杂 ZnO 纳米线的氧气传感性能研究[45]

(a)、(b) 未掺杂 ZnO 在 100℃、140℃工作温度下对氧气的传感特性；(c)、(d) 在 100℃、140℃的操作温度下，9% Fe 掺杂的 ZnO 纳米线对氧气的传感特性；R_g 为待测气氛中的电阻；R_a 为空气气氛中的电阻

以表现出两个以上的优异特征(见图 5.6)。同样地，Liu 等[47]将预先制备的均一 $ZnFe_2O_4$ 纳米粒子装饰到 ZnO 微纳米花的表面，所得样品对 50 ppm 丙酮也有接近 8.3 的传感响应，是纯 ZnO 微纳米花的 5.4 倍。此外，由三种不同组分构成的金属氧化物亦表明出优异的响应性能。例如，Wang 等[51]采用溶剂挥发诱导自组装技术合成了一系列不同 Ce 含量的有序介孔 Ni-Ce-Al 三元复合氧化物，该复合物也显示出多组分的优越性。再者，常见的稀土金属，如 La、Ce、Pr、Nd、Sm、Eu、Gd、Dy、Ho、Er 和 Yb，也可以作为辅助客体掺入 SMO 半导体中。Yang 等[52]成功地制备了由上述稀土金属掺杂的介孔 In_2O_3 材料，该材料在低浓度下对 NH_3 表现出优异的敏感性能，这主要是由于材料表面碱度、晶畴尺寸、氧空位浓度和电子云重叠的变化。综上，系列研究均表明掺杂不同元素能够调节材料的气体传感性能。

事实上，除了金属元素和贵金属元素，非金属元素也能够调节金属氧化物内部的电子结构或价态平衡。Righettoni 等[34]合成了 Si 掺杂的 ε-WO_3 纳米结构薄膜，并将其应用于丙酮的便携式传感，该薄膜展现出相对较低的检测限(约

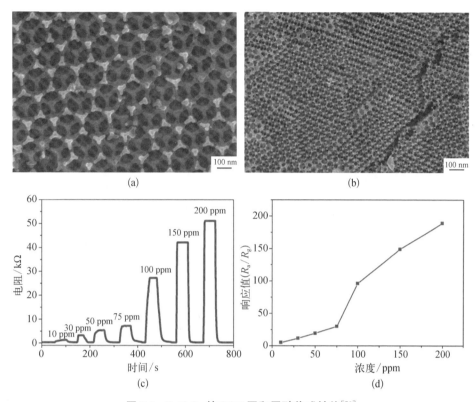

图 5.6 LaFeO₃ 的 SEM 图和甲醇传感性能[50]

(a)、(b) 大孔 LaFeO₃ 材料的 SEM 图；(c)、(d) 大孔 LaFeO₃ 材料对 10~200 ppm 甲醇的响应性（工作温度为 190℃）

20 ppb)、短的响应时间（10~15 s）和快速的恢复速度（35~70 s），特别是它还具有较高的稳定性且适应实际环境中高的相对湿度（80%~90%）。笔者课题组亦专注于非金属掺杂半导体对气体传感性能的调控研究。2018 年，Zhang 等[53]以自行设计的两亲性嵌段共聚物 PS-b-P4VP 为结构导向剂，成功设计出高活性的原位 N 掺杂有序介孔 TiO_2 材料，该材料具有较高的比表面积（119 m²/g）、大的孔容（0.16 cm³/g）和均一的孔径（约 6.2 nm），从而为气体扩散和传质提供了良好的微观结构。其在 250 ppm 下对丙酮表现出较高的高响应值（约 17.6）（见图 5.7）。尽管响应值不是非常突出，但这种响应相比纯的 TiO_2 传感器高出了约 10.4 倍（相同浓度下响应值仅为 1.54）。进一步的研究还发现引入"N"可以使 TiO_2 晶格内的晶体尺寸减小，产生丰富的内部缺陷，从而促进传质过程和增加氧空位浓度。

与此同时，Wang 等[54]采用 PEO-b-PS 为模板、酚醛树脂为碳源，以类似的溶剂挥发诱导共组装策略成功合成了有序介孔碳-氧化钴复合半导体材料，合成的 CoO_x/C 复合材料具有高度有序的面心立方（fcc）结构、超高的比表面积（394~

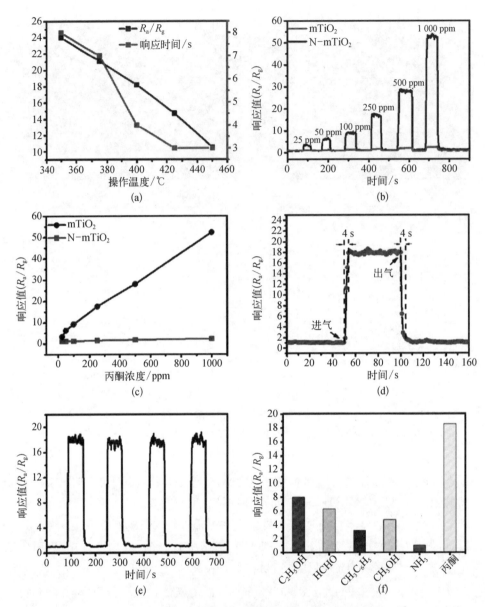

图 5.7 N 掺杂介孔 TiO_2 的丙酮传感性能研究[53]

(a) N 掺杂介孔 TiO_2 在不同操作温度下对 250 ppm 丙酮的传感响应;(b) 循环响应恢复曲线;(c) 丙酮浓度与气体响应之间的关系;(d) 响应-恢复曲线;(e) 响应恢复时间曲线;(f) N 掺杂 TiO_2 的气体传感选择性

483 m^2/g)、均一的大孔径(13.4～16.0 nm)、大的孔容(0.41～0.48 cm^3/g)及高度分散的超细 CoO_x 纳米颗粒(6.4～16.7 nm)。因其独特的微观结构和石墨化碳骨架的限域效应,该类材料制作的传感器对氢气具有优异的传感性能(见图 5.8)。作用机理研究表明,环境中的氧气分子易在 CoO_x/C 复合材料的导带捕获电子并

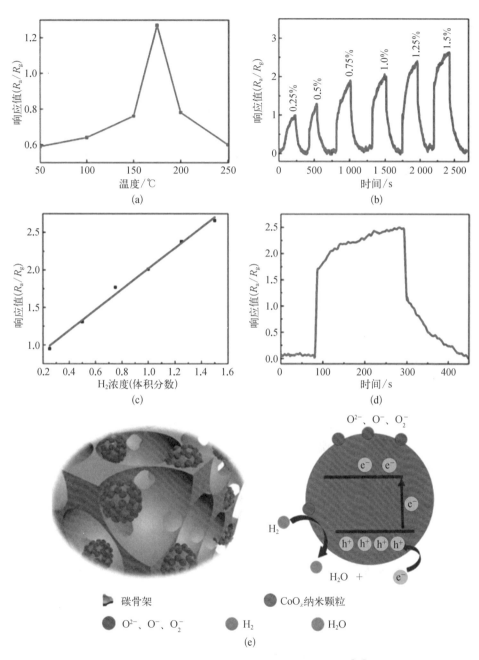

图 5.8 CoO_x/C 的传感性能研究及机理示意图[54]

(a) CoO_x/C 材料在不同温度下对 H_2 的响应;(b) 不同 H_2 浓度下的响应恢复曲线;(c) H_2 浓度与传感响应之间的关系;(d) 响应-恢复曲线;(e) CoO_x/C 材料对 H_2 的传感机理

相互反应，电子耗尽层则会在 CoO_x 纳米颗粒表面产生大量带负电的氧物种（如 O^{2-}、O^- 和 O_2^-），当接触到还原性气氛 H_2 时，带负电的氧物种将与 H_2 反应生成 H_2O，电子则返回至 CoO_x 纳米颗粒层并与空穴复合，从而导致电阻迅速增加。许多实验均表明，非金属的掺杂主要是基于混合效应和晶格匹配效应，这两种效应能够显著地调节电子结构和电子-空穴载流子浓度。

除引入单一元素外，衍生于基底材料与其他金属氧化物或聚合物复合的 SMO 材料往往具有多功能电子/光学效应。截至目前，众多学者已经对复合材料（如 ZnO/TiO_2、In_2O_3/SnO_2 和 ZnO/SnO_2）用于具有增强的气体传感性能的高性能传感器展开了系列研究[37,55-57]。Dong 等[58]通过非液相合成技术制备了 In_2O_3/ZnO 纳米复合材料（IZO）和 3.0% Ag 功能化的 IZO 纳米复合材料，两种材料均表现出优异的甲醛性能（见图 5.9）。此外，3.0% Ag 功能化的 IZO 纳米复合材料对 2000 ppm 甲醛的传感响应值达 842.9，远远高于纯的 IZO 材料，这表明高活性 Ag 可以显著提高材料的传感性能。此外，Choi 等[59]通过旋涂、热退火及水热处理三步法成功设计了涂有 Fe_2O_3 纳米粒子的新型 SnO_2 纳米线，结果表明，该纳米复合材料对 C_2H_5OH 的敏感性比纯 SnO_2 高 1.48～7.54 倍。该材料对乙醇传感性能增强的原因是表面耗尽层和晶界处潜在势垒的变化。

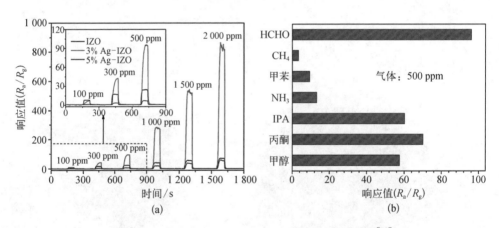

图 5.9 Ag 功能化 IZO 纳米复合材料的传感性能研究[58]

(a) 动态传感特性与不同含量（0%、3%和 5%）的 Ag 功能化 IZO 纳米复合材料的时间的关系（插图为 100 ppm、300 ppm 和 500 ppm 的放大曲线）；(b) 传感器对 VOC 气体的选择性响应（浓度为 500 ppm、温度为 300℃）

有趣的是，除了金属氧化物，聚合物也能增强 SMO 材料的传感性能。Su 和 Peng[36]使用聚吡咯（PPy）和 WO_3 纳米粒子作为原料，在氧化铝基底上生长出了均匀的纳米复合材料薄膜（定义为 PPy/WO_3 薄膜），其对 H_2S 的响应比任意单一组分都高（见图 5.10）。当 H_2S 的浓度从 100 ppb 增加至 1000 ppb 时，PPy/WO_3

薄膜对 H_2S 的响应变化显著,且具有较好的循环稳定性和重复实用性。此外,该研究还发现这种高效的传感性能归因于在 WO_3 薄膜表面、PPy 和 WO_3 薄膜界面处调节了两个耗尽层的宽度。Chen 等[60]制备了 Co_3O_4 插层的还原氧化石墨烯(rGO)基薄膜材料(定义为 Co_3O_4-rGO),该薄膜材料制作的纳米器件对 NO_2 的响应远高于纯 Co_3O_4 传感器和纯 rGO 传感器,这种典型的协同效应归因于 rGO 的大比表面积、小 Co_3O_4 纳米晶体的嵌入和 Co^{3+}-sp^2 C 偶联效应。Zhang 等[61]采用支化聚乙烯亚胺[PEI-(CH_2CHNH)]修饰多壁碳纳米管(MWCNT)并将其作为基质材料,在功能性碳纳米管表面构建 Co_3O_4 纳米粒子(定义为 Co_3O_4/PEI-CNTs)。该材料具有独特的电学性能、高表面积与体积比以及对 CO 的优异敏感特性,即使在 5 ppm 时也是如此,这是因为其具有特殊的一维结构,且在这些不同的组合物中产生了典型的协同效应(见图 5.11)。不论对于碳基材料还是硅基材

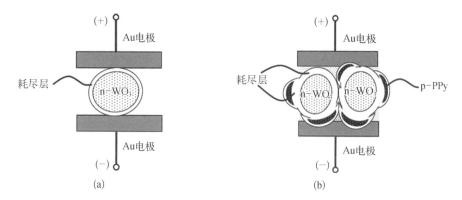

图 5.10 PPy/WO_3 纳米复合薄膜的传感机理示意图[36]

(a)、(b) WO_3 膜、异质结构的 PPy/WO_3 纳米复合薄膜(Au/p-PPy/n-WO_3/p-PPy/Au)的 H_2S 气体传感器的等效横截面图

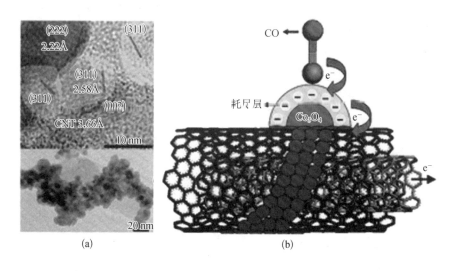

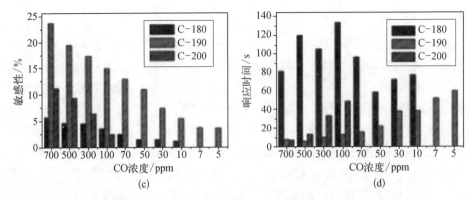

图 5.11 Co_3O_4/PEI-CNTs 复合物的微观形貌和对 CO 的传感性能研究[61]

(a) Co_3O_4 的 TEM 图;(b) Co_3O_4/PEI-CNTs 复合物的纳米结构形态和相应的传感机制示意图;(c)、(d) 室温下在空气中(湿度为 26%)于不同温度下合成的 Co_3O_4/PEI-CNTs 复合物的对 5~700 ppm CO 的气体敏感性、响应时间

料,都可以视为活性组分。Yang 等[62]选择商业化的 SBA-15 作为载体制备了纳米 SnO_2/SBA-15 复合材料,其在 1 000 ppm 下显示出对氢气相对高的灵敏度和敏感性。与纯的 SnO_2 材料相比,该复合材料对相同气体的响应值比纯 SnO_2 高接近 40 倍,传感性能优异。综上,在大多数情况下,聚合物的引入可以产生典型的协同效应并增强传感性能。

5.3 晶粒尺寸与多孔结构

在确定半导体的微相组成时,晶粒尺寸在调节电子结构和电阻中起重要作用。通常电子密度变化仅存在于金属氧化物颗粒的外层中,并且耗尽区的内部区域对电阻变化没有影响。早在 1991 年,Xu 等[63]就系统研究了 4~27 nm 范围内的微小 SnO_2 颗粒传感器的气体响应,发现当平均晶粒尺寸(D)小于 10 nm 时,材料对 H_2 或 CO 的响应和灵敏度会迅速增加;一旦晶粒尺寸超过 20 nm,材料晶粒尺寸与气体响应之间的关联性就变小,而响应则随着晶粒尺寸在 10~20 nm 减小而逐渐增加。因此,研究清楚地表明晶粒尺寸可以对气体响应产生显著影响,并为高效气体传感器的设计提供了有意义的参考。随后,为进一步理解晶粒尺寸对气体传感性能的影响,研究者们提出了一个经典的半定量理论模型来解释晶粒内部的变化。

理论上,由部分烧结的微小晶粒构成的气体传感器的颗粒界面通常相互渗透连接。这些相互连接的小颗粒可以聚集形成大晶粒,借助晶界连接周围的纳米颗粒[64-68]。当这些小晶粒暴露在空气或含氧氛围中时,其可以在微晶表面形成耗尽层。根据晶粒尺寸(D)与耗尽层宽度(L)之间的定量关系,可分为三种不同的情况。对于较大的晶粒($D \gg 2L$),晶粒的大部分体积不依赖于外部气体,并且外部气体对电导率的影响主要反映在晶界区域的势垒高度上[4,7,64-65]。此外,较高导电区域实际上涉及还原气体或氧化气体的放电-再充电反应,从而导致相对弱的响应和灵敏度。一方面,这些较大晶粒的传感机制属于晶界控制,晶界势垒与晶粒尺寸无关,导致传感性能与晶粒尺寸无关[66-68];另一方面,一些晶粒的 D 值接近但仍然大于 $2L(D>2L)$,环绕在每个颈部的耗尽区会形成受限制的诱导通道[69]。因此,材料的导电性取决于晶界阻挡层和相邻晶粒的横截面积,用于晶间电流传输的颈部的有效横截面对环境气体敏感。这些中粒子的传感机制属于颈部控制,由电流浓度和晶界势垒协同作用,与前一种情况相比,这些中粒子的电流收缩效应加上晶界屏障效应会使得灵敏度提高。材料的灵敏度随着晶粒尺寸的减小而逐渐增加。最后,当 $D<2L$ 时,微晶几乎完全占据耗尽层,这意味着整个晶粒都参与电荷转移与气体分子的相互作用,并诱导导电通道在周围消失,引发结晶性和导电率急剧下降[7]。相互连接的晶粒之间的能带结构变得平缓,意味着晶界势垒对晶粒间的载流子传输几乎没有影响,它只依赖于晶间电导率。因此,这些小晶粒的传感机制属于晶粒控制,该类材料将具有最佳的传感性能。Rothschild 和 Komen[69]通过模拟评估有效载流子浓度和表面态密度之间的数值关系解析了晶粒尺寸对半导体气体传感器灵敏度的影响,如图 5.12 所示。研究发现具有不同晶粒尺寸的 SnO_2(从 5.0 nm 到 80 nm)一旦形成完全耗尽层,其表面态密度具有一定的临界值,该临界值与粒径尺寸成正比。然而,在实际应用中,尽管越小的晶粒尺寸在理论上可以产生越高的灵敏度,但是纳米颗粒也往往因在较高工作温度下导电率急剧降低和增大不稳定性而受到应用限制,故合适的晶粒尺寸是设计和制造高效气体传感器需要考虑的重要因素。

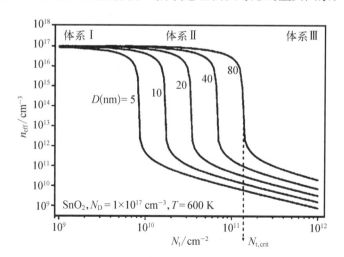

图 5.12　不同晶粒尺寸的 SnO_2 的有效载流子浓度(n_{eff})随表面态密度(N_t)的函数[69]

N_D 代表掺杂能级;T 代表温度;$N_{t,crit}$ 代表临界表面态密度除了这些理论研究,研究人员还设计和制备了具有不同晶粒尺寸的高效半导体金属氧化物,并系统地研究了晶粒尺寸对传感过程的影响。例如,1997年初,Gurlo等[70]通过改变煅烧温度成功制备了具有两种粒径的多晶In_2O_3薄膜。结果发现,粒径为5~6 nm的In_2O_3相比20 nm的较大晶粒具有更强的团聚效应、更高的传感器电阻和对NO_2的响应。同年,Ansari等[71]研究了SnO_2纳米粒子传感器对氢气的传感性能,结果亦表明较小的纳米粒子(20 nm)的气敏性能比25~40 nm的纳米粒子高约10倍。此外,就小晶粒和窄颈区域而言,一旦晶粒尺寸小于耗尽层宽度的两倍,晶粒将与空间电荷层完全连接,此时应该考虑自由电荷的流动对表面性质的影响。2005年,Vuong等[66]设计了具有不同晶粒尺寸(6~16 nm)的结晶SnO_2溶胶,研究了不同薄膜厚度(200~900 nm)制作的传感器件的性能,薄膜器件均表现出对H_2S气体优异的传感特性(见图5.13)。随着晶粒尺寸逐渐增加至16 nm,传感器响应亦逐渐增加,这与气体扩散反应理论是一致的。当采用预先制备的薄膜式气体传感器装置时,该传感器对H_2的响应随着晶粒尺寸D的增加而急剧增长。Xu等[72]系统研究了不同粒径的ZnO纳米材料对不同类型气体,如H_2、SF_6、C_4H_{10}、汽油、C_2H_5OH等的传感性能。结果一致表明ZnO传感器的气体传感灵敏度取决于晶粒尺寸。因此,晶粒尺寸可以对半导体金属氧化物的气体传感性能产生很大影响,并且合适的晶粒尺寸有利于实际应用。

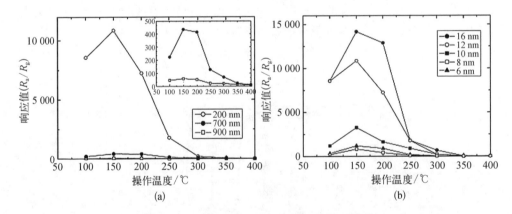

图5.13 薄膜器件对H_2S的传感性能研究[66]

(a) 薄膜器件(与图5.5中相同)在不同温度下对5 ppm H_2S的响应;(b) 与工作温度相关的微晶尺寸(厚度为200 nm)不同的薄膜器件传感器对5 ppm H_2S的响应

由于气体传感器是典型的气-固相互作用反应过程,因此传感特性取决于固相材料的孔隙率。更大的孔隙率可提供更丰富的吸附或反应位点和更大的电阻变化。同时,多孔结构在调节材料比表面积和气体扩散速率方面起着重要作用,平均孔径以及孔的互连程度可以影响气体扩散和传质行为。多孔传感材料可以促进薄膜深处的气体扩散并获得高的气体敏感性。根据 IUPAC 的分类,若材料具有丰富的孔,直径范围为 2.0～50 nm,则可以定义为介孔材料。当孔径大于 50 nm 时称为大孔,而小于 2.0 nm 则定义为微孔。这些丰富的多孔结构有利于许多应用,如分离、储能、催化、气体传感器等[73-75]。对气体传感器而言,大孔能促进气体的扩散但降低了比表面积,而小孔提供了更大的比表面积却缺乏合适的通道用于气体扩散至传感器氧化层。例如,笔者课题组设计了孔径为 29 nm 的有序介孔 n 型 ZnO 半导体材料,相比无孔 ZnO 材料,有序介孔 ZnO[76] 材料具有更优异的乙醇传感性能,如更快的响应时间(6 s)和恢复时间(7 s)、更高的灵敏度和选择性,这主要源于材料内部相互连通的双介孔结构、高的比表面积和结晶性骨架。2018 年,笔者课题组设计合成了孔壁高度晶化的 n 型二维六方介孔 SnO_2 材料(空间群为 $P6_3/mmc$),其对 H_2S 展现出优越的传感性能(选择性约为 170,50 ppm 和高循环稳定性)(见图 5.14),这种高传感性能主要归因于制备的 SnO_2 材料具有高的比表面积(98 m^2/g)、相互连通的介孔结构(孔径约为 18 nm)及晶化孔壁上丰富的活性位点的共同作用[13]。由此可见,丰富的介孔结构有利于气体扩散并暴露出更多的活性位点,从而为气体传感测试提供更多的气体-界面接触点。相似地,Tian 等[77]采用分级 SnO_2 介孔微纤维(孔径为 9.4～13.7 nm)作为甲醛传感器,该微纤维表现出独特的气体灵敏性,原因是合成的 SnO_2 介孔微纤维具有丰富的孔结构和大的比表面积,特别地,孔径及其所得的有效表面积而不是比表面积能显著提高材料的气体传感性能。他们同时发现较小的孔不利于气体转移到更活跃的位点,而较大的孔则允许大多数检测到的气体分子轻易地在材料的较深区域内扩散并且与吸附在这些有效表面上的大量氧化性物质反应。尽管如此,在大孔径材料中,孔径的进一步增加也会对气体传感器产生影响。例如,Xu 等[78]采用不同直径的 PS 微球作为硬模板(200 nm、750 nm 和 1 000 nm),成功制备出不同孔径的还原氧化石墨烯基有序介孔金属氧化物(如 rGO - SnO_2、rGO - Fe_2O_3 和 rGO - NiO)。相比之下,直径为 200 nm 的 PS 微球合成的金属氧化物对 C_2H_5OH 表现出更优异的响应性,这主要源于复合材料之间的异质结效应和更小孔径的结构优势。

除孔径大小和孔隙率外,孔结构的类型也会影响传感性能。Rossinyol 等[48]分别采用 SBA - 15(二维六方结构)和 KIT - 6(三维立方结构)作为硬模板,设计合成了两种具有多种多孔结构的 Cr 掺杂介孔 WO_3 材料。结果表明,由于不同的多孔结构,二维六方 $P6mm$ 型 WO_3 的气敏性低于三维立方 $Ia3d$ 型。KIT - 6 复制出的材料呈海绵状介孔结构,而另一种复合物则倾向于形成具有较弱有序性的聚集体。为了解释多孔系统中的气体扩散过程,研究人员提出了一系列用于各种多孔系统的扩散模型。在大多数介孔体系中,随着孔径的增加,气体扩散依次按以下顺序发生:表面扩散、Knudsen 扩散和分子扩散。当孔隙范围从 1.0 nm 或 2.0 nm

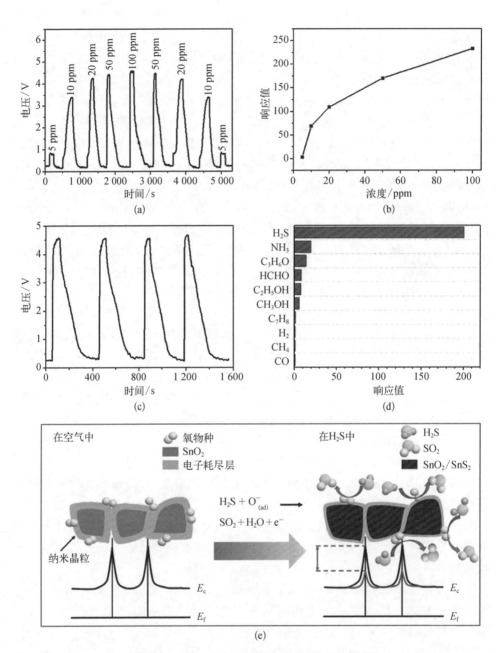

图 5.14 介孔 SnO_2 对 H_2S 的传感性能研究[13]

(a) 在 350℃ 的操作温度下,结晶性介孔 SnO_2-5-450-空气传感器对不同浓度(5~100 ppm)H_2S 的响应和恢复曲线;(b) 气体响应与 350℃ 下 H_2S 浓度之间的关系;(c) 结晶性介孔 SnO_2 基传感器对 50 ppmH_2S 的四次循环曲线;(d) 结晶性介孔 SnO_2 基传感器对 50 ppm 不同气体的响应;(e) 介孔 SnO_2 材料暴露于空气和 H_2S-空气混合物时的传感机制,其中 E_C:导电带边缘;E_F:费米能量

增加到 100 nm 时,通过 Knudsen 扩散引起气体扩散。根据模型,较大的孔径和较小的目标气体分子可以加速气体扩散并改善反应动力学。它假设可燃气体分子与氧化物颗粒相互作用,同时通过孔向内扩散以进入位于内部的颗粒。在稳态条件下,传感层内的气体浓度将随着扩散深度的增加而减小,导致气体浓度分布依赖于扩散速率和表面反应[65,67,77]。因此,对于较大的孔(通常直径大于 50 nm),气体分子的平均自由程小于孔径,这使得目标气体分子容易在框架内扩散。相反,较小的孔隙可以通过孔隙产生明显的气体扩散限制效应[48,78-79]。值得注意的是,气体分子的平均自由程也取决于环境温度和气固比。因此,可获得的平均孔径和介观结构可以有效地调节材料的理化性能和传感过程。

关于介孔金属氧化物的设计合成及其气敏性能的研究已经有很多相关报道。如 Zhu 等[80]发展了配体辅助嵌段共聚物诱导界面的自组装技术,以实验室合成的两亲性嵌段共聚物 PEO-b-PS 作为模板剂,巧妙地采用乙酰丙酮作为配位剂延缓前驱体水解交联的速度,合成了一系列孔道高度连通的高比表面积介孔 WO_3 材料,并首次将该材料用于构建高性能的气体传感器,快速选择性地检测食源性致病菌(见图 5.15)。其中,具有结晶骨架的介孔 WO_3 材料的孔径在 10.6~15.3 nm 范围内可调,比表面积可达 136 m^2/g。介孔 WO_3 材料独特的孔结构以及敏感特性使其对李氏特菌产生的特有气体 3-羟基-2-丁酮具有超快的响应速度(<10 s)、高灵敏度(R_a/R_g>50)和极高的选择性,有望用于快速有效检测食品、水体的微生物污染。此外,该研究团队利用 GC-MS 分析原位鉴定了 3-羟基-2-丁酮敏感材料在反应过程中产物的成分,发现气敏检测反应过程的最终产物是乙酸,而非传统观念中认知的水和二氧化碳,为揭示敏感机理提供了直接的证据,也为优化气体传感器的性能提供了新的思路。

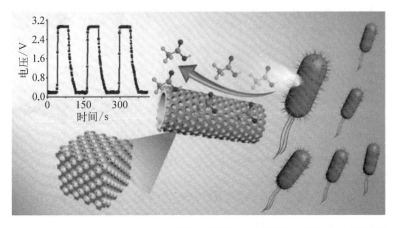

图 5.15　基于有序介孔 WO_3 材料的 3-羟基-2-丁酮的气体传感示意图[80]

大多数报道的介孔氧化钨都具有单介孔结构,这不利于物质扩散和主客体相互作用。到目前为止,合成具有双峰或分级孔和晶体框架的有序介孔 WO_3 仍然是一个巨大的挑战。Li 等[81]提出了一种造孔工程策略,用于合成具有良好连接双峰孔和结晶孔壁的有序介孔 WO_3,其使用亲水性酚醛树脂作为牺牲碳源,可优先与聚环氧乙烷(PEO)相互作用,并用作连接钨物种和聚环氧乙烷-聚苯乙烯嵌段共聚物的黏合剂。所获得的有序介孔氧化钨材料具有双介孔尺寸(5.8 nm 和 15.8 nm)、高比表面积(128 m^2/g)、大窗口尺寸(7.7 nm)和高度结晶的介观结构。双介孔 WO_3 基气体传感器对 H_2S 具有十分优异的气敏性能,即使在低浓度(0.2 ppm)下,也具有快速响应时间(3 s)和恢复时间(14 s)以及高选择性,远优于先前报道的 WO_3 基传感器(见图 5.16)。

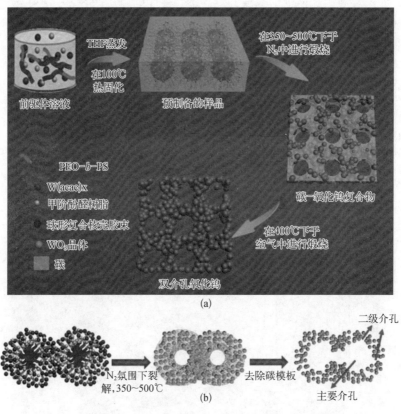

图 5.16 基于 Resol 辅助的双介孔 WO_3 材料的合成示意图[81]

此外,Zhao 等[82]开发了一种受限界面胶束聚集组装方法,用于使用三维 SiO_2 光子晶体(PCs)作为钨前驱体和 PEO-b-PS 模板的受限组装的纳米反应器来合成有序的大孔-介孔 WO_3(OMMW)纳米结构。经过热处理和蚀刻工艺后,所获得的大孔-介孔 WO_3 具有紧密堆积的球形介孔的层次有序多孔纳米结构(约 34.1 nm)、互连空腔(约 420 nm)、高的比表面积(约 78 m^2/g)和高度结晶的骨架。由于双模板的保护,将 OMMW 纳米结构组装到用于检测 H_2S 的气体传感器中时,得到的传感器表现出良好的综合传感性能,包括快速响应-恢复动力学、高选择性和长期稳定性,明显优于之前报道的基于 WO_3 的传感器(见图 5.17)。

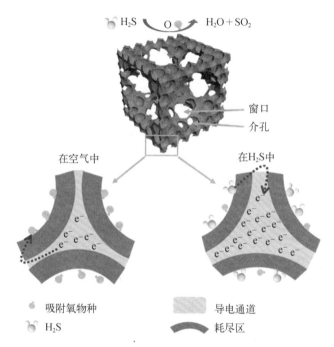

图 5.17　OMMW 暴露于空气(左)和 H_2S-空气混合物(右)中时 OMMW 的 H_2S 传感机理示意图[82]

5.4　比表面与异质结界面

材料的比表面积对于气体传感器也很重要,较大的表面积可以提供足够的接触面积和基于表面氧物种和目标气体之间存在的主要反应的反应界面[83-84]。对于多孔材料,表面积的影响与孔隙率一致,而对于非多孔材料,表面积将对气体传感器显示出独特的影响。例如,Kawi 课题组[85-86]设计了一系列高表面积 SnO_2 半导体,结果表明,较高的表面积有助于使材料对氢气具有较高的灵敏度。根据上述可知,较小的晶粒尺寸可以基于颗粒堆积原理提供较大的表面积,因此,可通过类似的策略得到可用的晶粒尺寸和表面积。此外,近似的线性关系清楚地表明灵敏度对传感器表面积的线性依赖性。同样,Li 等[85]发现 SnO_2 的表面积与气体(如氢气和CO)的敏感性之间具有线性关系。丰富的表面活性位点，如表面吸附的氧物种，通过较大的表面积提供，可引起传感器较大的电导率变化。然而，当孔径增加直到超过临界值时，表面积将随着孔径的增加而减小。因此，为了获得最佳的综合传感性能，必须要求达到孔径和表面积的平衡[83]。

就微结构而言，异质界面是决定基于高效电子结构和有源界面的半导体特性的最重要因素。一旦在两种不同的半导体金属氧化物或其他材料之间的界面处存在紧密的电接触或碰撞，界面中的费米能级将平衡到相同的能量，诱导电荷转移并产生电荷耗尽层[8,87-88]。此外，两种不同的半导体材料通常会发生额外的异常，两种材料暴露在大气中的两个界面非常接近，而这些异质结构在气体传感器中显示出巨大的潜力，因为它具有高的表面积和协同效应[16,87]。根据参与半导体的类型，异质结构由四个经典模型组成，即 p-n 异质结、n-n 异质结、p-p 异质结和 n-p-n 异质结。p-n 异质结是多学科领域中使用最广泛的界面效应，p-n 异质结附近的电子和空穴复合可定义为"费米能级介导的电荷转移"[89-90]。例如，用 p 型 Co_3O_4 纳米颗粒修饰 n 型 ZnO 纳米线，这种复合纳米线异质结材料对 NO_2 和 C_2H_5OH 均表现出优异的传感性能[91]。纳米线在空气中的正常环境电阻甚至高于没有异质结的单体材料，通过延伸到 ZnO 纳米线的异质结界面的耗尽区，减小了电荷传导通道的宽度。此外，较小的横截面积可用于纳米线中的电荷传导并引起较高的电阻，跨越 p-n 界面的电荷传导可以进一步促进电阻的增加。

早在20世纪中期，许多科学家就已经对设计和制造具有异质结界面的 SMO 传感器产生了极大的关注。如1979年，Waldrop 和 Grant[92] 首先研究了半导体异质结界面的能带不连续性，发现它是非传递性的，为异质结的应用提供了理论参考。之后，Zhang 等[89] 通过结合共沉淀技术和水热生长法设计了花状 p-CuO/n-ZnO 异质结纳米棒，该纳米棒对 C_2H_5OH 表现出优异的气敏特性(见图5.18)，其中 p-CuO/n-ZnO(1∶4)传感器对 100 ppm C_2H_5OH 的响应达到 98.8，是纯 n-ZnO 的 2.5 倍，且响应恢复时间达到 7 s 和 9 s。这种快速响应和高灵敏度取决于材料内部较宽的耗尽层和 p-CuO/n-ZnO 异质结界面效应，其为电子传输提供了足够的能量。同样，Ju 等[93] 通过化学沉积技术设计了由 n 型 SnO_2 空心球和 p 型 NiO 纳米粒子结合而成的 p-n 异质结(见图5.18)，该 NiO/SnO_2 异质结对三甲胺(TEA)气体具有高响应(10 ppm 响应值高达 48.6)和选择性，以及较低的检测限(2 ppm)，且最佳工作温度相比纯 SnO_2 空心球材料低 40℃。这种优异的气体传感特性可能是因为在 p-n 异质结界面处形成了耗尽层，从而导致材料在接触空气和 TEA 气体时电阻产生显著变化。

n-n 异质结和 p-p 异质结也可以改变电子结构并调整传感特性，其特殊界面都是由两个均匀半导体组合而成的。例如，Xu 等[87] 成功制备了 TiO_2 纳米颗粒/

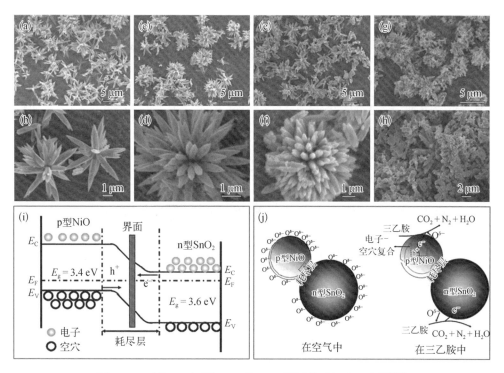

图 5.18 p 型 NiO/n 型 SnO$_2$ 的 SEM 图及传感机理示意图[93]

(a)、(b) 纯 ZnO 的低倍率和高倍率 FESEM 图；CuO/ZnO(0.125∶1)的低倍率和高倍率 FESEM 图；(e)、(f) CuO/ZnO(0.25∶1)的低倍率和高倍率 FESEM 图；(g)、(h) CuO/ZnO(0.5∶1)的低倍率和高倍率 FESEM 图；(i) p 型 NiO/n 型 SnO$_2$ 异质接触的能带结构图；(j) 暴露于 TEA 气体时 p 型 NiO/n 型 SnO$_2$ 异质结传感器的示意模型

SnO$_2$ 纳米片 n-n 异质结复合材料，该材料表现出对 TEA 气体的高传感特性，对 100 ppm 的 TEA 气体响应值高达 52.3（最佳工作温度为 260℃），传感性能显著高于纯的 SnO$_2$ 纳米片（响应值约为 3，最佳工作温度为 320℃）。这种特殊的 n-n 异质结界面形成的耗尽层能够有效降低材料在 TEA 气体中的电阻并增加其在空气氛围中的电阻。Ma 等[94]通过简单的声化学合成技术制备了 n-ZnO/n-In$_2$O$_3$ 异质结。该异质结界面可以增强材料对 HCHO 的气体响应，并通过调节电子转移的速率以降低检测限（见图 5.19）。In$_2$O$_3$ 纳米结构表面上的 ZnO 聚结层促进了 ZnO/In$_2$O$_3$ 界面上 n-n 异质结的形成，其被认为是电子转移的桥梁，以增强电阻的变化。此外，n-n 异质结界面容易将电子转移到较低能量导带中，产生"累积层"而不是耗尽层[26,93,95]。形成的累积层将通过在主基板表面上的后续氧吸附而耗尽，这可以进一步改善界面处的势垒并提高响应灵敏度。除 n-n 异质结外，p-p 异质结也被设计并应用于许多领域，相互作用过程类似于 n-n 异质结。最后一种典型的异质结是 n-p-n 异质结，它由两个以上的半导体相结合。Huang 等[9]

在 SnO_2@ZnO 核壳纳米棒中设计了典型的 n-p-n 三元异质结构,该结构分别由 n 型 ZnO、p 型 Zn-O-Sn 及 n 型 SnO_2 异质结构构成,这种复合结构具有独特的理化性能,对氢气具有较高的选择性和灵敏度[9]。研究结果证实在界面处产生了混合的 Zn-O-Sn 相,具有 Sn^{4+} 位点上 Zn^{2+} 的受体型掺杂的本征 p 型半导体特征。此外,界面处主要的载流子形式从电子主导型变为空穴主导型,这是由表面吸附产生电子耗尽层及 Zn^{2+} 置换 Sn^{4+} 产生大量空穴而导致的。因此,合理的设计和制造具有异质结的 SMO 材料对于气体传感器是非常重要的,且我们应该选择有效的基底材料和额外的异质部件。

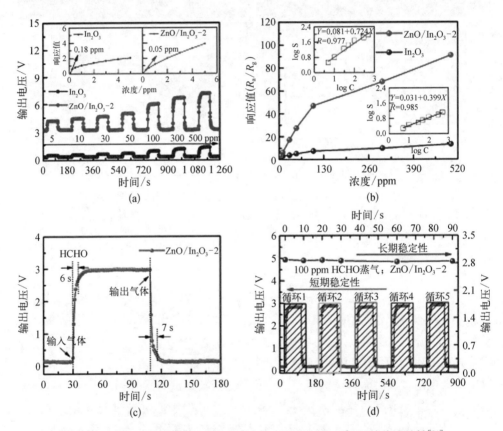

图 5.19 In_2O_3 纳米结构和 ZnO/In_2O_3-2 异质结在 300℃下的传感特性[94]

(a) 传感器对 HCHO 的响应随浓度(5~500 ppm)的变化;(b) 气体传感器对 HCHO 浓度的敏感性响应;(c) ZnO/In_2O_3-2 暴露于 100 ppm HCHO 的响应/恢复时间;(d) ZnO/In_2O_3-2 传感器暴露于 100 ppm HCHO 的稳定性

Yang 等[96]报道了具有丰富的表面氧物种和氧空位的多孔 n-n 异质结 CeO_2/SnO_2 纳米片,该纳米片显示出优异的 3-羟基-2-丁酮(3H-2B)的气敏性能(见图 5.20)。值得一提的是,结晶金属氧化物纳米片因大的表面体积比和量子

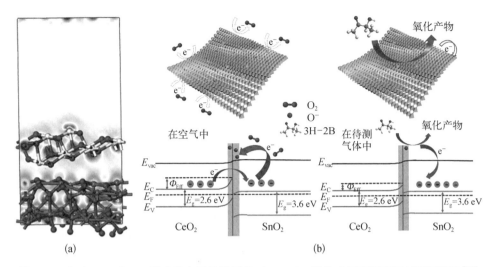

图 5.20 多孔 CeO_2/SnO_2 纳米的电子结构及其对 3H-2B 的传感机理示意图（后附彩图）[96]

(a) CeO_2/SnO_2 纳米片中电荷密度差异的侧视图，其中红色和蓝色区域分别表示电子积累和耗尽，Sn、Ce 和 O 原子分别用灰色、白色和红色标记；(b) 多孔 CeO_2/SnO_2 纳米片对 3H-2B 的表面反应示意图以及相应的传感机理图

限制效应显示出优异的催化性能。然而，开发一种简便、通用的方法来合成金属氧化物纳米片仍然是一个挑战。上述研究中，研究人员采用一种共晶诱导的空间自约束方法来合成金属氧化物纳米片。以 SnO_2 的合成为例，KCl 和 $SnCl_2$ 溶液中的溶剂蒸发导致 KCl 和 K_2SnCl_6 共结晶，所获得的包覆 K_2SnCl_6 的复合材料可原位转化为限制在 KCl 基质中的 SnO_2 纳米片，经水洗去除 KCl 后，可获得多孔 SnO_2 纳米片。这种策略具有非常广泛的普适性，可以通过这种通用且有效的绿色路线获得一系列金属氧化物纳米片。金属氧化物的显著表面/界面效应在表面催化反应中起着重要作用，这也有助于增强传感性能和改善吸附、解吸过程。因此，通过这种策略合成的 2D 多孔 CeO_2/SnO_2 纳米片暴露于 3H-2B 时能够显著催化 3H-2B 的氧化。根据 GC-MS 测试结果，以 CeO_2/SnO_2 为敏感材料时，检测过程中可以检测到 3-甲基-2,4-戊二酮和乙酸作为 3H-2B 气敏催化过程中的主要产物，而对于 SnO_2 样品中仅生成乙酸，表明 CeO_2/SnO_2 纳米片能够显著促进 3H-2B 的催化氧化。此外，该研究还分析了在 3H-2B 中暴露后的传感材料的 XPS 结果，发现晶格氧浓度和 Ce^{4+} 浓度均下降，揭示了目标分子可以通过 Mars-van Krevelen 机制与 CeO_2/SnO_2 界面中的晶格氧反应。此外，这项研究还对多孔纳米片的结构对传感性能的影响进行了探究，通过制备具有不规则形态的晶体 CeO_2/SnO_2 复合材料，在 160℃对 50 ppm 3H-2B 的气敏性能进行测试。结果显示，CeO_2/SnO_2 块体复合材料的传感响应值（R_a/R_g）为 164，远低于 2D 多孔 CeO_2/SnO_2 纳米片（$R_a/R_g=623$）。此外，CeO_2/SnO_2 复合材料的响应和恢复时间分别为 259 s 和

1 186 s，比 2D 多孔 CeO_2/SnO_2 纳米片（59 s 和 172 s）慢得多，这表明多孔纳米片结构可以显著提高传感性能。一方面，CeO_2 可以均匀地分散在多孔 2D SnO_2 纳米片上，使其具有更多的表面 n-n 异质结，并提供大量可用的 O^-（ads）物种，得到更好的传感性能；另一方面，多孔纳米片结构赋予材料高的表面积和短的扩散路径，显著提高气体的吸附和解吸速率。因此，具有纳米片形态的多孔结构是设计高性能气体传感器的理想选择。

5.5 晶体结构和内部缺陷

众所周知，SMO 传感器的化学活性强烈依赖于结晶度（即单晶、多晶和混晶等）[97-99]，晶形（即 α、β、γ 等）[100] 和暴露的晶面［即（110）、（101）和（100）对应于 SnO_2，（0001）、(10-10) 和 (10-11) 对应于 ZnO 等］[101-102]。本节将介绍晶体结构和内部缺陷对材料实际传感应用的相关影响。

在大多数情况下，所制备的或天然的金属氧化物半导体以多晶或混晶的形式存在，这是由于普通介质下的高表面能。2016 年，笔者课题组[54]设计了有序介孔碳-钴氧化物纳米复合材料，它具有高结晶度，同时属于混晶，通过改变退火温度可以很容易地调节结晶度。该纳米复合材料显示出优异的氢气传感性能。随后，笔者课题组[80]报道了具有高结晶骨架（混合晶体）的介孔 WO_3，其对 3-羟基-2-丁酮的检测性能优于低结晶性的 WO_3。除了普通的混晶之外，多晶半导体在实际生产中也很重要。例如，Gurlo 等[103]报道了多晶立方和六方 In_2O_3 在低水平下对臭氧表现出高灵敏度。Wang 等[104]通过溶液相反应制备了多晶金红石相 SnO_2 纳米线，该材料对 CO 和氢气有类似的传感行为。此外，Zhou 等认为通过气体分子的吸附和解吸实现电阻的变化是由于晶粒尺寸小、结晶性高和表面体积比大[7,67,69]。同样，Wang 等[105]描述了具有可控直径的多晶 WO_3 纳米纤维（见图 5.21），该纳米纤维显示出在一系列浓度下对氨的快速响应。此外，将微晶尺寸减小到超过"纳米"尺寸时可以得到更大比例的微晶表面原子，更容易与周围环境反应或表现出不寻常的结构特征，表明多晶结构有利于调整材料内部微观结构。有趣的是，Fan 等[98]研究了在紫外光下一维和二维多晶 ZnO 对氢气的气体传感行为，发现多晶结构保持了较高的稳定性和化学活性。

虽然单晶材料难以形成和生长，但许多科学家仍成功地设计并制造了一系列具有优异的传感性能和完美晶体结构的单晶金属氧化物。许多常见的单晶 SMO 材料（即 ZnO、SnO、Fe_2O_3、In_2O_3 和 WO_3）已经应用于多个传感装置[99,106-108]。例如，Rai 等[106]合成了沿 c 轴生长的单晶 ZnO 纳米粒子，该材料具有六方纤锌矿结构，在低温下对乙醛表现出高选择性和灵敏度。如图 5.22 所示，气体传感机制可归因于目标气体和 ZnO 纳米颗粒之间的化学/电子相互作用引发的电阻变化。化

图 5.21 WO$_3$/PVAc 复合纳米纤维的 SEM 图[105]

(a)~(d) WO$_3$/PVAc 复合纳米纤维包含 W(iPr)$_6$ 体积百分比分别为 20%、25%、30% 和 35% 的 SEM 图；(e)~(h) 对应材料煅烧后的相应产物

学相互作用涉及目标气体在 ZnO 表面上的吸附，随后与表现吸附氧反应从而实现对目标气体的检测。此外，Rai 等[107]通过传统的微波辅助水热法设计了单晶 ZnO 纳米棒，该纳米棒对乙醇(250~50 ppm)和 CO(1 000~200 ppm)均具有较好的检测性能。Liu 等[97]则报道了具有丰富多孔结构的新型单晶 ZnO 纳米片，其显示出对甲醛和氨较高的气敏响应、高选择性、短的响应/恢复时间，以及用于检测甲醛和氨的长期稳定性。因此，单晶结构可以改善材料的半导体性能并提高材料的长期稳定性。

相比之下，单晶 SnO$_2$ 材料也得到了关注和研究，Cheng 等[109]首次通过溶液反应合成了单晶 SnO$_2$ 纳米棒，其显示出可调节的电学、光学、磁学和化学性质。此外，Chen 等[99]也制造了具有窄直径(4~15 nm)的单晶 SnO$_2$ 纳米棒，其对 300 ppm C$_2$H$_5$OH 显示出高灵敏度(约 31.4)。除了纯单晶 SnO$_2$ 半导体外，Wan 和 Wang[33]制备了单晶 Sb 掺杂的 SnO$_2$ 纳米线，并将其应用于 C$_2$H$_5$OH 的检

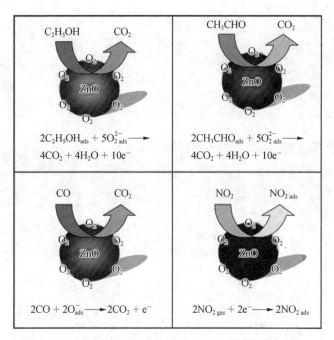

图 5.22 ZnO 纳米颗粒对不同气体的气敏机理示意图

测,具有 Sb 掺杂的单晶结构可以提高基于取向纹理和晶格畸变的传感性能。Li 等[110]报道了高长径比的单晶多孔 In_2O_3 纳米带,该纳米带在 ppb 水平下检测各种气体(如甲醇、乙醇和丙酮)方面显示出巨大的传感潜力。在某种程度上,单晶 SMO 材料表现出优于其他晶体的传感和理化性能。然而,这些晶体的载流子迁移率(幅度和温度依赖性)几乎相同并且对晶粒边界不敏感,此外,潜在的载流子迁移率强烈依赖于晶格的取向和暴露的晶面。一旦晶粒边界对载流子迁移率形成较大的限制,多晶材料中的霍尔迁移率将显著小于单晶材料[98-100]。

除了材料的结晶度可以影响基于 SMO 的传感器的传感性能,晶体形式也可以改变传感反应的基本特征。最常见的表现形式是具有丰富晶型的 Fe_2O_3 基材料,如 $\alpha\text{-}Fe_2O_3$、$\beta\text{-}Fe_2O_3$、$\gamma\text{-}Fe_2O_3$、$\delta\text{-}Fe_2O_3$、$\eta\text{-}Fe_2O_3$ 和 $\varepsilon\text{-}Fe_2O_3$,这些不同的形式将表现出不同的物理和化学性质[111-113]。因此,以 Fe_2O_3 基材料为例,详细讨论晶形在传感过程中的作用。$\alpha\text{-}Fe_2O_3$ 也被称为赤铁矿,其与最稳定的氧化铁一起存在,属于 n 型半导体。基于其低成本、高稳定性和高化学活性等特征,该类材料对于还原性分子或气体(如丙酮、C_2H_5OH 和 H_2S 等)具有巨大的传感应用潜力[111-112,114]。相比之下,$\beta\text{-}Fe_2O_3$ 是 Fe_2O_3 的亚稳态方铁锰矿相,是一种极为稀有的金属氧化物,$\beta\text{-}Fe_2O_3$ 在气体传感器中的应用很少有报道,这可能是由于其化学稳定性较弱。但是，Carraro 等[115]首次设计了贵金属(Au 和 Ag)纳米粒子功能性 β-Fe_2O_3，并将其用于气体传感器领域。合成的贵金属敏化的多功能 β-Fe_2O_3 材料可以检测多种气体，如氢气、C_2H_5OH 和丙酮等。γ-Fe_2O_3 也称为磁赤铁矿，是典型的缺陷型氧化物，具有晶胞$(3/4)[Fe^{3+}[Fe_{5/3}^{3+}\square_{1/3}]O_4]$（"$\square$"代表空位）。许多研究发现 γ-Fe_2O_3 不仅可以检测普通的挥发性有毒气体，而且可以用于检测常见的可燃气体，包括液化石油气、氢气和有毒性可燃气体。Biswal[116]通过声化学方法合成的纳米 γ-Fe_2O_3 对丙酮具有高灵敏度和优异的检测性能。截至目前，暂无关于 δ-Fe_2O_3 及其衍生物的报道，但该类材料亦可能具有良好的气敏特性。对于 η-Fe_2O_3，人们发现 η-Fe_2O_3 微球具有特殊的空芯-壳结构，且对 0.5 ppm 丙酮和 C_2H_5OH 均表现出优异的灵敏度（15.69 和 8.15），而纯 η-Fe_2O_3 粉体材料往往因其组成不唯一或均匀的微观形态而在传感过程中没有明显的应用价值。就 ε-Fe_2O_3 而言，它是一种稀有的亚稳相，Peeters 等[117]合成了 Au 掺杂的 ε-Fe_2O_3 材料，该材料在中等工作温度下可以选择性检测气态 NO_2，并展现出非常优异的检测性和灵敏度。最重要的是，研究还发现各种晶体形式将在传感过程中具有产生明显的多样性。在这些 Fe_2O_3 基传感材料中，即便不采用任何手段进行掺杂、功能化或修饰，仅仅使用纯的单体材料，α-Fe_2O_3 和 γ-Fe_2O_3 也具有优异的传感潜力，这归因于其特有的表面缺陷和电子结构。

在实际应用中，SMO 材料的裸露晶面可以在气体传感器中发挥另一个重要作用。最常见的是 TiO_2、SnO_2 和 ZnO 类传感材料，并且基于 TiO_2 的材料暴露的晶面不仅在气体传感器中，而且在其他各种应用中也能够产生强烈的性能变化。不同的晶面/表面通常具有不同的几何和电子结构，这导致不同的功能特性。许多研究明确指出 SMO 半导体的气体传感活性取决于暴露在环境气体中表面的性质[19,101,118]。因此，控制和调节暴露的晶面对于提高气体传感的选择性和灵敏度是必要的。宏观 SnO_2 被(110)、(101)和(100)晶面包围，其具有金红石型微结构（空间群 $P4_2mnm$，Nr. 136，$a=4.734$ Å，$c=3.185$ Å，$Z=2$）。这些晶面通常具有不相等的势能面，这将改变材料的化学活性和表面稳定性[119]。通常，(110)晶面由于氧势能、温度和晶体尺寸而具有最稳定的界面能[120]。Han 等[121]首先提供了许多证据并成功地制备了具有(221)和(110)晶面的不同比例形状控制的 SnO_2 晶体，其中独特的(221)晶面是均匀的四方双锥（见图 5.23）。研究发现，这些不同比例的(221)和(110)晶面暴露于 VOC 中时，SnO_2 晶体对 C_2H_5OH 的传感器响应产生了明显的变化[122]。(221)晶面的较高感应活性主要是由于这些表面增强的能量促进了反应性的提高，且(221)晶面含有的不饱和阳离子往往被视为表面活性位点。另一种典型的是六方 ZnO，其具有纤锌矿型结构（空间群 $P6_3mc$，Nr. 186，$a=3.249$ Å，$c=5.205$ Å，$Z=2$）。共同的晶面由(0001)、(10-10)和(10-11)组成，研究发现(0001)晶面具有高感应活性，容易对 C_2H_5OH 产生高的敏感性和选择性,而不是(10-10)或(10-11)晶面[101]。有趣的是,非极性(00-10)晶面含有丰富的悬空键和Zn-O表面二聚体,这使得它们对NO_2和氧的吸附具有高反应性和选择性,而极性(0001)晶面的反应性,无论是Zn-终止还是O-终止,均取决于其极性。实际上,除SnO_2和ZnO之外,In_2O_3和WO_3也具有相似的晶面特性。大部分研究都集中在用一组对称(hkl)平面封闭的明确定义的形状的可用性,以及暴露的晶面与气体传感特性之间的关系。

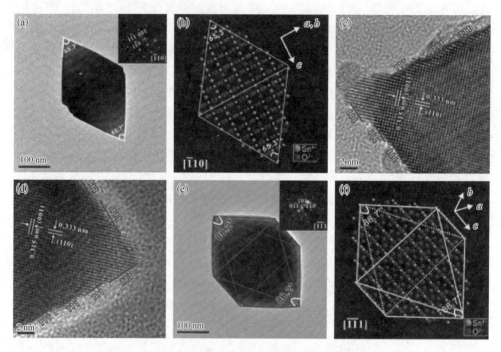

图5.23 八面体SnO_2颗粒的TEM图[121]

(a)沿[110]方向观察的八面体SnO_2颗粒的典型低倍放大TEM图,其中插图为相应的SEAD模式;(b)沿[110]方向投影的(221)小平面内的理想SnO_2八面体的示意模型;(c)从八面体的顶部顶点取得的HRTEM图;(d)从八面体的底部顶点取得的HRTEM图;(e)沿[111]方向观察的相同SnO_2八面体颗粒的低倍放大TEM图,其中插图为相应的SAED模式;(f)在[111]方向上投影的理想SnO_2八面体的示意模型,投影在[111]方向上

另一方面,由晶体在晶格位置处的平衡振动而形成的材料缺陷,如点缺陷、线缺陷和平面缺陷,对SMO材料上的气体靶的吸附过程具有显著影响[123-125]。通过诱导电荷转移和调节其表面反应性,这些独特缺陷对电子结构和光学性质具有经典的实质性影响。在气体传感器中,基于SMO的材料的响应是通过氧与离子和电子点缺陷的平衡产生的,并且在大多数金属氧化物中,固有的离子点缺陷是典型的氧空位(如肖特基缺陷)。在相对低的氧分压下,主要的点缺陷是氧空位和电子;而在相对高的氧分压下,主要的点缺陷是金属空位和空穴[14,126]。此外,

氧分压取决于电子传导率，并且这种相互依赖的相关性与供体掺杂和元素比例相关。体晶粒内的点缺陷需要更长的扩散到表面的距离，并且它可以增加传感器响应和界面活动的速度。固体表面或晶面由分离的杂质、充当电子源和吸收剂的吸附气体以及相关的空间电荷区组成[1,124-125]。因此，缺陷和相关电荷载体的控制在利用基于 SMO 的材料的广泛性质的应用中是至关重要的。以 ZnO 为例，ZnO 的带隙内存在丰富的缺陷态，包括供体缺陷和受体缺陷，缺陷电离能为 0.05~2.8 eV。这种变化将导致各种缺陷的相对浓度强烈依赖于环境温度，并且缺陷控制的 ZnO 纳米线表现出对 NO_2 的快速响应和恢复行为[46,72,127]。研究发现气体灵敏度与制造的和缺陷控制的气体传感器中的氧空位相关缺陷的光致发光强度成线性比例。

同时，线缺陷也被定义为一维缺陷或简单的位错，与二维和三维缺陷相比，线缺陷是最简单的扩展缺陷，具有较少的断裂键，因此具有较低的形成能[128-129]。它还可以对 SMO 材料的传感性能产生很大的影响，例如，Adepalli 等[129]合成了单晶 TiO_2 并讨论了线缺陷的影响。研究发现，一旦材料从主要的电子传导向离子传导，离子和电子浓度都会局部增加，故线缺陷对调节活性和空间分布很重要。尽管如此，由于线缺陷的形成过程复杂，SMO 材料的线缺陷很少在气体传感器中产生或使用。

参考文献

[1] Basu S, Bhattacharyya P. Recent developments on graphene and graphene oxide based solid state gas sensors[J]. Sens. Actuators B, 2012, 173: 1-21.

[2] Comini E, Faglia G, Sberveglieri G, et al. Stable and highly sensitive gas sensors based on semiconducting oxide nanobelts[J]. Appl. Phys. Lett., 2002, 81: 1869-1871.

[3] Natale C D, Paolesse R, Martinelli E, et al. Solid-state gas sensors for breath analysis: a review[J]. Anal. Chim. Acta, 2014, 824: 1-17.

[4] Fergus J W. Solid electrolyte based sensors for the measurement of CO and hydrocarbon gases[J]. Sens. Actuators B, 2007, 123: 1169-1179.

[5] Fine G F, Cavanagh L M, Afonja A, et al. Metal oxide semi-conductor gas sensors in environmental monitoring[J]. Sensors, 2010, 10: 5469-5502.

[6] Zhou X, Lee S, Xu Z, et al. Recent progress on the development of chemosensors for gases[J]. Chem. Rev., 2015, 115: 7944-8000.

[7] Zhou X, Cheng X, Zhu Y, et al. Ordered porous metal oxide semiconductors for gas sensing[J]. Chin. Chem. Lett. 2017, 29: 405-416.

[8] Miller D R, Akbar S A, Morris P A. Nanoscalemetal oxide-based heterojunctions for gas sensing: A review[J]. Sens. Actuators B, 2014, 204: 250-272.

[9] Huang H, Gong H, Chow C L, et al. Low-temperature growth of SnO_2 nanorod arrays and tunable n-p-n sensing response of a ZnO/SnO_2 heterojunction for exclusive hydrogen sensors[J]. Adv. Funct. Mater., 2011, 21: 2680-2686.

[10] Fu D, Zhu C, Zhang X, et al. Two-dimensional net-like SnO_2/ZnO heteronanostructures for high-performance H_2S gas sensor[J]. J. Mater. Chem. A, 2016, 4: 1390-1398.

[11] Dhawale D S, Salunkhe R R, Patil U M, et al. Room temperature liquefied petroleum gas (LPG) sensor based on p-polyaniline/n-TiO_2 heterojunction[J]. Sens. Actuators B, 2008, 134: 988-992.

[12] Ma J, Ren Y, Zhou X, et al. Pt nanoparticles sensitized ordered mesoporous WO_3 semiconductor: gas sensing performance and mechanism study[J]. Adv. Funct. Mater., 2017, 28: 1705268-1705279.

[13] Xiao X, Liu L, Ma J, et al. Ordered mesoporous tin oxide semiconductors with large pores and crystallized walls for high-performance gas sensing[J]. ACS Appl. Mater. Interfaces, 2018, 10: 1871-1880.

[14] Ahn M W, Park K S, Heo J H, et al. Gas sensing properties of defect-controlled ZnO-nanowire gas sensor[J]. Appl. Phys. Lett., 2008, 93: 263103-263106.

[15] Espid E, Taghipour F. Development of highly sensitive ZnO/In_2O_3 composite gas sensor activated by UV-LED[J]. Sens. Actuators B, 2017, 241: 828-839.

[16] Wu H, Kan K, Wang L, et al. Electrospinning of mesoporous p-type In_2O_3/TiO_2 composite nanofibers for enhancing NO_x gas sensing properties at room temperature[J]. CrystEngComm, 2014, 16: 9116-9124.

[17] Wen Z, Zhu L, Mei W, et al. Rhombus-shaped Co_3O_4 nanorod arrays for high-performance gas sensor[J]. Sens. Actuators B, 2013, 186: 172-179.

[18] Wang J, Wei L, Zhang L, et al. Zinc-doped nickel oxide dendritic crystals with fast response and self-recovery for ammonia detection at room temperature[J]. J. Mater. Chem., 2012, 22: 20038-20047.

[19] Wang C, Yin L, Zhang L, et al. Metal oxide gas sensors: sensitivity and influencing factors[J]. Sensors, 2010, 10: 2088-2106.

[20] Li Y, Luo W, Qin N, et al. Highly ordered mesoporous tungsten oxides with a large pore size and crystalline framework for H_2S sensing[J]. Angew. Chem. Int. Ed., 2014, 53: 9035-9040.

[21] Ren Y, Zhou X, Luo W, et al. Amphiphilic block copolymer templated synthesis of mesoporous indium oxides with nanosheet-assembled pore walls[J]. Chem. Mater., 2016, 28: 7997-8005.

[22] Wen Z, Zhu L, Li Y, et al. Mesoporous Co_3O_4 nanoneedle arrays for high-performance gas sensor[J]. Sens. Actuators B, 2014, 203: 873-879.

[23] Nguyen H, El-Safty S A. Meso- and macroporous Co_3O_4 nanorods for effective VOC gas sensors[J]. J. Phys. Chem. C, 2011, 115: 8466-8474.

[24] Han D, Zhai L, Gu F, et al. Highly sensitive NO_2 gas sensor of ppb-level detection based on In_2O_3 nanobricks at low temperature[J]. Sens. Actuators B, 2018, 262: 655-663.

[25] Garcia-Sanchez R F, Ahmido T, Casimir D, et al. Thermal effects associated with the Daman spectroscopy of WO_3 gas-sensor materials[J]. J. Phys. Chem. A, 2013, 117: 13825-13831.

[26] Dandeneau C S, Jeon Y H, Shelton C T, et al. Thin film chemical sensors based on p-

CuO/n‑ZnO heterocontacts[J]. Thin Solid Films, 2009, 517: 4448‑4454.

[27] Rahman M M, Ahammad A J, Jin J H, et al. A comprehensive review of glucose biosensors based on nanostructured metal-oxides[J]. Sensors, 2010, 10: 4855‑4886.

[28] Mirzaei A, Leonardi S G, Neri G. Detection of hazardous volatile organic compounds (VOCs) by metal oxide nanostructures-based gas sensors: A review[J]. Ceram. Int., 2016, 42: 15119‑15141.

[29] Eranna G, Joshi B C, Runthala D P, et al. Oxide materials for development of integrated gas sensors—A comprehensive review[J]. Solid State Mater. Sci., 2010, 29: 111‑188.

[30] Arafat M M, Dinan B, Akbar S A, et al. Gas sensors based on one dimensional nanostructured metal-oxides: A review[J]. Sensors, 2012, 12: 7207‑7258.

[31] Wei B Y, Hsu M C, Su P G, et al. A novel SnO_2 gas sensor doped with carbon nanotubes operating at room temperature[J]. Sens. Actuators B, 2004, 101: 81‑89.

[32] Shishiyanu S T, Shishiyanu T S, Lupan O I. Sensing characteristics of tin-doped ZnO thin films as NO_2 gas sensor[J]. Sens. Actuators B, 2005, 107: 379‑386.

[33] Wan Q, Wang T H. Single-crystalline Sb-doped SnO_2 nanowires: synthesis and gas sensor application[J]. Chem. Commun., 2005, 30: 3841‑3843.

[34] Righettoni M, Tricoli A, Gass S, et al. Breath acetone monitoring by portable Si: WO_3 gas sensors[J]. Anal. Chim. Acta, 2012, 738: 69‑75.

[35] Zhang Y, He W, Zhao H, et al. Template-free to fabricate highly sensitive and selective acetone gas sensor based on WO_3 microspheres[J]. Vacuum, 2013, 95: 30‑34.

[36] Su P G, Peng Y T. Fabrication of a room-temperature H_2S gas sensor based on PPy/WO_3 nanocomposite films by in-situ photopolymerization[J]. Sens. Actuators B, 2014, 193: 637‑643.

[37] Ling Z, Leach C. The effect of relative humidity on the NO_2 sensitivity of a SnO_2/WO_3 heterojunction gas sensor[J]. Sens. Actuators B, 2004, 102: 102‑106.

[38] Kukkola J, Mohl M, Leino A R, et al. Inkjet-printed gas sensors: metal decorated WO_3 nanoparticles and their gas sensing properties[J]. J. Mater. Chem., 2012, 22: 17878‑17886.

[39] Adami A, Lorenzelli L, Guarnieri V, et al. A WO_3‑based gas sensor array with linear temperature gradient for wine quality monitoring[J]. Sens. Actuators B, 2006, 117: 115‑122.

[40] Lee I, Choi S J, Park K M, et al. The stability, sensitivity and response transients of ZnO, SnO_2 and WO_3 sensors under acetone, toluene and H_2S environments[J]. Sens. Actuators B, 2014, 197: 300‑307.

[41] Kruefu V, Wisitsoraat A, Tuantranont A, et al. Ultra-sensitive H_2S sensors based on hydrothermal/impregnation-made Ru-functionalized WO_3 nanorods[J]. Sens. Actuators B, 2015, 215: 630‑636.

[42] Kim S J, Hwang I S, Na C W, et al. Ultrasensitive and selective C_2H_5OH sensors using Rh-loaded In_2O_3 hollow spheres[J]. J. Mater. Chem., 2011, 21: 18560‑18567.

[43] Wang S, Xiao B, Yang T, et al. Enhanced HCHO gas sensing properties by Ag-loaded sunflower-like In_2O_3 hierarchical nanostructures[J]. J. Mater. Chem. A, 2014, 2: 6598‑

6604.

[44] Zhao J, Yang T, Liu Y, et al. Enhancement of NO_2 gas sensing response based on ordered mesoporous Fe-doped In_2O_3[J]. Sens. Actuators B, 2014, 191: 806-812.

[45] Vyas R, Kumar P, Dwivedi J, et al. Probing luminescent Fe-doped ZnO nanowires for high-performance oxygen gas sensing application[J]. RSC Adv., 2014, 4: 54953-54959.

[46] Schmidt-Mende L, MacManus-Driscoll J L. ZnO-nanostructures, defects, and devices[J]. Mater. Today, 2007, 10: 40-48.

[47] Liu C, Wang B, Wang T, et al. Enhanced gas sensing characteristics of the flower-like $ZnFe_2O_4/ZnO$ microstructures[J]. Sens. Actuators B, 2017, 248: 902-909.

[48] Rossinyol E, Prim A, Pellicer E, et al. Synthesis and characterization of chromium-doped mesoporous tungsten oxide for gas sensing applications[J]. Adv. Funct. Mater., 2007, 17: 1801-1806.

[49] Tabassum R, Mishra S K, Gupta B D. Surface plasmon resonance-based fiber optic hydrogen sulphide gas sensor utilizing Cu-ZnO thin films[J]. Phys. Chem. Chem. Phys., 2013, 15: 11868-11874.

[50] Qin J, Cui Z, Yang X, et al. Synthesis of three-dimensionally ordered macroporous $LaFeO_3$ with enhanced methanol gas sensing properties[J]. Sens. Actuators B, 2015, 209: 706-713.

[51] Wang N, Shen K, Huang L, et al. Facile route for synthesizing ordered mesoporous Ni-Ce-Al oxide materials and their catalytic performance for methane dry reforming to hydrogen and syngas[J]. ACS Catal., 2013, 3: 1638-1651.

[52] Yang H, Zhang X, Li J, et al. Synthesis of mesostructured indium oxide doped with rare earth metals for gas detection[J]. Microporous Mesoporous Mater., 2014, 200: 140-144.

[53] Zhang Y, Yang Q, Yang X, et al. One-step synthesis of in-situ N-doped ordered mesoporous titania for enhanced gas sensing performance[J]. Microporous Mesoporous Mater., 2018, 270: 75-81.

[54] Wang Z, Zhu Y, Luo W, et al. Controlled synthesis of ordered mesoporous carbon-cobalt oxide nanocomposites with large mesopores and graphitic walls[J]. Chem. Mater., 2016, 28: 7773-7780.

[55] Liu J, Wang T, Wang B, et al. Highly sensitive and low detection limit of ethanol gas sensor based on hollow ZnO/SnO_2 spheres composite material[J]. Sens. Actuators B, 2017, 245: 551-559.

[56] Zhou X, Cao Q, Huang H, et al. Study on sensing mechanism of $CuO-SnO_2$ gas sensors [J]. Mater. Sci. Eng. B, 2003, 99: 44-47.

[57] Hu Y, Zhou X, Han Q, et al. Sensing properties of CuO-ZnO heterojunction gas sensors [J]. Mater. Sci. Eng. B, 2003, 99: 41-43.

[58] Dong C, Liu X, Han B, et al. Nonaqueous synthesis of Ag-functionalized In_2O_3/ZnO nanocomposites for highly sensitive formaldehyde sensor[J]. Sens. Actuators B, 2016, 224: 193-200.

[59] Choi K S, Park S, Chang S P. Enhanced ethanol sensing properties based on SnO_2 nanowires coated with Fe_2O_3 nanoparticles[J]. Sens. Actuators B, 2017, 238: 871-879.

[60] Chen N, Li X, Wang X, et al. Enhanced room temperature sensing of Co_3O_4-intercalated reduced graphene oxide based gas sensors[J]. Sens. Actuators B, 2013, 188: 902-908.

[61] Zhang G, Dang L, Li L, et al. Design and construction of Co_3O_4/PEI-CNTs composite exhibiting fast responding CO sensor at room temperature[J]. CrystEngComm, 2013, 15: 4730-4738.

[62] Yang J, Hidajat K, Kawi S. Synthesis of nano-SnO_2/SBA-15 composite as a highly sensitive semiconductor oxide gas sensor[J]. Mater. Lett., 2008, 62: 1441-1443.

[63] Xu C, Tamaki J, Miura N, et al. Grain size effects on gas sensitivity of porous SnO_2-based elements[J]. Sens. Actuators B, 1991, 3: 147-155.

[64] Korotcenkov G. Metal oxides for solid-state gas sensors: What determines our choice? [J]. Mater. Sci. Eng. B, 2007, 139: 1-23.

[65] Tiemann M. Porous metal oxides as gas sensors[J]. Chem. Eur. J., 2007, 13: 8376-8388.

[66] Vuong D D, Sakai G, Shimanoe K, et al. Hydrogen sulfide gas sensing properties of thin films derived from SnO_2 sols different in grain size[J]. Sens. Actuators B, 2005, 105: 437-442.

[67] Vuong D D, Sakai G, Shimanoe K, et al. Preparation of grain size-controlled tin oxide sols by hydrothermal treatment for thin film sensor application[J]. Sens. Actuators B, 2004, 103: 386-391.

[68] Korotcenkov G, Han S D, Cho B K, et al. Grain size effects in sensor response of nanostructured SnO_2-and In_2O_3-based conductometric thin film gas sensor[J]. Solid State Mater. Sci., 2009, 34: 1-17.

[69] Rothschild A, Komem Y. The effect of grain size on the sensitivity of nanocrystalline metal-oxide gas sensors[J]. J. Appl. Phys., 2004, 95: 6374-6380.

[70] Gurlo A, Ivanovskaya M, Pfau A, et al. Sol-gel prepared In_2O_3 thin films[J]. Thin Solid Films, 1997, 307: 288-293.

[71] Ansari S G, Boroojerdian P, Sainkar S R, et al. Grain size effects on H_2 gas sensitivity of thick film resistor using SnO_2 nanoparticles[J]. Thin Solid Films, 1997, 295: 271-276.

[72] Xu J, Pan Q, Shun Y, et al. Grain size control and gas sensing properties of ZnO gas sensor[J]. Sens. Actuators B, 2000, 66: 277-279.

[73] Wan Y, Zhao D. On the controllable soft-templating approach to mesoporous silicates[J]. Chem. Rev., 2007, 107: 2822-2861.

[74] Suib S L. A review of recent developments of mesoporous materials[J]. Chem. Rec., 2017, 17: 1169-1183.

[75] Ren Y, Ma Z, Bruce P G. Ordered mesoporous metal oxides: synthesis and applications [J]. Chem. Soc. Rev., 2012, 41: 4909-4927.

[76] Zhou X, Zhu Y, Luo W, et al. Chelation-assisted soft-template synthesis of ordered mesoporous zinc oxides for low concentration gas sensing[J]. J. Mater. Chem. A, 2016, 4: 15064-15071.

[77] Tian S, Ding X, Zeng D, et al. Pore-size-dependent sensing property of hierarchical SnO_2 mesoporous microfibers as formaldehyde sensors[J]. Sens. Actuators B, 2013, 186: 640-647.

[78] Xu S, Sun F, Pan Z, et al. Reduced graphene oxide-based ordered macroporous films on a curved surface: general fabrication and application in gas sensors[J]. ACS Appl. Mater. Interfaces, 2016, 8: 3428-3437.

[79] Rossinyol E, Arbiol J, Peiró F, et al. Nanostructured metal oxides synthesized by hard template method for gas sensing applications[J]. Sens. Actuators B, 2005, 109: 57-63.

[80] Zhu Y, Zhao Y, Ma J, et al. Mesoporous tungsten oxides with crystalline framework for highly sensitive and selective detection of foodborne pathogens[J]. J. Am. Chem. Soc., 2017, 139: 10365-10373.

[81] Li Y, Zhou X, Luo W, et al. Pore engineering of mesoporous tungsten oxides for ultrasensitive gas sensing[J]. Adv. Mater. Interfaces, 2019, 6: 1801269.

[82] Zhao T, Fan Y, Sun Z, et al. Confined interfacial micelle aggregating assembly of ordered macro-mesoporous tungsten oxides for H_2S sensing[J]. Nanoscale, 2020, 12: 20811-20819.

[83] Shen Y, Yamazaki T, Liu Z, et al. Influence of effective surface area on gas sensing properties of WO_3 sputtered thin films[J]. Thin Solid Films, 2009, 517: 2069-2072.

[84] Dey A. Semiconductor metal oxide gas sensors: A review[J]. Mater. Sci. Eng. B, 2018, 229: 206-217.

[85] Li G J, Zhang X H, Kawi S. Relationships between sensitivity, catalytic activity, and surface areas of SnO_2 gas sensors[J]. Sens. Actuators B, 1999, 60: 64-70.

[86] Li G J, Kawi S. High-surface-area SnO_2: a novel semiconductor-oxide gas sensor[J]. Mater. Lett., 1998, 34: 99-102.

[87] Xu H, Ju J, Li W, et al. Superior triethylamine-sensing properties based on TiO_2/SnO_2 n-n heterojunction nanosheets directly grown on ceramic tubes[J]. Sens. Actuators B, 2016, 228: 634-642.

[88] Xing X, Xiao X, Wang L, et al. Highly sensitive formaldehyde gas sensor based on hierarchically porous Ag-loaded ZnO heterojunction nanocomposites[J]. Sens. Actuators B, 2017, 247: 797-806.

[89] Zhang Y B, Yin J, Li L, et al. Enhanced ethanol gas-sensing properties of flower-like p-CuO/n-ZnO heterojunction nanorods[J]. Sens. Actuators B, 2014, 202: 500-507.

[90] Xu Z, Duan G, Li Y, et al. CuO-ZnO micro/nanoporous array-film-based chemosensors: new sensing properties to H_2S[J]. Chem. Eur. J., 2014, 20: 6040-6046.

[91] Na C W, Woo H S, Kim I D, et al. Selective detection of NO_2 and C_2H_5OH using a Co_3O_4-decorated ZnO nanowire network sensor[J]. Chem. Commun., 2011, 47: 5148-5150.

[92] Waldrop J R, Grant R W. Semiconductor heterojunction interfaces: nontransitivity of energy-band discontiuities[J]. Phys. Rev. Lett., 1979, 43: 1686-1689.

[93] Ju D, Xu H, Xu Q, et al. High triethylamine-sensing properties of NiO/SnO_2 hollow sphere P-N heterojunction sensors[J]. Sens. Actuators B, 2015, 215: 39-44.

[94] Ma L, Fan H, Tian H, et al. The n-ZnO/n-In_2O_3 heterojunction formed by a surface-modification and their potential barrier-control in methanal gas sensing[J]. Sens. Actuators B, 2016, 222: 508-516.

[95] Langer J M, Heinrich H. Deep-level impurities: a possible guide to prediction of band-edge discontinuities in semiconductor heterojunctions[J]. Phys. Rev. Lett., 1985, 55: 1414 - 1417.

[96] Yang X, Shi Y, Xie K, et al. Cocrystallization enabled spatial self-confinement approach to synthesize crystalline porous metal oxide nanosheets for gas sensing[J]. Angew. Chem. Int. Ed. 2022, 134, e202207816.

[97] Liu J, Guo Z, Meng F, et al. Novel porous single-crystalline ZnO nanosheets fabricated by annealing ZnS(en)$_{0.5}$ (en = ethylenediamine) precursor. Application in a gas sensor for indoor air contaminant detection[J]. Nanotechnology, 2009, 20: 125501 - 125508.

[98] Fan S W, Srivastava A K, Dravid V P. UV-activated room-temperature gas sensing mechanism of polycrystalline ZnO[J]. Appl. Phys. Lett., 2009, 95: 142106 - 142108.

[99] Chen Y J, Xue X Y, Wang Y G, et al. Synthesis and ethanol sensing characteristics of single crystalline SnO$_2$ nanorods[J]. Appl. Phys. Lett., 2005, 87: 233503 - 233505.

[100] Li X, Wei W, Wang S, et al. Single-crystalline α - Fe$_2$O$_3$ oblique nanoparallelepipeds: High-yield synthesis, growth mechanism and structure enhanced gas-sensing properties [J]. Nanoscale, 2011, 3: 718 - 724.

[101] Tian S, Yang F, Zeng D, et al. Solution-processed gas sensors based on ZnO nanorods array with an exposed (0001) facet for enhanced gas-sensing properties[J]. J. Phys. Chem. C, 2012, 116: 10586 - 10591.

[102] Kim H J, Lee J H. Highly sensitive and selective gas sensors using p-type oxide semiconductors: Overview[J]. Sens. Actuators B, 2014, 192: 607 - 627.

[103] Gurlo A, Barsan N, Weimar U, et al. Polycrystalline well-shaped blocks of indium oxide obtained by the sol - gel method and their gas-sensing properties[J]. Chem. Mater., 2003, 15: 4377 - 4383.

[104] Wang Y, Jiang X, Xia Y. A solution-phase, precursor route to polycrystalline SnO$_2$ nanowires that can be used for gas sensing under ambient conditions[J]. J. Am. Chem. Soc., 2003, 125: 16176 - 16177.

[105] Wang G, Yi Y, Huang X, et al. Fabrication and characterization of polycrystalline WO$_3$ nanofibers and their application for ammonia sensing[J]. J. Phys. Chem. B, 2006, 110: 23777 - 23782.

[106] Rai P, Yu Y T. Citrate-assisted hydrothermal synthesis of single crystalline ZnO nanoparticles for gas sensor application[J]. Sens. Actuators B, 2012, 173: 58 - 65.

[107] Rai P, Song H M, Kim Y S, et al. Microwave assisted hydrothermal synthesis of single crystalline ZnO nanorods for gas sensor application[J]. Mater. Lett., 2012, 68: 90 - 93.

[108] Hwang S, Kwon H, Chhajed S, et al. A near single crystalline TiO$_2$ nanohelix array: enhanced gas sensing performance and its application as a monolithically integrated electronic nose[J]. Analyst, 2013, 138: 443 - 450.

[109] Cheng B, Russell J, Shi W, et al. Large-scale, solution-phase growth of single-crystalline SnO$_2$ nanorods[J]. J. Am. Chem. Soc., 2004, 126: 5972 - 5973.

[110] Li Y, Xu J, Chao J, et al. High-aspect-ratio single-crystalline porous In$_2$O$_3$ nanobelts with enhanced gas sensing properties[J]. J. Mater. Chem., 2011, 21: 12852 - 12857.

[111] Xu Z, Duan G, Kong M, et al. Fabrication of α-Fe$_2$O$_3$ porous array film and its crystallization effect on its H$_2$S sensing properties[J]. ChemistrySelect, 2016, 1: 2377-2382.

[112] Wang S, Zhang H, Wang Y, et al. Facile one-pot synthesis of Au nanoparticles decorated porous α-Fe$_2$O$_3$ nanorods for in situ detection of VOCs[J]. RSC Adv., 2014, 4: 369-373.

[113] Wang B, Chen J S, Wu H B, et al. Quasiemulsion-templated formation of α-Fe$_2$O$_3$ hollow spheres with enhanced lithium storage properties[J]. J Am. Chem. Soc., 2011, 133: 17146-17148.

[114] Sun X, Ji H, Li X, et al. Open-system nanocasting synthesis of nanoscale α-Fe$_2$O$_3$ porous structure with enhanced acetone-sensing properties[J]. J. Alloys Compd., 2014, 600: 111-117.

[115] Carraro G, Barreca D, Comini E, et al. Controlled synthesis and properties of β-Fe$_2$O$_3$ nanosystems functionalized with Ag or Pt nanoparticles[J]. CrystEngComm, 2012, 14: 6469-6476.

[116] Biswal R C. Pure and Pt-loaded gamma iron oxide as sensor for detection of sub ppm level of acetone[J]. Sens. Actuators B, 2011, 157: 183-188.

[117] Peeters D, Barreca D, Carraro G, et al. Au/ε-Fe$_2$O$_3$ nanocomposites as selective NO$_2$ gas sensors[J]. J. Phys. Chem. C, 2014, 118: 11813-11819.

[118] Walcarius A. Mesoporous materials-based electrochemical sensors[J]. Electroanalysis, 2015, 27: 1303-1340.

[119] Gurlo A. Nanosensors: towards morphological control of gas sensing activity. SnO$_2$, In$_2$O$_3$, ZnO and WO$_3$ case studies[J]. Nanoscale, 2011, 3: 154-165.

[120] Batzill M, Katsiev K, Burst J M, et al. Gas-phase-dependent properties of SnO$_2$ (110), (100), and (101) single-crystal surfaces: Structure, composition, and electronic properties[J]. Phys. Rev. B, 2005, 72: 165414-165433.

[121] Han X, Jin M, Xie S, et al. Synthesis of tin dioxide octahedral nanoparticles with exposed high-energy {221} facets and enhanced gas-sensing properties[J]. Angew. Chem. Int. Ed., 2009, 48: 9180-9183.

[122] Wang C, Du G, Ståhl K, et al. Ultrathin SnO$_2$ nanosheets: oriented attachment mechanism, nonstoichiometric defects, and enhanced lithium-ion battery performances [J]. J. Phys. Chem. C, 2012, 116: 4000-4011.

[123] Zakrzewska K. Gas sensing mechanism of TiO$_2$-based thin films[J]. Vacuum, 2004, 74: 335-338.

[124] Nisar J, Topalian Z, Sarkar A D, et al. TiO$_2$-based gas sensor: a possible application to SO$_2$[J]. ACS Appl. Mater. Interfaces, 2013, 5: 8516-8522.

[125] Jiménez I, Arbiol J, Dezanneau G, et al. Crystalline structure, defects and gas sensor response to NO$_2$ and H$_2$S of tungsten trioxide nanopowders[J]. Sens. Actuators B, 2003, 93: 475-485.

[126] Lupan O, Ursaki V V, Chai G, et al. Selective hydrogen gas nanosensor using individual ZnO nanowire with fast response at room temperature[J]. Sens. Actuators B, 2010, 144: 56-66.

[127] Liu J, Huang H, Zhao H, et al. Enhanced gas sensitivity and selectivity on aperture-controllable 3D interconnected macro‐mesoporous ZnO nanostructures[J]. ACS Appl. Mater. Interfaces, 2016, 8: 8583-8590.

[128] Kim K, Lee H B, Johnson R W, et al. Selective metal deposition at graphene line defects by atomic layer deposition[J]. Nat. Commun., 2014, 5: 4781-4789.

[129] Adepalli K K, Kelsch M, Merkle R, et al. Influence of line defects on the electrical properties of single crystal TiO_2[J]. Adv. Funct. Mater., 2013, 23: 1798-1806.

第 6 章
气体分子和金属氧化物半导体界面相互作用模型

随着物联网(internet of things,IoT)和人工智能技术的发展,基于金属氧化物半导体(metal oxide semiconductors,MOS)的传感材料越来越受到基础研究和实际应用的关注。MOS材料具有优良的内在物理化学性质、可调的组成以及电子结构,因此特别适用于集成和小型化开发化学电阻气体传感器。在气体传感过程中,气-固界面的相互作用在提升传感器性能中扮演关键角色,而大多数研究却着眼于MOS材料与气体分子间的化学反应。因此,从一个崭新的视角,即更多地考虑气体传感过程中物理的气-固相互作用出发,本章将对气体分子所经历的动态过程,即吸附、脱附和扩散的基础理论和最新进展进行系统性的概括和阐述,并从多个角度详细介绍独特的电子传感机理,包括分子界面模型、气体扩散机理和界面反应行为,其中重点介绍传感器的结构-活性关系以及气体分子的扩散行为。特别地,本章将讨论并评估表面吸脱附动态过程,而它们对于气体传感性能的潜在效应通过气-固界面位点调节的假定来阐述。本章内容可为将来在MOS气体传感器中提升气体分子吸脱附和扩散动力学的进一步研究提出展望,为高性能MOS气体传感器的开发进行补充。

近来,随着纳米技术的迅速发展,各种新式传感器在各项领域,包括环境保护[1-2]、工业制造[3]、军事[4],以及公共安全和健康[5]都发挥了无可替代的作用。气体传感器的原理主要是基于传感材料对于不同气体分子所展现出的化学电阻特性,从而将测试气体的浓度值转化为相应的信号[6-18]。值得注意的是,金属氧化物半导体(MOS)已经成为检测气体分子时最广泛使用的传感材料之一,因为它们具有优良的物理化学性质、可调的组成和独特的微/纳米结构,以及多元的形貌结构。

自MOS材料在气体传感中的应用被开创以来[19-21],大量对于不同形貌结构和组成的气体传感材料的合成和应用的研究已经被开展[22-31]。然而,迄今为止,一些重要的科学问题,如传感性能的调控原则,还没有被充分阐述,因此也限制了高性能气体传感器的应用与开发。尽管研究者们确实已经提出了一些传感材料的优化技术,包括表面改性(如多孔以及低维材料)[23-24,29-30]、异质结(掺杂其他MOS材料)[22,27-28,31]、负载贵金属(溢流效应)[25-27]、光能激发[9,31],但对于其他重要因素,如框架微环境以及界面相互作用的影响,仍没有得到足够的关注,因此应被考虑并深入探讨。

实际上,在一个典型的 MOS 气体传感器的传感过程中,注入的气体分子首先快速弥漫并充满整个气室,然后一部分吸附到气敏材料表面,使得 MOS 材料的电阻或电导率发生变化,便产生了"响应"[32]。换句话说,测试气体分子的吸附标志着响应过程的"开始",这是气体响应的关键步骤。此外,气体分子的脱附作为吸附的逆过程,也会同时发生,将部分吸附的气体分子重新释放到空气中。剩余的吸附分子接下来会经历一个耦合的扩散-反应(diffusion-reaction,DR)过程[33],在此过程中,气体分子扩散到材料的深处并在其暴露的活性位点处发生反应,从而改变电阻。气体分子的消耗和/或转化会导致电阻发生快速变化,从而相应提升响应值。在达到一段时间的平衡后,排出测试气体,传感材料的导电性会重新恢复到初始状态,在这个过程中测试气体与衍生物分子从 MOS 材料表面脱附并释放到空气中。因此,为了了解传感机理,有必要对整个测试过程中气体的吸附、脱附与扩散进行深入研究。然而,通过分析传感过程中气体吸附、脱附和扩散来进行机理解释需要构建大量的数学和物理模型[34-40],这些内容常常在过去大多数传感材料设计与合成的研究中被忽略。虽然先前已有一些气体传感研究涉及这些方面,但很少有报道能够系统地串联 MOS 材料传感器中气体吸附、脱附与扩散和气体响应值之间的内在联系。

综上,本章着眼于气体分子的动态过程与传感机理之间的紧密联系,详细讨论气体传感过程中吸附与脱附的影响。根据不同模型,MOS 上的吸附机理主要分为三种,即氧吸附、化学吸附和物理吸附。幂律规则和响应/恢复过程与气体的吸脱附有紧密联系,因此同样作为气体传感机理进行进一步的阐述。此外,本章同时引入了一些有代表性的扩散模型,如 Gardner 模型、Yamazoe 模型以及其他修改过的模型。相关的经典数学与物理理论以及详尽的推导和推论也一并给出,这样可以为清晰地描述机理打下坚实的基础。本章主要的研究内容如图 6.1 所示,能够将以传统材料为主的传感研究与以理论为主的数学或物理推导高度结合起来,从而

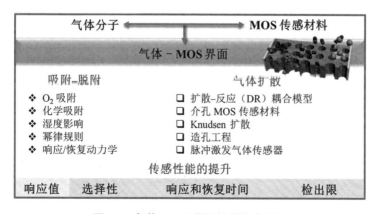

图 6.1 气体-MOS 界面处的动态过程

将气体传感领域扩展到更广阔的范围。

6.1 吸附和脱附

6.1.1 经典的吸附与脱附理论

总体而言,吸附定义为流体中的微粒附着到表面的过程。根据国际纯粹与应用化学联合会(international union of pure and applied chemistry, IUPAC)的定义,吸附是指由于表面力的作用,物质在一密实层与一液相或气体层的交界处的浓度增加,而脱附则是吸附的逆过程。传统上吸附可分为物理吸附与化学吸附,两者主要通过活化能大小来区分,一般认为物理吸附的活化能低于 25.116 kJ·mol^{-1},而化学吸附高于 62.79 kJ·mol^{-1}。在物理吸附过程中,通过 van der Waals(范德瓦尔斯)力与极化相互作用,粒子的电子结构在吸附后几乎不受影响;而当吸附质与吸附剂之间形成化学键时,则发生化学吸附,此时包含了载流子的交换[41]。

粒子的吸附,即吸附剂表面吸附质的量,可通过固定温度下基于各种模型的等温式来精确描述。常见的等温式主要包括 Henry 吸附等温式、Freundlich 方程、Langmuir 方程以及 Brunauer-Emmett-Teller(BET)理论。Henry 吸附等温式是最简单的模型,由英国化学家 William Henry 命名,它主要基于 Henry 规则,即溶液中溶解的气体的量与其气相中的分压成正比。因此,在 Henry 吸附等温式中,表面吸附的气体分子的量也与气体的分压成正比。如式(6.1)所示,X、K_H 和 P 分别代表表面覆盖度、Henry 吸附常数和分压。事实上,上述提到的所有等温线在低压下均遵循线性关联,而 Henry 吸附等温式只对低表面覆盖度的吸附有效。

$$X = K_H P \tag{6.1}$$

Freundlich 方程由德国化学家 Herbert Freundlich 命名,是 Henry 吸附等温式的扩展。通过经验估计可得,单位质量吸附剂上吸附质的质量与平衡压力的幂成正比。Freundlich 吸附等温式可表述为式(6.2),其中 x、M 和 p 分别代表吸附质质量、吸附剂质量和吸附质的平衡压力,而 K 和 n 在某一温度下对于特定吸附剂保持为常数。在高温下,n 接近 1,式(6.2)就转化成了式(6.1)。通常,由于研究者经常在理论研究中通过经验估计来拟合等温式,因此最终推导得出的方程中存在一定的局限性。Freundlich 方程最明显的局限性在于其不适用于高压。在高压下,根据该模型会产生吸附饱和,而这与实际实验结果完全不同。例如,H_2S 吸附到活性炭上时,在高压和低压下从等温式中可获得不同的推导结果[42]。

$$\frac{x}{M} = KP^{\frac{1}{n}} \tag{6.2}$$

1918 年,美国物理学家和化学家 Irving Langmuir 基于动力学假设和统计热力学理论,提出了一个半经验型的吸附等温式,即 Langmuir 方程。由于 Langmuir 方程的精简性以及在多种吸附条件下的有效性,至今它已成为最常见的等温式方程之一。Langmuir 方程主要基于以下四大假设[43]:

(1) 所有的吸附位点都是等价的,且每个位点只能由一个气体分子吸附;
(2) 吸附剂表面是均相的,并且吸附质分子之间没有相互作用存在;
(3) 吸附过程中没有相转换;
(4) 吸附过程中气体分子只形成单一吸附层。

基于动力学、热力学或统计力学推导,Langmuir 方程通常可表示成式(6.3),其中 θ、K 和 P 分别代表表面覆盖度、吸脱附反应的平衡常数以及气体分压。在低压下,$\theta \approx KP$;而在高压下,$\theta \approx 1$。Burke 等[44]指出在选择以上的等温式模型时应考虑到气体分子吸附焓的存在,并且仅仅根据最佳拟合数据来选择吸附模型是一个常见的错觉。

$$\theta = \frac{KP}{1+KP} \tag{6.3}$$

虽然 Langmuir 方程应用范围很广,它仍然无法应用于多层吸附条件。多层吸附指的是气体分子在已经吸附的单分子层上发生进一步吸附。1938 年,美国化学家 Stephen Brunauer、Paul Emmett 和物理学家 Edward Teller 提出了一个新的等温式(BET 理论),该理论考虑了多层吸附,并基于以下四大假设[45]:

(1) 气体分子的物理吸附发生在一个层数无穷大的固体上;
(2) 只有相邻层的气体分子间有相互作用,并且 Langmuir 方程可应用于每一个单分子层;
(3) 第一层气体分子的吸附焓是常数,并且比更高层的吸附焓高得多;
(4) 更高层的吸附焓大小与气体分子的液化焓相同。

BET 理论可最终表述成式(6.4),其中 x 代表吸附质分子的分压,v 代表标准状况(standard conditions for temperature and pressure,STP)下的吸附质体积,v_{mon} 代表 STP 下形成单层吸附所需的吸附质体积,而 c 代表吸脱附反应的平衡常数。

$$\frac{x}{v(1-x)} = \frac{1+x(c-1)}{v_{mon}c} \tag{6.4}$$

虽然 BET 理论考虑了多层吸附,但吸附实验通常在 N_2 的沸点(77 K)下开展,而该条件无法应用到传统气体传感的研究中,其工作温度为几百开尔文。因此,气体分子在更高层中的吸附对于气体响应值的贡献很少,在实际气体传感研究中,Langmuir 方程是最常用的吸附模型[46-48]。此外,吸脱附机理也能应用于气体分子的表面反应中。Irving Langmuir(1921 年)和英国化学家 Cyril Hinshelwood(1926 年)提出了一种反应机理(Langmuir - Hinshelwood 模型,简称 L - H 模型)。该模型指出两个分子先吸附到相邻位点处,然后经历一个双分子反应[49-51],最终的平衡常数与材料上反应物分子的表面覆盖度以及吸脱附过程的速率常数有关。1938

年,英国化学家 Dan Eley 和 Eric Rideal 提出了另一种机理(Eley-Rideal 模型,简称 E-R 模型)。该模型指出一个已经吸附到材料表面的分子(A 分子)与另一个气相中未吸附的分子发生反应[52-54],最终平衡常数与材料上吸附的 A 分子的表面覆盖度有关。1954 年,荷兰化学家 Mars 和 Van Krevelen 进一步提出了一个新的基于吸附的催化机理(MvK 机理[55],也叫氧化还原/再生机理[56])。该机理指出催化剂的晶格成分与反应物反应并消耗,导致表面空位产生,此空位可能会随着之后的重构而被填补。通常,在 MvK 机理中,在催化剂的晶格成分释放到气相中并消耗前,其表面会产生基团(如 CH_4 的自由基偶联反应)[57]。

进一步而言,Langmuir 等温线已被认为可用于电子吸附理论框架中。这一观点来源于 Wolkenstein,其指出式(6.3)中的 K 不仅与温度有关,也与材料中的 Fermi 能级有关,进而对该公式进行了升级[58]。换言之,这意味着吸附受到固体材料及其表面的电子性质的影响,主要体现为材料的禁带宽度。Wolkenstein 等温式可能是最完整解释了吸附质-吸附剂相互作用的模型。Rothschild[59]指出,在部分条件下电子吸附模型可以观察到 Henry、Freundlich 和对数等温线等各种形式(见图 6.2)。

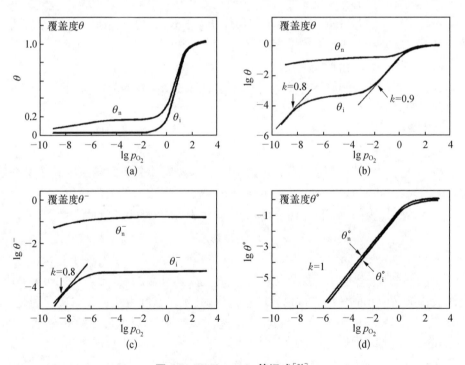

图 6.2　Wolkenstein 等温式[61]

根据 O_2 分压不同,n 型 ZnO 薄膜上化学吸附的受体物种覆盖度可展示为高补偿情况(用 i 表示)和无补偿情况(用 n 表示)
(a)、(b) 总覆盖度 $\theta(p)$ 的 θ-$\lg p$ 关系和 $\lg \theta$-$\lg p$ 关系;(c)、(d) 带电强化学吸附与中性弱化学吸附物种的覆盖度的对数关系

例如,应用该理论,通过数值研究可以正确解释实验中观测到的气体响应值(S)和晶粒大小(D)之间的 S-$1/D$ 关系曲线[59]。此外,将吸附的物种视为金属氧化物表面给体或受体的缺陷使得研究者可以计算材料暴露到还原性或氧化性气体中后电阻的变化情况,至少在 Fermi 能级被固定在相应状态时确实如此[60]。

1920 年,法国化学家 Paul Sabatier 提出了一个原理,指出催化反应中催化剂与反应物的相互作用应该适中(Sabatier 均衡),否则,如果相互作用较弱,反应物很难活化;而当相互作用较强时,产物则很难脱附[62]。1969 年,Balandin[63] 进一步指出反应速率与催化剂成键能力之间的关系可通过有一个最大值峰的火山形曲线来描述(见图 6.3)。Sabatier – Balandin 原理在定性理解如何设计合适的催化剂中有广泛应用。

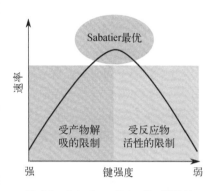

图 6.3　Sabatier – Balandin 原理的定性示意图[62]

6.1.2　MOS 材料气体传感中吸附与脱附的影响

众所周知,MOS 气体传感材料通过电阻变化在不同气体中产生响应。在实际气体传感实验中,物理吸附通常很弱且易受实验条件(如湿度和温度)影响,而化学吸脱附在传感过程中占主导地位,且由于产生了强化学键相互作用而更稳定。换句话说,在大多数情况下,由于吸附的气体分子与 MOS 材料之间形成了化学键,电阻快速变化。然而,由于在测试环境中氧广泛存在,O_2 的吸附是一个非常重要且不可忽略的过程,因而常常被从化学吸附中单独列出[64]。此外,如果将从氧吸附模型中推导出的耗尽层理论与气体分子本身的吸附和反应过程相结合,可以进一步推导得出 MOS 材料气体传感的幂律规则,即材料电阻与气体分子的分压有幂函数关系[40]。除此之外,在实际的气体传感实验中,氧和测试气体的化学吸附以及产物分子的脱附对气体传感器的响应和恢复有重要影响,并可进一步用来提升传感材料的响应活性和针对某些特定气体的选择性。

1. 氧吸附模型

当 MOS 材料暴露于干燥空气中时,N_2 和 O_2 必然会在气-固界面上产生影响[65]。虽然 N_2 在空气中的体积分数达到 78%,但其具有三键的分子具有高键能(946 kJ·mol^{-1}),使其对 MOS 材料产生高惰性,导致电阻几乎没有任何变化。Hoa 等[66] 指出在 p 型半导体 CuO 薄膜中氮气有化学惰性。因此 N_2 通常在研究测试气体与 MOS 材料之间的化学动力学时被用来作为载气[65]。除此之外,O_2 在空气中的体积分数达到 21%,是空气中浓度第二高的气体,仅次于 N_2,它可以吸附到 MOS 材料上而对气体传感器产生巨大影响。n 型半导体的载流子是电子(e^-)。当该类型半导体材料暴露于空气中时，O_2 分子吸附到其表面并从中"攫取"电子，形成活性氧负离子，使得 MOS 材料的表面上方会产生一个包含绝缘区域的电子核-壳结构，通常称为电子耗尽层[67]，进而使得 MOS 材料的电阻增大[见图 6.4(a)]。EDL 内的电子分布范围为表面上方 Debye 长度(λ_D,几个纳米)的有限深度，并且其几乎不会受到 O_2 分子吸附的影响[68]。Debye 长度(也叫 Debye 半径)是一个讨论溶液中载流子的净静电效应的参数，反映了等离子体的静电屏蔽效应。当讨论的尺度大于 Debye 长度时，等离子体可视作电中性；否则则视作带电。对于 n 型半导体，Debye 长度由式(6.5)定义，其中 ε、k_B、T、e 和 N_d 分别代表介电常数、Boltzmann 常数、温度、元电荷以及掺杂物的数密度[69]。

$$\lambda_D = \sqrt{\frac{\varepsilon k_B T}{e^2 N_d}} \qquad (6.5)$$

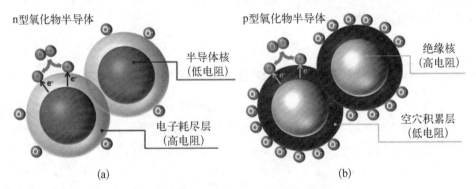

图 6.4 不同类型半导体的电子核-壳结构[70]
(a) n 型半导体；(b) p 型半导体

Li 等[71]使用 n 型半导体 SnO_2 制备了一个简单但有效的啤酒检测传感器。该研究指出拥有化学计量比的 SnO_2 在高温下通入惰性或还原性气体时表现出高掺杂的半导体性质，这主要归功于表面 O 原子脱离而形成的氧空位[见式(6.5)，图 6.5(a)]。当 O_2 分子吸附到 MOS 材料表面时，它们首先占据之前形成的氧空位并从材料的导带中"攫取"电子，从而形成各种活性氧物种(O_2^-、O^- 和 O^{2-})，因而材料表面上方形成了 EDL。总体而言，O_2^-、O^-(解离氧)和 O^{2-}(晶格氧)分别在<150℃、150～400℃、>400℃的温度范围内占主导地位[67-68]。通过提高工作温度、掺杂以及降低晶粒尺寸等方法可以进一步提升 O_2 分子的吸附和解离能力[72]。此外，如果 MOS 材料的厚度小于 λ_D 的两倍，EDL 可以扩展到整个材料中，使得材料电阻值产生最大响应。实验结果表明单晶 SnO_2 纳米带中的 EDL 在几纳米范围内被金属催化剂纳米点和表面吸附的物种所改变[73]。如图 6.5(b)所示，O_2 分子的吸附会降低 Fermi 能级,并导致导带向更高能级处弯曲,从而进一步增加材料的电阻。对于 n 型半导体,当其暴露在还原性气体中时,MOS 材料表面发生了氧化还原反应[见图 6.5(c)],从而降低表面吸附氧的浓度以及 EDL 深度,因此电阻随着 Fermi 能级和导带恢复到正常水平而得以恢复。

$$O_0 \leftrightarrow V_o^{\cdot\cdot} + 2e' + \frac{1}{2}O_2(g) \tag{6.6}$$

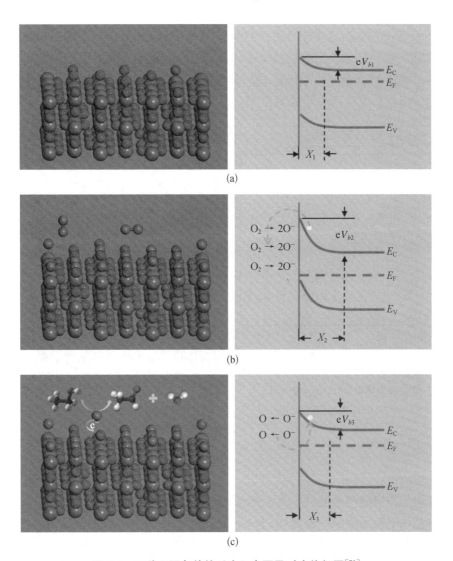

图 6.5 三种不同条件的反应示意图及对应能级图[71]

(a) 拥有氧空位的非化学计量 SnO_2;(b) 部分重新填充吸附氧的 SnO_2;(c) C_2H_5OH 和预吸附 O 原子的反应

当关注吸附的还原性气体的反应过程时,可以发现它们被各种氧物种氧化并随之消耗。在较低温度下,相对更活泼的吸附氧优先与还原性气体发生反应。然而,在较高温度下,由于材料表面活性提升,同时自由吸附的氧物种被消耗,MOS 材料晶格表面将发生氧负离子的传输,在这一过程中晶格氧将直接与还原性气体反应,并产生表面氧空位。此时材料表面的催化活性通过遵循典型的 MvK 机理得到进一步提升,并可通过同位素交换法得到验证[74]。材料表面的氧空位能进一步被空气中的自由氧固化而重新填充。

在 p 型半导体中,空穴充当载流子,表面氧分子的吸附同样也会降低电子密度,但此时形成了一个空穴积累层[hole accumulation layer,HAL,见图 6.4(b)][75]。研究者已发现,电子核-壳结构在 n 型和 p 型半导体中都适用,但 p 型半导体中核-壳结构的内核是绝缘体而外部的 HAL 则为导体,这一点与 n 型半导体完全不同。因此,当 p 型半导体暴露于空气中时,氧分子将首先吸附到氧空位中。然而,在这种情况下,氧负离子也能够和作为载流子的空穴复合[式(6.7)][76]。Iwamoto 等[77]使用程序升温脱附(temperature-programmed desorption,TPD)方法研究了 560℃ 下不同 MOS 材料表面氧脱附的总量(V_{560},见图 6.6)。研究发现,p 型半导体(如 CuO、Co_3O_4、MnO_2、NiO 和 Cr_2O_3)的 V_{560} 值比 n 型半导体更高。此外,尽管有一些偏差,但 V_{560} 值大致随着生成焓的增加而降低。总体而言,由于 p 型半导体相比 n 型半导体热力学更不稳定,有多种氧化态,能促进氧化还原反应,因此前者氧吸附的程度相比后者更高。

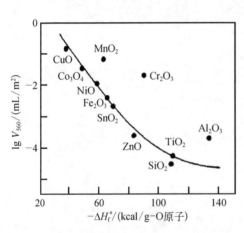

图 6.6 脱附氧的量(V_{560})与每克氧气对应的氧化物生成焓($-\Delta H_f^\circ$)的关系

$$O_2 + V_O^{\cdot\cdot} + e' \leftrightarrow (O_2-V_O^{\cdot\cdot})_{ad} \tag{6.7}$$

2. 化学吸脱附

一般而言,气体分子吸附到 MOS 材料表面,使得材料的电子分布发生变化,从而产生了特定的电阻变化。严格地说,上述讨论的 MOS 材料上发生的氧吸附过程也是化学吸脱附的一种,但由于 O_2 在空气中广泛存在,因而氧吸附模型应该被单独列出和讨论。尽管氧吸附模型似乎在解释传统 MOS 材料机理时显得万能,但在很多更复杂的条件下,上述理论不可避免地有所不足,而在这些条件下测试气体的化学吸脱附对响应值有主要影响。

传统的氧吸附模型的一个例外是无氧体系。Moradi 等[42]使用 Pd/Pt 负载的

SnO_2 作为传感材料,成功使其在低 O_2 浓度(20 ppm)下对 CO 产生非常好的响应,并且该响应值随着 O_2 浓度的升高而降低。CO 的无氧传感测试通常在超高真空条件下进行,此时吸附的 CO 分子直接与材料表面的晶格氧(O_{lat})反应,遵循 MvK 机理。在低 O_2 浓度下,由于吸附氧占据的氧空位数量很少,占主导的机理和无氧条件类似。随着 O_2 浓度提升(25~50 ppm),CO 分子占据的活性位点数减少,机理逐渐转化为氧吸附模型。进一步的研究探讨了 SnO_2 材料中不同 O_2 背景浓度对 H_2 和 CO 响应的影响[78-80]。Henrich 和 Cox[81] 提出了一个可行的机理,能够解释无氧条件下 H_2 的响应。该机理包括 H_2 分子的解离、H 原子吸附到表面晶格氧离子处,以及"扎根的"羟基(OH_O^+)的形成。这种基团的电子亲和力比晶格氧离子更弱,因此可以作为电子给体。在这种情况下,导带中电子数量增加,而材料的电阻下降。与此同时,材料表面形成了一个电子积累层(electron accumulation layer,EAL),使得能带向上弯曲,这一点和 p 型半导体暴露于空气中时产生的 HAL 十分类似。然而,虽然材料导电性随着 H_2 浓度的增大而升高,但当电子给体的能级与 Fermi 能级交叉时会达到饱和,因为电子无法越过 Fermi 能级。简而言之,只要材料表面有足够的氧离子,传感机理将由表面电子受体的浓度所支配;而对那些拥有电子受体数量很少的材料而言,表面电子给体支配主要机理,使得 n 型半导体表面产生 EAL。在含氧条件下,上述提到的两种机理[见图 6.7(a)、图 6.7(b)]平行发生,但最终的响应值比无氧条件下只发生 H_2 吸附的情况要低。这是因为在含氧条件下,电子给体的电离不会导致 SnO_2 表面载流子浓度增加。相反地,电子给体释放出的电子最终被受体所捕获,进一步发生氧的电离吸附。由于表面电荷没有变化,能带也不会弯曲。事实上,在含氧条件下,只有氧吸附模型对响应有主要贡献,因为它降低了吸附氧离子的浓度并因此增加了半导体内部自由载流子的浓度。因此,这两种机理不能简单叠加,相似的结果在 CO 的实验中也得到了验证[80]。Zhu 等[82] 使用原子尺度材料模拟的计算机程序包(Vienna Ab initio Simulation Package,VASP)进一步分析了含氧和无氧条件下 SnO_2 表面 H_2 分子的相对吸附能以及最佳活性位点。第一性计算结果表明,SnO_2 的热力学最稳定的(110)晶面上 O_2 的吸附能相对较低(-0.38 eV),表明其吸附不稳定。此外,H_2 分子在吸附氧上的吸附能更低(-0.018 eV)。此外,在不同的吸附位点中,端氧处 H_2 分子的吸附能最高[-0.029 eV,见图 6.7(c)~图 6.7(f)],表明 H_2 分子更倾向于在无氧条件下直接吸附到 SnO_2 表面而非在预吸附的氧离子上发生进一步吸附,这与实验结果一致。类似地,研究者们还使用了密度泛函理论(density-functional theory,DFT)对 SnO_2 晶粒不同晶面上的 CO 吸附与氧化机理进行了研究。例如,Lu 等[83] 指出 SnO_2(110)晶面上 CO 的氧化遵循 MvK 机理而非 L-H 模型,在这一过程中吸附氧通过将 e^- 传输给化学吸附氧而转化为各种氧负离子。Zakaryan 和 Aroutiounian[84] 进一步指出在别的晶面如(101)晶面和(001)晶面上,

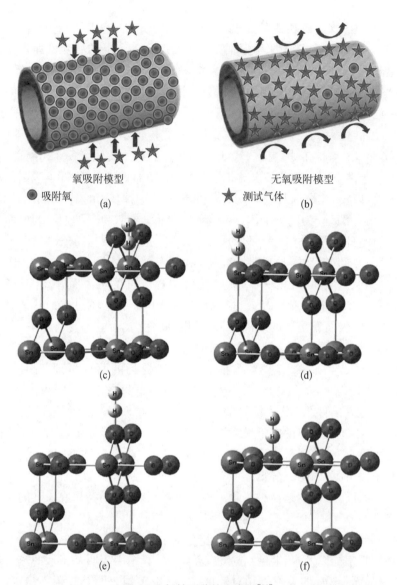

图 6.7 气敏测试的示意图[82]

(a) 氧吸附模型;(b) 无氧吸附模型;(c)~(f) 金红石相 SnO_2(110)晶面不同位置的 H_2 吸附模型,依次表示为 H_1、H_2、H_3 和 H_4

CO 分子通过 C 原子与晶格氧成键而吸附,这使得 MvK 机理无效。

化学吸脱附占主导的另一种典型情况是对能够与晶粒直接接触且发生化学反应的气体分子进行测试,最终导致材料电阻变化[64]。这一过程通常和氧吸附模型同时发生,但和后者相比,该过程能够对提升气体响应产生更多的贡献。Xu 等[85]利用自组装方法制备了单层胶体粒子组成的 CuO 薄膜,其对 H_2S 有很好的响应。

CuO 传感器的可恢复性依赖于 H_2S 的浓度。具体而言,低浓度 H_2S 下能产生很高的可恢复性[见图 6.8(a)],而高浓度 H_2S 下则难以恢复[见图 6.8(b)],因此该研究提出了两种不同的机理来解释这一气体传感行为[见图 6.8(c)～图 6.8(e)]。

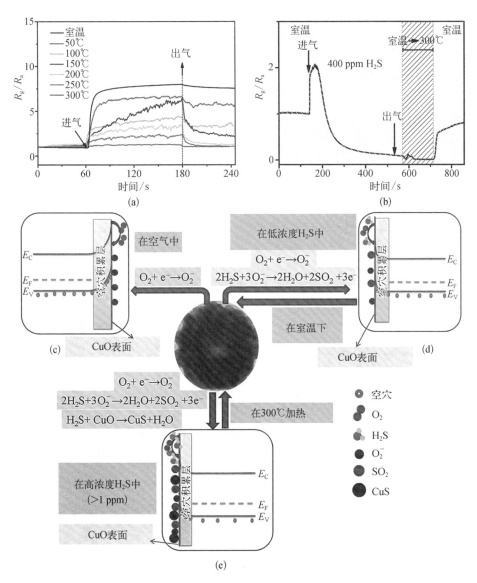

图 6.8　CuO 对 H_2S 的传感性能和机理(后附彩图)[85]

(a) CuO 传感器在 10 ppm H_2S 下响应值与工作温度的关系;(b) CuO 传感器在 400 ppm H_2S 下响应值随时间的变化;(c)～(e) CuO 传感器与不同浓度的 H_2S 产生的不同传感机理,其中(c)为在空气中,O_2 分子吸附到 CuO 表面形成氧负离子,产生 HAL,(d)表示注入低浓度 H_2S 时,H_2S 与吸附在 CuO 表面的 O_2^- 反应,减少了 HAL,响应值提升,(e)表示注入高浓度 H_2S 时,除了 H_2S 氧化反应,H_2S 还与 CuO 反应而在表面产生了一层 CuS,响应值下降

作为一种 p 型半导体,当 CuO 暴露于空气中时,表面发生 O_2 吸附,形成了 HAL。当接触到低浓度 H_2S(<500 ppb)时,后者将吸附于 CuO 表面预吸附的氧负离子之上并与之发生反应,释放电子并使作为载流子的空穴失效,从而增加电阻[见图 6.8(c)]。当移走 H_2S 并使材料重新暴露于空气中时,吸附的氧负离子将得到补充,先前失效的空穴可重新形成并使得材料电阻恢复[86]。对于更高浓度 H_2S(>1 ppm),由于没有足够的氧负离子占据 CuO 的活性位点,H_2S 分子倾向于直接吸附于 CuO 表面并与之反应[见图 6.8(d)][87]。由于产物 CuS 的能带宽度比 CuO 窄,因此材料电阻会迅速下降,这与传统 p 型半导体的传感机理相反。传感器从室温到 300℃ 的温度范围内均无法完全恢复,这主要归结于 CuS 重新还原的速率很低[87]。对于超高浓度 H_2S(>100 ppm),以上两种机理都对最终的响应产生贡献[见图 6.8(e)]。如图 6.8(b)所示,一开始,当注入 H_2S 后,氧吸附模型占主导,使得电阻上升。但即使如此,不久后由于 H_2S 在 CuO 表面发生化学吸附并直接发生反应,仍使得材料电阻下降。随着越来越多 CuS 生成,电阻最终快速下降,表明此时 CuS 在气敏机理中起决定性作用。对于 n 型 MOS 传感器,H_2S 的化学吸附会直接增强响应。例如,笔者课题组已经开展了基于有序介孔 MOS 材料(包括 Fe_2O_3、SnO_2 和 WO_3)的 H_2S 传感器制备的研究[88-93]。H_2S 在 MOS 传感层上的化学吸附导致了金属硫化物(SnS_2、WS_2 等)的产生,与原始的氧化物材料之间形成了异质结[91-93]。在这种情况下,拥有更窄能带宽度的金属硫化物能在 n 型半导体中进一步促使电阻下降,同时不会出现 CuO 中的那种响应反转。然而,其恢复时间会不可避免地延长,这是因为生成的硫化物必须要重新氧化才能回到初始状态。

还有一种常见的化学吸附形式是 H_2O 分子的吸附。虽然 H_2O 分子的吸附一般被认为归属于物理吸附,但实际情况下化学吸附同样存在。H_2O 分子的吸附机理由 H_2O 分子占据的活性位点比例来决定。在低湿度下,单层化学吸附起主导作用;而在高湿度下,由于材料表面活性位点被占据,多层物理吸附可能发生。在较低温度下进行的气体传感研究中,湿度对气体响应也有副作用。在相对湿度(relative humidity,RH)低于 20% 的干燥空气中[65],由上述提到的传统化学吸脱附模型(尤其是氧吸附)对最终的响应值占主导作用。而在潮湿条件下,密集的 H_2O 分子在材料表面发生解离吸附,并电离产生 H^+ 和 OH^-。

Heiland 等[94]提出了 SnO_2 表面两种不同的 H_2O 分子的吸附机理。一方面,对于一个 H_2O 分子与两个金属位点反应的情况[见式(6.8),图 6.9(a)],解离的 OH^- 直接与 Sn 原子形成化学键,而剩下的 H^+ 则与一个晶格氧结合,从而形成两个 Sn—OH 偶极并释放出两个自由电子。另一方面,对于一个 H_2O 分子与一个金属位点反应的情况[见式(6.9),图 6.9(b)],H^+ 扩散到材料深处并与晶格氧结合,而剩余的 OH^- 也与 Sn 位点结合。"扎根"固定住的—OH 由于电子亲和力和电离度较弱,它们主要充当电子给体。简而言之,H_2O 分子的吸附会导致响应值发生变化,可能是由于材料电导率发生变化,也可能是由于占据了活性位点。然而,以上理论完全不足以解释所有包含 H_2O 分子吸附的条件下的机理,因为它们中的大部分忽略了下述将提到的 H_2O 分子的物理吸附。

$$H_2O(g) + 2(Sn_{Sn} + O_O) \longleftrightarrow 2(Sn_{Sn}^{\delta+} - OH^{\delta-}) + V_O^{2+} + 2e^- \quad (6.8)$$

$$H_2O(g) + (Sn_{Sn} + O_O) \longleftrightarrow (Sn_{Sn}^{\delta+} - OH^{\delta-}) + (OH)_O^+ + e^- \quad (6.9)$$

总而言之,化学吸脱附模型与气体传感过程有密切联系,但在很多研究中被忽视。因此,化学吸脱附模型对于传统的氧吸附模型而言也是重要的补充。

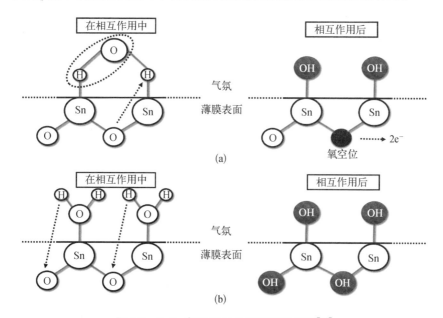

图 6.9 SnO_2 表面 H_2O 分子吸附的机理[65]

(a) 两个金属位点对应一个 H_2O 分子;(b) 每个金属位点对应一个 H_2O 分子

3. 物理吸脱附与湿度影响

物理吸附是传感过程中的另一种常见吸附过程。然而,与氧吸附与化学吸附不同,物理吸附理论很少用于解释气敏机理。这是因为在大多数条件下,物理吸附造成的 MOS 材料的导电性变化可以忽略不计[64]。如前所述,物理吸附是 H_2O 分子吸附的主要形式,因此,最常见的基于物理吸附机理的 MOS 材料传感器是湿度传感器。即便如此,湿度传感器的响应仍然受到化学吸附,包括氧吸附的影响[95]。

除了此前讨论的低湿度下的化学吸附,在高湿度下,H_2O 分子的吸附主要以物理吸附形式存在;而在更高湿度下,Morrison 进一步提出了"共吸附"的概念[96]。这时 H_2O 分子将驱除其他已经占据活性位点的分子,因而将在材料表面同时发生单层的化学吸附和多层的物理吸附,从而产生共吸附。在这种情况下,根据

Grotthuss 质子跳跃机理,物理吸附层中的 H^+ 可以通过氢键作用在 H_2O 分子间自由流动(见图 6.10)[97]。特别地,由于跳跃机理与吸附以及扩散之间有很深的联系,6.2 节气体扩散中将对其机理进行详细介绍。材料电阻的下降速率随着吸附形式的不同而不同,通常在化学吸附中最快、物理吸附中适中,而在共吸附中几乎为零。在某些情况下,湿度对于材料的内在性质也会产生影响。例如,Deng 等[98]制备的 p 型半导体 $CuScO_2$ 对 NH_3 产生了"伪 n 型"的性质。由于 $CuScO_2$ 本身电阻很高,在潮湿的室温下,材料的导电性主要体现在表面吸附的 H_2O 层,这时 NH_3 分子溶解并电离,为 H_2O 层提供了 NH_4^+ 和 OH^-。此外,遵循 Grotthuss 质子跳跃机理,H^+ 的流动提升了材料的导电性,从而建立起伪 n 型机理。之后,由于 NH_3 的 H^+ 亲和力比 H_2O 强[99],流动的 H^+ 可被 NH_3 分子所捕获,产生伪 p 型机理[见图 6.11(a)]。随着 H_2O 的蒸发,H_2O 分子的脱附削弱了伪 p 型响应[见图 6.11(b)]。而在高温下(>100℃),所有的 H_2O 分子均发生脱附,因此只剩下 $CuScO_2$ 自身的 p 型半导体性质[见图 6.11(c)]。

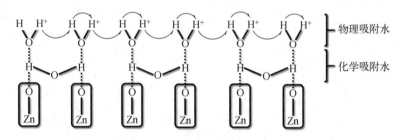

图 6.10 ZnO 中湿度传感的示意图[100]

除了 H_2O 分子以外,室温下 O_2 分子也能发生物理吸附。Hong 等[101]使用场效应管(field-effect transistor,FET)制备的 O_2 传感器来观察室温下 O_2 分子的物理吸附。在大多数情况下,这一效应被认为对响应值影响很小。然而,即使在氧吸附模型中,低温下材料表面也是由 O_2 分子的物理吸附占主导。虽然这种物理吸附对电子没有直接影响,但研究者认为该效应会对材料表面载流子的直接传输产生干扰,通过降低迁移率而使得电阻略微上升。通常来说,这一效应看上去在低噪声(准)二维(two-dimensional,2D)材料如石墨烯或 Mxene 层中占主导,而体相甚至介孔金属氧化物中化学电阻的形成则主要来自化学吸附造成的自由载流子浓度变化[102]。

4. MOS 气体传感器的幂律规则

关于 MOS 气体传感器的电阻与气体浓度的关系已有深入的研究。1987 年,Morrison 提出了一个 SnO_2 材料上还原性气体的质量作用规律,即电阻在低 P_R 下随着 $P_R^{-0.5}$ 变化,其中 P_R 是还原性气体的分压[103-104]。在之后的研究中,研究者们获得了一些经验结论[33,105],但仍然没有系统性地解释幂律规则的来源。之后,

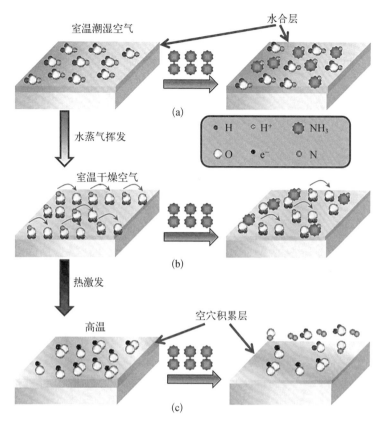

图 6.11 CuScO$_2$ 在不同条件下的传感机理示意图[98]
(a) 潮湿的室温；(b) 干燥的室温；(c) 高温

基于前述提到的气体吸附理论，Yamazoe 和 Shimanoe[40]进一步为 MOS 材料气体传感器的幂律规则提供了理论基础，即 MOS 材料暴露于分压为 P 的气体中时，其电阻正比于 P^n，其中 n 是对该气体而言特定的常数。传统上，一般认为材料的电阻与气体浓度大致呈线性关系，在低浓度下尤为如此，这一点被进一步应用于扩散理论[36]、检出限(limit of detection, LoD)的计算等方面[63]。进一步而言，幂律规则可以扩大气体扩散的研究范围[104,106-108]。幂律规则可以通过将电子给体的功能(气体分子的吸附与反应)与传感器功能(表面电势的变化)相结合得到。换言之，即将氧化物表面的化学过程与 MOS 材料表面的物理过程相结合。作者应用了一些传统半导体和晶粒影响的基础理论，如双 Schottky 层和隧道模型[68,109]。之后作者进一步讨论了吸附不同气体分子后产生的化学变化。为了简化模型，Ghosh 假定氧吸附后形成的氧负离子为 O$^-$。O$^-$ 的累积可以表示为式(6.10)，其中 k_1 和 k_{-1} 分别表示正反应和逆反应的速率常数。另一方面，对于 n 型半导体表面的物理变化，表面的导电电子密度可表示为式(6.11)，其中 N_d 和 m 分别代表电子给体的数密度和减少的耗尽层深度。将上述提到的化学过程与物理过程相结合,幂指数 n 可最终表示为式(6.12)。当 m 足够大时,n 等于 0.5,这是幂律规则的极限值。Yamazoe 等[110]之后将幂律规则的推导拓展到其他氧负离子中,并指出如果吸附的氧负离子完全是 O^{2-},则幂指数 n 为 0.25。

$$\frac{d[O^-]}{dt}=k_1 P_{O_2}[e^-]^2-k_{-1}[O^-]^2 \tag{6.10}$$

$$[e^-]=N_d e^{-\frac{m^2}{2}} \tag{6.11}$$

$$n=\frac{d\lg R}{d\lg P_{O_2}}=\frac{m^2}{2(m^2+1)} \tag{6.12}$$

如上所述,不同工作温度下 MOS 材料表面形成的氧负离子物种有所不同,这也导致幂律规则的幂指数值发生变化。实际操作中,气体传感器的响应值通常定义为 $S=R_a/R_g$ 或 $S=(R_a-R_g)/R_a\times 100\%$。特别地,Windischman 和 Mark[111]研究了 CO 吸附的典型案例,并指出 SnO_2 表面的氧化还原相互作用可以用来解释化学电阻响应值 S[式(6.13)],其中 e、μ、d、α、γ 和 P_R 分别代表元电荷、电子迁移率、传感层厚度或特征晶粒尺寸、表征还原性气体 R 黏附到表面的系数、表面重组系数,以及气体 R 的分压。根据众多实验结果,如果简单气体分子与氧化物表面发生相互作用,并且所得的简单吸附物种提供一个自由电子(到材料的导带中),此时式(6.13)中的幂指数 β 值为 0.5。对此进行扩展,如果考虑更大的气体分子,如挥发性有机化合物,产生了一连串的表面反应,提供了超过一个电子,因而幂指数的值会偏离 0.5。除了改变气体分子,Scott 等[112]还研究了 MOS 表面改性对 β 值的影响,其中反蛋白石结构的 SnO_2 充当了球形"聚集体"。不规则的微结构以及表面聚集的存在会使得 β 值偏离 0.5 而有所提升,但不超过 1[113],具体大小由表面带电物种的状态,尤其是氧负离子所决定[112,114]。具体而言,O^{2-} 的高表面覆盖度将使得 β 值低于 0.5,而丰富的 O^- 则有利于提升 β 值至接近 1,并提升响应值[115]。Bai 等[115]通过在 In_2O_3 纳米管(nanotube, NT)上掺杂不同稀土(rare earth, RE)元素制备了一系列有丰富氧空位的 C_2H_5OH 传感器,之后进一步研究了气体响应值和不同 RE 元素掺杂的传感器之间的幂律规则。未掺杂的纯 In_2O_3 NT 的 β 值十分接近 0.5[见图 6.12(a)],这与先前的模型结论一致。当掺杂不同 RE 元素后,β 值有所提升,同时在各种氧负离子中 O^- 所占比例也有所提升,这是因为表面氧空位增加。气体响应值也随之一并提升。$Tb-In_2O_3$ 中得到了超高的 β 值[1.073 56,见图 6.12(f)],超越了传统的极限值 1,这主要是由于材料表面出现了极高比例的 O^-。

$$S=e\mu d\left(\frac{\alpha}{\gamma}P_R\right)^\beta \tag{6.13}$$

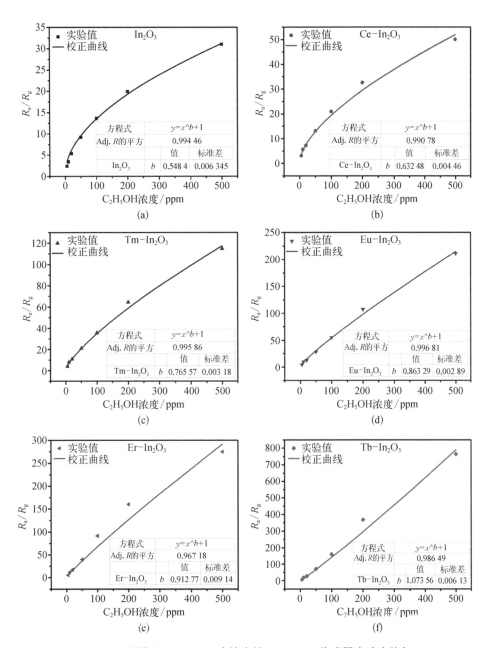

图 6.12　纯的和不同 RE 元素掺杂的 In_2O_3 NT 传感器中响应值与 C_2H_5OH 浓度(5~500 ppm)的校准曲线[115]

值得注意的是,在气体浓度较高时,从线性关系扩展到幂律规则十分关键。一般而言,高浓度气体的测试通常发生在有重要实际应用价值(如 H_2)[116],或测试气体本身过于稳定而难以在较低浓度下产生合适的响应(如 CH_4)[117]的情况下。此外,线性关系足够符合大多数情况下的传统测试浓度范围。例如，LoD 和测定限（limit of quantitation，LoQ）的概念对应了一个极低浓度，在此浓度下响应值和背景噪声大小可比拟[118-119]。在这种情况下，实际计算时幂律规则当然可以退化为线性关系[106-107]。

5. 气体吸脱附对响应/恢复过程的影响

当切换到某一测试气体后，MOS 材料传感器的响应可表示为电阻或电导率的变化。当切换回空气时，材料电阻恢复到初始状态。传统上，一方面，如果传感材料的电导率在接触某测试气体时增加，例如 n 型半导体接触到还原性气体，它的响应值可定义为 $S=R_a/R_g$ 或 $S=(R_a-R_g)/R_a \times 100\%$，其中 R_a 和 R_g 分别表示材料暴露于空气中与测试气体中的电阻值。另一方面，如果电导率降低，则响应值可定义为 $S=R_g/R_a$ 或 $S=(R_g-R_a)/R_a \times 100\%$。经验指出，响应/恢复速率可由一个完整的响应/恢复过程的 90% 所需的时间（τ）来衡量[32]。在实际操作中，由于传感器的响应值由前后两个平衡态所决定，中间的响应/恢复瞬态常常被人们所忽略；并且需要引入动态测量体系才能对该瞬态进行研究[120]。但事实上，传感器的响应瞬态很大程度上由传感层表面和内部发生的化学反应所决定[34]，因此在传感研究中起到重要作用。

Lundström[34] 讨论了固体传感材料中的响应/恢复瞬态，其通过应用类 Langmuir 等温方程并假定一级动力学过程，引入了两个时间常数（τ_f 和 τ_r）来分别代表响应过程的正反应和恢复过程的逆反应。响应和恢复的初始速率可进一步通过式（6.14）和式（6.15）进行计算，其中 θ 代表响应值。式（6.14）的值恒大于式（6.15），表明在一级动力学过程中，恢复时间长于响应时间。之后对二级动力学过程进行了进一步推导（如 H_2 在 Pd 上的吸附[121]，形成了 PdH_x），发现响应速率与恢复速率大小接近。Korotcenkov 等[39] 讨论了 SnO_2 传感中响应/恢复的动力学，并测量了不同工作温度范围下相应的活化能（见图 6.13）。进一步的实验表明晶格氧对于 H_2O 分子的形成不起作用，因此在无氧条件下传感器的响应动力学由材料自身决定；而在含氧条件下响应和恢复时间的差异被削弱。此外，在 150℃ 以上，响应时间接近或长于恢复时间；而在 150℃ 以下，响应时间短于恢复时间。前者归属于一级动力学，其决速步骤为气体分子的吸脱附过程，没有进一步的解离；而后者归属于二级动力学，其决速步骤为表面物质（尤其是 OH^- 和 O 原子）的解离吸附。

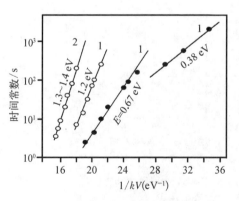

图 6.13 潮湿气氛中（相对湿度为 30%～50%）测定的 SnO_2 薄膜气体响应的瞬时特性的时间常数的温度依赖性[39]

1 表示 d 为 30～60 nm；2 表示 d 约为 200 nm

$$\left(\frac{\mathrm{d}\theta}{\mathrm{d}t}\right)_0 = \frac{1}{\tau_\mathrm{f}} \tag{6.14}$$

$$\left(\frac{\mathrm{d}\theta}{\mathrm{d}t}\right)_0 = -\frac{\theta_\mathrm{s}}{\tau_\mathrm{r}} = \frac{1}{\tau_\mathrm{f} + \tau_\mathrm{r}} \tag{6.15}$$

Korotcenkov 等[122]进一步的研究探讨了 In_2O_3 传感器的动力学。SnO_2 的瞬态变化主要与时间成幂函数关系[39],而 In_2O_3 中有更复杂的关系,如幂函数、平方根、正比例关系,这主要是因为其表面有大量由不同氧物种(化学吸附氧、氧空位和晶格氧)产生的活性位点。In_2O_3 接触到不同气体时会展现出不同的行为:对于 CO 和 H_2,在低于 250℃时表现出受体性质,而在高于 250℃时表现出给体性质;对于 O_3,恢复过程的时间常数(τ_rec)远长于响应过程的时间常数(τ_res)。此后,Korotcenkov 等进一步提出了 In_2O_3 的响应和恢复过程的机理,主要分为两个过程:吸脱附过程和氧化还原过程。前者主要由气体分子的吸脱附与解离、表面氧扩散、表面反应及其产物脱附构成;而后者除了以上过程外,还包括气体分子(O_2、H_2O 和测试气体)与 In_2O_3 晶格的相互作用(还原或重新氧化)、表面重构以及氧物种或氧空位的体相扩散。例如,CO 和 H_2 的恢复过程是 In_2O_3 的表面重新氧化,而 O_3 的恢复过程则仅仅是脱附。在氧化还原过程中,响应时间总体上与恢复时间接近;而在吸脱附过程中恢复时间更长。此外,虽然还原性气体主要遵循氧化还原过程,但 O_2 分子的吸脱附也起到重要作用。例如,CO 与不同氧物种之间发生的氧化反应有不同的结果,其中晶格氧主要起给体作用,而吸附氧则为受体,并且后者的响应时间小于前者。除了纯 MOS 材料以外,研究者们也对更复杂的材料进行了探讨,如高分子和异质结材料[123-124]。Hu 等[123]分析了 NH_3 在聚苯胺薄膜上的吸附动力学,并同时应用了 Langmuir 和 Freundlich 模型,最终发现 NH_3 的响应主要遵循后者。不过,基于最佳拟合数据来选择模型是一种常见的误解[44],因此该结论的可靠性可能要打折扣。

Mukherjee 和 Majumder[120]进一步使用了 $ZnFe_2O_4$ 传感器来研究 H_2 的响应和恢复动力学。一方面,如果考虑一级动力学过程,可以确定的是响应过程的决速步骤是化学吸附氧和测试气体的反应,而氧化产物的脱附则是恢复过程的决速步骤[108]。另一方面,时间常数的温度依赖性遵循 Arrhenius 方程[式(6.16)和式(6.17)],其中 E_A 和 E_D 分别代表测试气体吸附和反应产物脱附的活化能;τ_0 和 τ_0' 为指前因子,只由反应本身决定[117]。基于以上理论,Ghosh 和 Majumder[108]在 Co 掺杂的 ZnO 上进一步进行了 CO 和 H_2 传感测试。如图 6.14(a)所示,G_0 和 G_0' 几乎不随 CO 浓度变化,而 $G/(1-G)$ 与 CO 浓度间的线性关系[见图 6.14(b)]表明遵循 Langmuir 吸附机理[123]。Li 等[117]合成了 PdPt 合金负载的 SnO_2 纳米片,并讨论了响应过程的动态变化,该纳米片在不同温度下对 CO 和 CH_4 产生了双重选择性(见图 6.15)。对于相对活泼的 CO 分子,其动态过程出现了一个折点[见图

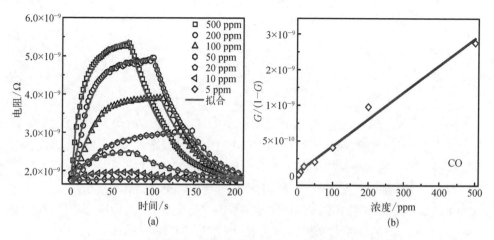

图 6.14 Co 掺杂 ZnO 薄膜传感器对 CO 传感性能研究[108]

(a) 300℃下 Co 掺杂 ZnO 薄膜传感器接触到不同浓度 CO(5~500 ppm)时的响应和恢复瞬态(点标志)以及相应的拟合数据(实线);(b) $G/(1-G)$ 对 CO 浓度的线性拟合直线

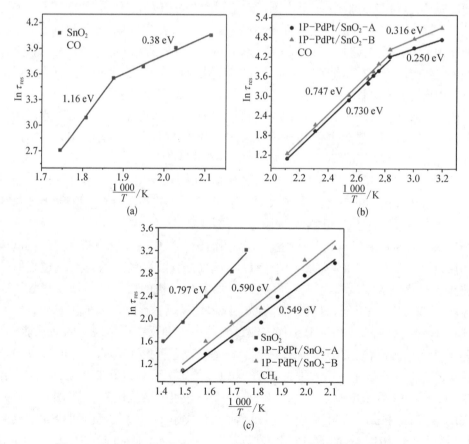

图 6.15 SnO_2、$1P-PdPt/SnO_2-A$ 和 $1P-PdPt/SnO_2-B$ 为传感材料时对 CO 和 CH_4 气体的 $\ln \tau_{res}$ 拟合直线与温度的关系[117]

6.15(a)和图 6.15(b)],并且高温下的活化能比低温下的活化能高,表明存在两种不同的 CO 氧化路径。在低温下,只有表面吸附氧参与氧化;而在高温下,表面吸附氧和晶格氧同时参与,从而使平均活化能升高。而由于 CH_4 分子具有非常稳定的正四面体结构[见图 6.15(c)],需要更高温度来活化,因此只能观察到一种氧化路径,即表面吸附氧和晶格氧同时参与。此外,在不同的传感材料中,PdPt 合金负载的 SnO_2(表示为 1P-PdPt/SnO_2-A)活化能最低(见图 6.15 蓝线),主要归因于溢流效应,在这一过程中贵金属充当活性位点,加强了其周围氧的解离吸附,从而提高了气体氧化效率[72,125]。

$$\tau_{res} = \tau_0 e^{\frac{E_A}{2k_B T}} \tag{6.16}$$

$$\tau_{rec} = \tau'_0 e^{\frac{E_D}{2k_B T}} \tag{6.17}$$

6.2 气体扩散

6.2.1 经典扩散理论和模型

1. Fick 扩散定律

在热力学中,输运过程指热力学系统从非平衡态向平衡态转换时所经历的过程,如热传导、黏性现象和扩散现象。如果系统中的粒子密度不均匀,粒子将通过热运动从高浓度向低浓度方向迁移,这一过程称为扩散。由于实际体系的复杂性,扩散过程总是与其他宏观过程一并发生,这使得扩散本身变得复杂。特别地,在一个恒温恒压的单组分系统中,仅由浓度差异造成的纯扩散被称为自扩散;而对于含有相近扩散能力的不同粒子的体系(如 CO 和 N_2),其扩散过程被称为互扩散。1855 年,A. Fick 提出了两条探讨扩散本质的定律。Fick 第一定律将扩散通量与浓度梯度联系起来,指出扩散的流量从高浓度区域向低浓度区域移动。在不同环境下,Fick 第一定律可表示为不同形式,在 1D 条件下最常见的形式是式(6.18),其中 J 代表扩散通量,用来衡量参与扩散过程的物质的量;D 代表扩散系数或扩散率;φ 代表物质的浓度;x 代表位置,即 1D 扩散路径上的长度。Fick 第二定律指出扩散如何随着时间影响浓度变化,其最常见的形式是式(6.19),其中 t 代表时间。事实上,Fick 第二定律可进一步通过 Fick 第一定律和能量守恒定律推导得出。

$$J = D \frac{d\varphi}{dx} \tag{6.18}$$

$$\frac{\partial \varphi}{\partial t} = D \frac{\partial^2 \varphi}{\partial x^2} \tag{6.19}$$

2. 硬球模型和扩散系数的推导

基于碰撞理论和 Newton 第一定律,研究者提出了硬球模型,该模型在分子动力学(molecular dynamics,MD)模拟中得到广泛应用[126-127]。在这种情况下,如果一个分子被简化为一个直径为 d(分子动力学直径)的球[128],它在空间中将保持匀速直线运动,直到与另一个分子相互碰撞。该分子总的运动路径是一条折线,因此在空间中,它的碰撞范围可简化为一个底面半径为 d 的弯折圆柱体。该分子碰撞前沿着弯折圆柱体行进的总路程定义为自由程(λ)。某一特定气体分子的平均自由程可通过 Maxwell 分布定律计算得到[式(6.20)],其中 $\bar{\lambda}$、n 和 σ 分别代表分子的平均自由程、数密度和有效碰撞面积。对于理想气体,式(6.20)可表示为式(6.21),其中 k_B、T 和 p 分别代表 Boltzmann 常数、温度和压强。平均自由程 $\bar{\lambda}$ 和温度 T 的关系是一个关键问题,而根据式(6.21),似乎 $\bar{\lambda}$ 与 T 成正比。然而,参数 T 没有出现在式(6.20)中,因此一般认为在体积一定时,$\bar{\lambda}$ 与 T 无关。事实上,一方面,随着温度升高,虽然平均碰撞时间减少,但碰撞前所需的平均运动距离不受影响。另一方面,随着温度升高,$\bar{\lambda}$ 将略微减小。这一现象主要源于温度升高后分子的振动加剧,从而使 σ 值略微增加,最终降低了平均自由程。事实上,通过调节温度来控制平均自由程是提升气体响应的一个关键方法[129]。扩散系数(D)可以进一步由式(6.22)定义[127],其中 $\overline{u_A}$、R 和 M 分别代表气体分子的方均根(root-mean-square,RMS)速率、普适气体常数和分子量[130]。对于理想气体,式(6.22)可进一步表示为式(6.23),其中 N_A 代表 Avogadro 常数。可以发现,D 与 M 的平方根以及 d 的平方分别成反比。因此,分子量和动力学直径更小的气体分子优先发生扩散,而留下更大的分子[128]。

$$\bar{\lambda} = \frac{1}{\sqrt{2}\,n\sigma} \tag{6.20}$$

$$\bar{\lambda} = \frac{k_B T}{\sqrt{2}\,p\sigma} \tag{6.21}$$

$$D = \frac{1}{3}\overline{u_A}\bar{\lambda} = \frac{1}{3}\bar{\lambda}\sqrt{\frac{8RT}{\pi M}} \tag{6.22}$$

$$D = \frac{2}{3\sqrt{M}\,d^2 p N_A}\left(\frac{RT}{\pi}\right)^{\frac{3}{2}} \tag{6.23}$$

3. 约束性扩散以及多孔材料中的扩散机理

除了气体分子自由扩散并相互碰撞产生的普通 Fick 扩散以外[131],气体分子也会在一个外在物理尺度范围内(如气体容器的大小或者多孔材料中孔道的直径/半径,见图 6.16)发生约束性扩散[132-133]。在流体力学中,平均自由程和物理尺度的关系可由一个量纲为 1 的量衡量,称为 Knudsen 数(Kn),由式(6.24)定义,其中 L 代表有代表性的物理尺度。高 Kn 值表示流体的平均自由程大于物理尺度,使得流体可以更自由地通过相应的容器。典型地,对于不同的 Kn 值,根据经验可定义如下四种不同的流[134]。① $Kn<0.01$:连续流或 Poiseuille 流,此时流体的平均自由程比物

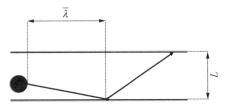

图 6.16 粒子的平均自由程和扩散的物理尺度的示意图

理尺度小得多,因此流体会以连续传质(体相流体)而非离散粒子的形式通过孔道[134]。在宏观体系中,流体的浓度非常高,使得黏度增加。② $Kn>10$:自由分子流或 Knudsen 流,此时粒子的流动可认为互相独立,因而每个粒子都沿直线行进[134]。在宏观体系中,流体浓度极低(通常低于 0.1 Pa),处于高真空或超高真空状态,因而黏度趋近于零。③ $0.01<Kn<0.1$:滑移流,此时若干个流体粒子也一并连续移动,但规模不如连续流的连续传质过程大,粒子之间仍然有相互联系[134]。在宏观体系中,流体压强相对较高。④ $0.1<Kn<10$:过渡流,此时平均自由程与物理尺度相当,属于连续流与自由分子流的叠加。流体同时具有连续流与自由分子流的性质[134]。虽然流和扩散很相似,并且在一些体系中同时发生,但它们来源不同。前者由压强梯度驱动,后者则由浓度梯度驱动。换言之,在颗粒与分子均匀分布的体系中,由于没有梯度驱动的传质过程,不会产生流和扩散。

$$Kn=\frac{\bar{\lambda}}{L} \tag{6.24}$$

气体分子首先扩散到传感材料的内部,并在活性位点处发生反应。对于多孔材料,孔径大小会导致 Kn 值的变化,从而产生不同形式的扩散。Li 等[135]研究了膜材料中粒子的扩散,并绘制了不同孔径大小的膜材料中的不同气体扩散机理,如图 6.17 所示。当孔径大于 $\bar{\lambda}$ 时,发生体相 Poiseuille 流[见图 6.17(a)],也叫对流性流。此时,不同的气体分子将以体相流体的形式自由通过孔道,因而选择性较低。当扩散体系的尺度接近或略小于 $\bar{\lambda}$ 时,发生 Knudsen 扩散[见图 6.17(b)],此时 Kn 接近或略大于 1。在这种情况下,气体分子可能与传感材料(膜材料、MOS 材料等)发生碰撞。对于多孔材料,孔道半径可以作为扩散尺度($\bar{\lambda}/2$)看待。因而,Knudsen 扩散系数(D_K)可表示为式(6.25),从而将分子内在的参数(M)和外在的扩散尺度(r)组合起来。类似地,D_K 与 M 的平方根成反比的关系使得 Knudsen 扩散相比体相 Poiseuille 流有一定的选择性,但仍然不够。当孔径大小介于两种不同的气体分子之间时,发生尺寸限制的扩散[见图 6.17(c)],也叫分子筛扩散。此时更大的分子(绿球)将无法扩散进入孔道,而只有更小的分子(红球)能够通过,从而产生更高的选择性。尺寸限制的扩散可进一步分为两类:表面模型[见

图 6.17(c) 左小图]和气体平移模型[见图 6.17(c) 右小图][87]。固态扩散[见图 6.17(d)]是膜材料中的一种特殊扩散形式,此时膜中几乎没有孔。在这种情况下,气体分子吸附到膜的表面,其中一部分溶解到膜材料中。在膜材料内部经历一系列过程后,气体分子从膜表面的另一侧脱附并释放。

$$D_K = \frac{4r}{3}\sqrt{\frac{2RT}{\pi M}} \quad (6.25)$$

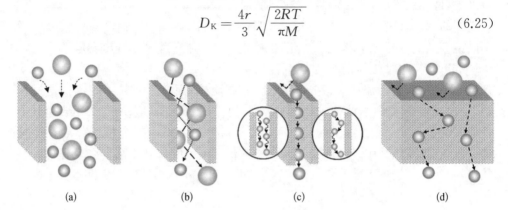

图 6.17 气体扩散机理的示意图(后附彩图)[135]

(a) 体相 Poiseuille 流;(b) Knudsen 扩散;(c) 尺寸限制的扩散(左小图:表面模型,右小图:气体平移模型);(d) 固态扩散机理

值得注意的是以上四种扩散机理主要适用于膜材料的扩散。对于其他材料,如多孔 MOS 材料,其孔径大小可调控的范围不如膜材料那么宽,因而可发生的扩散种类也相应减少。特别地,如果孔径小到气体分子之间无法相互通过而形成 1D 通道时,会发生尺寸限制的扩散,被称为单档扩散或构型扩散[136-137]。研究者对于多孔材料中的限制性扩散也提出了一些其他概念(如非通量扩散和悬浮效应)[138-141],但它们的内在机理仍然不够明确[142]。另一种对于不同扩散的经验性分类包括分子扩散、Knudsen 扩散、构型扩散以及表面扩散[136]。根据 IUPAC 的分类,孔径小于 2 nm 的为微孔,大于 50 nm 的为大孔,而介于 2~50 nm 的为介孔。随着孔径大小增加,气体分子进入孔道的扩散率增大[36],从而依次经历表面扩散、Knudsen 扩散和体相分子扩散的过程[86,143]。孔径大小为 1~100 nm 时,气体分子的平均自由程 λ 值与孔径大小相当,此时 Knudsen 扩散占据主导。换言之,大孔径将促使气体分子自由地扩散通过,而小孔径对于扩散的限制更显著[69]。因此,介孔材料在气体传感中的广泛应用,不仅是因为其自身的优点[88,144-147],也包括其促进了气体分子发生 Knudsen 扩散[69,93,145,147-151],从而进一步提升了传感性能。一些研究者也对多孔材料中 Fick 扩散和约束性扩散的区别进行了研究。Raccis 等[131]以反蛋白石结构作为模型(见图 6.18),该模型可通过调节粒径(a)、孔径(L)和腔体半径(R)等参数而得到各种腔体大小和开口,并在此基础上开展 Brown 动

力学(Brownian dynamics,BD)研究。研究结果表明在各种条件下都能预测到传统 Fick 扩散的存在,但非 Fick 指数被归因于空腔开口的多分散性。Malek 和 Coppens[152]使用动态 Monte Carlo 模拟对纳米多孔介质的表面粗糙度对于 Knudsen 自扩散与 Fick 扩散的影响进行了研究。由于 Knudsen 自扩散系数与分子停留时间有关,导致其受到表面粗糙度的影响,而 Knudsen Fick 扩散则不受影响。

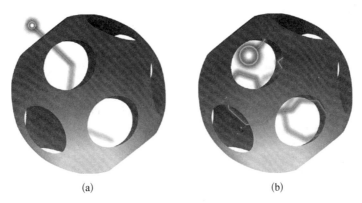

图 6.18 示踪粒子在逃到邻近空腔前与空腔壁发生多次碰撞的过程简图[131]

(a) 对于小比例的 a/L,少量反弹即足够;(b) 对于大比例的 a/L,颗粒逃逸之前要与腔壁发生大量的碰撞

4. 吸附影响的扩散

除了以上讨论的含有外部物理尺度的约束性扩散,在很多情况下扩散都与吸附影响有关。例如,通过对比气体分子是否吸附到吸附剂表面,可区分表面扩散与体相分子扩散(见图 6.19)。在这种情况下,对于气体分子而言,垂直于吸附晶面的维度方向将变得无法进入[130]。此外,在孔径小于 1 nm 的多孔材料中,气体分子的表面扩散占据主导,导致孔内扩散通量较低。和其他几种扩散相比,研究者认为表面扩散由吸附剂表面的化学势梯度驱动,而扩散势垒则与化学吸附能有关。

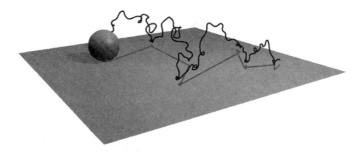

图 6.19 分子表面扩散的示意图(后附彩图)[153]

该图将分子固定在表面的时间与在表面上方发生体相扩散的时间结合起来,黑色曲线是真实的三维分子轨迹,而品红色折线是穿越表面的有效二维轨迹

跳跃模型是被广泛应用于分析表面扩散的模型之一。在跳跃模型中,吸附的分子在吸附剂表面相邻的吸附位点间跳跃[154]。特别地,在亲 H^+ 溶剂中 H^+ 迁移的机理可因此通过 Grotthuss H^+ 跳跃模型来解释[97]。H^+ 不断地吸附到溶剂分子上,再跳跃到相邻的位点处。在这个过程中 H^+ 离子簇(H_3O^+、$H_9O_4^+$、$H_5O_2^+$ 等)不断产生和离解。Agmon[97]指出 H^+ 迁移是非相干的 H^+ 跳跃过程,并且决速步骤为氢键的裂解而非 H^+ 运动或离子簇的产生。除了 H^+ 跳跃,研究者也针对不同粒子在不同界面上的跳跃过程进行了建模和规整化。例如,碳基催化中氧的溢流起到关键作用,因此 Radovic 等[155]将 DFT 计算应用于石墨烯表面氧跳跃的研究,并着眼于原型集群和周期性结构。当石墨烯包含空余边位点以及氧官能团,导致电子密度增加时,可观察到跳跃能垒下降。在最佳条件下的氧迁移情况与微孔中气相 O_2 分子的扩散情况相同,这与研究者通常认为的在 sp^2 杂化碳材料参与的各种吸附与反应过程中基面氧所起到的主导(而非旁观者)作用一致。在气-固界面以外,Skaug 等[153]进一步将单分子追踪应用于固-液界面,并发现一系列不同分子经历了具有非 Gauss 位移的间歇性随机行走,这与通常基于随机行走与 Gauss 统计的分子表面扩散的假定形成对比。进一步地,研究者假定间歇性跳跃是分子在固-液界面发生表面扩散的主要形式。总体而言,对于大多数气-固体系,气体吸附能远超气体碰撞能,因此气体吸附能将随着气体压力的增大而减小,从而提升气体表面扩散能力。此外,如果气体吸附能和气体碰撞能大小接近,气体分子的表面扩散率和气体压力间将呈现相反趋势。在这种情况下,将发生 2D 气体行为,即由吸附分子之间的碰撞支配、以表面 λ 为特征的形式。由于表面 λ 远超相邻吸附位点的空间距离,跳跃模型将不再可用[130]。

Sun 和 Bai[130]研究了石墨烯上不同气体分子的表面扩散机理。石墨烯的单层结构使其与物理吸附于其上的气体分子间作用较弱,导致传统的跳跃模型可靠性存疑。研究者使用 MD 手段对 CH_4 和 CO_2 分子建模,模拟和计算了不同气体压力下石墨烯表面的扩散系数[见图 6.20(a)]。根据 Einstein 方程[式(6.26)],表面扩散系数通过均方位移(mean-square-displacement, MSD)和时间的线性关系的斜率计算得到[130],其中 x 和 y 分别代表 t 时刻下的分子坐标,而 x_0 和 y_0 分别代表初始吸附时刻 t_0 时的分子坐标。CH_4 和 CO_2 分子的表面扩散系数都随气体压力增大而减小[见图 6.20(b)离散点],表明石墨烯表面的气体扩散为 2D 气体行为,并主要通过吸附的分子间的碰撞而非跳跃模型来支配。随后研究者进一步比较了表面扩散系数和对应的体相分子扩散系数[见图 6.20(b)断续线],后者通过硬球模型预测[由式(6.22)计算]得到。结果表明表面扩散系数和体相分子扩散系数之间有定性关联,并且随气压升高而降低。然而,由于其他气体分子之间以及气体分子和石墨烯层之间的相互作用导致的限制,表面扩散系数的值低于对应的体相分子扩散系数,对于吸附程度较高的 CO_2 分子而言尤为如此。此后研究者又对多层石

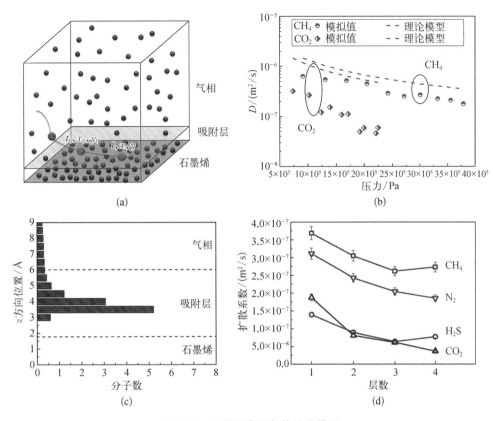

图 6.20 石墨烯表面气体扩散模拟

(a) 使用 Einstein 方程计算扩散系数的模型;(b) 不同压力下 CH_4 和 CO_2 分子的表面扩散系数,断续线代表使用硬球模型基于理想气体动力学预测得到的体相扩散系数的理论值[83];(c) 吸附在单层石墨烯表面的 CH_4 沿 z 方向的分子数密度分布;(d) CH_4、H_2S、CO_2 和 N_2 的扩散系数随石墨烯层数的变化关系,其中扩散系数的误差棒基于 5% 不确定度绘制[128]

墨烯进行了进一步的研究[81]。由于气体分子和石墨烯层之间的较强相互作用,在石墨烯表面上方形成了一个单原子厚的高气体密度的区域[即吸附层,图 6.20(c)],表明石墨烯表面发生的是单层气体吸附。随着石墨烯层数的增加,石墨烯与气体分子之间的相互作用增强,从而提升了吸附程度。然而,当石墨烯层数超过 2 时,吸附程度达到饱和,这主要是因为石墨烯上能够实际与气体分子产生相互作用的有效 C 原子数量几乎不变。一方面,随着石墨烯层数增加,气体吸附能和气体碰撞能一直保持接近状态,表明发生 2D 气体行为;另一方面,气体吸附能的值增大,对气体分子产生更多限制,因而气体的扩散系数值减小[见图 6.20(d)]。此外,当石墨烯层数较大时,由于相互作用距离很有限,扩散系数将不再显著受到石墨烯层数的影响。相同石墨烯层数上不同气体分子的扩散能力[见图 6.20(d)]可以进一步通过硬球模型来解释。CH_4 和 N_2 分子的分子量和动力学直径较小,因此扩散系数较大;而 H_2S 和 CO_2 分子的分子量和动力学直径较大,因此扩散受约束。此外,H_2S 和 CO_2 分子在石墨烯表面的吸附程度更高,因而石墨烯上方的气相分子数更少,压力更低。

$$D = \frac{\langle (x-x_0)^2 + (y-y_0)^2 \rangle}{4(t-t_0)} = \frac{\langle (x-x_0)^2 + (y-y_0)^2 \rangle}{4\Delta t} \quad (6.26)$$

5. 界面传质与 DR 耦合理论

作为一种典型的由浓度梯度驱动的传质过程,扩散在质量传递以及将体系从当前的未平衡态导向一个新的平衡态的过程中起到重要作用。对于多相体系,界面处某一特定组分的扩散将使其质量传递到另一相中,这一点在加强两相之间的异相连接中十分关键。最著名的界面传质过程之一是 Kirkendall 效应(以美国化学家与冶金学家 Ernest Oliver Kirkendall 命名)。Kirkendall 效应阐述了两种具有不同扩散率的金属间因发生传质而导致金属-金属界面发生移动[156-157]。在此之前,研究者们曾一度认为金属原子的扩散机理只是扩散的原子与其他原子或其他空穴之间发生位置交换,而在扩散过程中没有界面移动。但 Kirkendall 效应指出不同金属具有不同扩散率,进一步而言,扩散过程中金属-金属界面将向具有高扩散率的那一相移动,并产生孔洞(Kirkendall 孔洞)[158]。

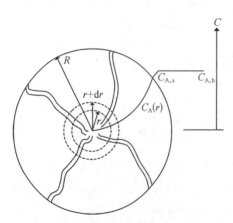

图 6.21 一个半径为 R 的球形多孔固相颗粒的示意图[160]

物质 A 的浓度从体相($C_{A,b}$)到表面($C_{A,s}$)再到孔道内部[$C_A(r)$]中不断降低

一个更被广泛接受的界面传质形式是发生在含有一个或多个流体相的多相体系中的情况,即气-固界面或气-液界面。在多相催化体系和气体传感等领域中,这是一个常见的现象。以气-固界面为例,图 6.21 基于滞膜模型展示了一个多孔固相材料中的传质过程。一般来说,研究者会假定气体的压强梯度促使其在固体颗粒周围流动,然后在体相气体和固相外表面之间产生一个边界层,并且气体在边界层中的流动相比体相中静止得多,导致产生一个穿越边界层的浓度梯度。气体分子将在边界层中发生扩散而非流动,直到到达气-固界面,在此之后它们吸附到固体表面并经历进一步的过程。这一过程使得体相的流体相中的流体分子传递并聚集到固体表面上,这被称为外扩散过程[159]。当气体分子接触到固相表面后,它们可能经历多个平行发生的过程,即耦合的 DR 过程。在此过程中,气体分子可能进一步扩散到固相体系的更深处并与内部的固体颗粒发生接触,尤其是在多孔材料中发生约

束性扩散,这被称为内扩散过程[159];而另一些分子将在固相体系内部某处或者直接在表面上发生反应而被消耗。在大多数情况下,气体分子在体相气氛中经历的整个过程如图6.21所示,这是一个由浓度梯度驱动同时包含外扩散与内扩散的组合的过程。

一方面,在实际DR耦合条件下,由于扩散和反应同时发生,研究者通常把扩散的作用归属为它与最终反应之间的比较。研究者们提出了一些假定,如Mears假定(C_m)和Weisz-Prater假定(C_{WP}),来分别衡量气-固界面下整个DR耦合过程中外扩散和内扩散的作用[159-164]。另一方面,气体分子在实际固相体系内部的扩散率小于拥有均匀分布孔道结构的理想条件,因此,气体A在固相体系中的有效扩散率($D_{A,e}$)由式(6.27)定义,其中ε、D_A、σ和τ分别代表固相体系的孔隙率、气体A在固相体系中的扩散系数、构效因子和曲率[162]。在实际条件下,研究者采用伪连续孔模型将固相体系的颗粒简化为球体,因而气体分子从体相气氛中扩散到球体表面并进一步扩散到孔道深处[165]。

$$D_{A,e} = \frac{\varepsilon D_A \sigma}{\tau} \quad (6.27)$$

此外,对于边界层很薄以至于可忽略的多孔材料,可以认为流体分子在其中只有内扩散。如果假定发生一级反应,A分子的浓度与r的相关性可表示为式(6.28),其中量纲为1的量ϕ是反应速率与内扩散速率的比值。该参数被称为Thiele模数(以美国学者E. W. Thiele命名),对于一级反应($n=1$),它可通过式(6.29)计算得到,其中ρ_s代表表观密度[159,162,166]。高Thiele模数代表反应速率远高于内扩散速率,因此大部分A分子在固体表面附近的一个浅层区域内反应并被消耗,使得固体球形颗粒的大部分区域没有得到利用。反言之,低Thiele模数意味着内扩散速率远高于反应速率,此时A分子能够扩散到固体球形颗粒内的几乎所有地方并发生反应,从而提高固相的利用率[162]。因此,研究者进一步引入了另一个衡量内扩散阻力重要性的准则,即内扩散效用因子η。对于固体球形颗粒内部发生的一级反应,η由式(6.30)定义[162,167]。内扩散效用因子在不同反应条件下均与Thiele模数保持负相关(见图6.22),因而能很好地详细阐明固体球形颗粒的利用率。当η趋近于1时,固体球形颗粒中的内扩散没有阻力,

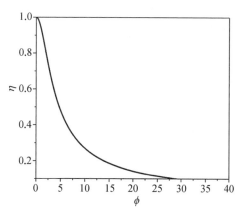

图6.22 固体球形颗粒中发生一级动力学过程时η与Thiele模数的关系[162]

因而 DR 耦合过程由表面反应控制；当 η 趋近于 0 时，该过程为内扩散控制[162]。

$$C_A(r) = C_{A,s} \frac{\sinh\left(\phi \frac{r}{R}\right)}{\frac{r}{R}\sinh\phi} \tag{6.28}$$

$$\phi = R\sqrt{\frac{k\rho_s}{D_{A,e}}} \tag{6.29}$$

$$\eta = \frac{3}{\phi}\left(\coth\phi - \frac{1}{\varphi}\right) \tag{6.30}$$

另一个值得讨论的传质情形是气-液界面的传质。与气-固界面相比，气-液界面的两相均为流体，因此气体分子通过外扩散和吸附到液体表面后，其接下来在新的流体系中很可能发生动态的传质。研究者针对这种情况已经提出了几种描述气-液界面传质的模型，包括滞膜模型和渗透模型[168]。气-液界面中的滞膜模型和气-固界面的类似，唯一的区别在于，在气-液界面中，边界层不仅存在于气体一侧，也存在于液体一侧。在这种情况下，气体分子穿过边界到达液相后的传质过程将遵循一个新的扩散路径，如同其在气相一侧的边界层中一样，因而最终形成了一个双膜模型，其中两相边界薄膜层的厚度分别用水动力学参数 $\delta_{1,\text{int}}$ 和 $\delta_{2,\text{int}}$ 表示[见图 6.23(a)][169-170]。尽管双膜模型较为简易，水动力学参数的预测仍然受到材料几何结构、液体搅动以及物理性质的影响[168]。渗透模型则认为液相体系没那么固定，因此气体分子进入液相体系中的流量并不稳定。事实上，与气相界面接触的液体会在一段时间（暴露时间）后更换，在此过程中它被体相新鲜的液体所取代[171]。暴露时间（θ）可通过水动力学参数（如界面的速率以及长度）来衡量[168]。既然气-液界面的表面不断变化，渗透模型也因此被称为表面更新模型[172]。Toor 和 Marchello[173] 最终把滞膜模型和渗透模型组合成了一个新的薄膜-渗透模型[见图 6.23(b)]。在动态系统中，渗透模型看上去比滞膜模型在物理上更可靠，而在实际条件下以上模型通常与流体流动模型组合起来衡量总体传质效果[168]，因此衍生出进一步修正的模型研究[174-175]。

和式（6.28）类似，如果假定发生一级反应，气体分子 A 在液相距离表面深度 x 处的浓度可表示为式（6.31），其中 $C_{A,i}$ 和 L 分别代表 A 在气-液界面处的浓度（与气-固界面中的 $C_{A,s}$ 类似）和液相薄膜的总厚度。研究者引入了另一个量纲为 1 的量，即 Hatta 数[表示为式（6.32）]来简化式子[168,170]。和 Thiele 模数类似，Hatta 数对比了薄膜内部的反应速率与穿过薄膜的扩散速率[176]。如果 $Ha > 3$，反应速率很快，大部分气体分子将在靠近边界的液体薄膜中反应，整个过程由扩散控制；如果 $Ha < 0.3$，气体分子的扩散速率足够使其到达体相液相并反应，整个过程由反应控制[168]。为了进一步衡量气-液界面液体薄膜层的扩散影响以及利用率，研究

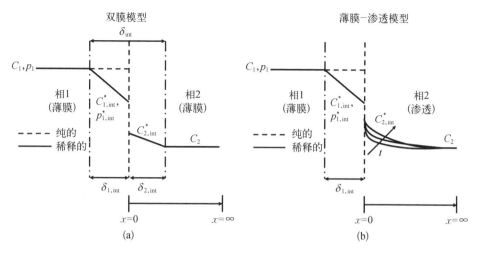

图 6.23 使用双膜模型和薄膜-渗透组合模型来描述气液体系和液液体系中气体分子浓度的示意图[168]

(a) 双膜模型；(b) 薄膜-渗透模型

者又定义了液体效用因子(η_L)。液体效用因子是实际反应速率与没有浓度梯度的理想条件下的反应速率的比值，可表示为式(6.33)，其中 k_c 代表传质系数。

$$C_A(x) = \frac{C_{A,i}\sinh\left[Ha\left(1-\dfrac{x}{L}\right)\right] + C_{A,b}\sinh\left(\dfrac{x}{L}\right)}{\sinh Ha} \tag{6.31}$$

$$Ha = x\sqrt{\dfrac{k}{D_{A,e}}} \tag{6.32}$$

$$\eta_L = \dfrac{D_{A,e}}{Hak_cL\tanh Ha}\left(1 - \dfrac{C_{A,b}}{C_{A,i}\cosh Ha}\right) \tag{6.33}$$

6.2.2 MOS 气体传感器的扩散模型

气体扩散的过程可视为对常见吸脱附理论的补充，后者侧重于 MOS 材料的物理化学性质。虽然气体扩散机理主要起到辅助作用，但对于气体扩散的操控将对响应值产生很大影响。此外，气体分子的外扩散发生在其吸附到 MOS 材料表面之前，因而该过程对传感器的响应值几乎没有贡献，也通常被忽略。当气体分子吸附到 MOS 气体传感器以后，大量的气体分子将在 MOS 材料内部发生内扩散并遵循 DR 耦合过程。在此过程中，一部分分子因吸附或反应而被固定。这里假定分子的固定速率大于其扩散速率，故最终产生两类不同分子：固定的分子和自由的分子。化学反应固定了一部分分子，从而减慢了扩散速率。基于 Fick 第二定律

[式(6.19)], Crank 对扩散方程进行了修改[式(6.34)], 其中 C_A 和 S_A 分别代表自由和固定的 A 分子的浓度[177]。

$$\frac{\partial C_A}{\partial t} = \nabla(D \nabla C) - \frac{\partial S_A}{\partial t} \tag{6.34}$$

1. Gardner 的线性与非线性 DR 模型

1989 年, Gardner 提出了一个线性 DR 模型[33]。在该模型中, 假定 S_A 与 C_A 成正比, 因而式(6.34)可简化为式(6.35)。气体浓度与扩散深度的关系(见图 6.24)表明在低 T 值下, 浓度曲线随着距离的增加而显著下降; 而在高 T 值下, 曲线逐步变得平缓, 表明浓度梯度与扩散速率的同时降低。但该模型假定扩散速率低于反应速率, 因此只在较厚的多孔层中有效, 进而限制了其应用范围。

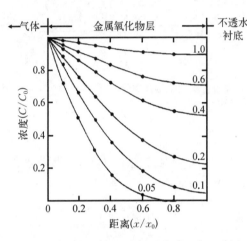

图 6.24 DR 方程的分析解(整体曲线)与数值解(点)[33]

显示了均质氧化层在不同时刻的浓度分布(以扩散时间常数的一部分表示)

$$\frac{\partial C_A}{\partial t} = D_{A,e} \nabla^2 C_A \tag{6.35}$$

1990 年, Gardner 又提出了一个非线性 DR 模型[105]。和线性 DR 模型相比, 这里假定 S_A 与 C_A 的 r 次方成正比[式(6.36)], 其中 B 和 r 分别为传感材料和测试气体的常数。在这种情况下, 扩散方程可简化为式(6.37)。非线性关系的假定扩展了该模型的应用范围, 1996 年 Vilanova 等[35]基于此对瞬态电导进行了建模。传统的选择性研究侧重于对不同气体分子产生不同响应值, 但该研究引入了基于响应时间来评判选择性的方法。20 ppm 苯和 50 ppm 邻二甲苯的平衡态电导变化 $\Delta G(\infty)$ 值接近, 而它们的归一化响应时间 τ' 则不同, 主要是因为两者的扩散能力不同, 因此, 这两种气体可以被区分开来。此外, Lu 等[178]指出 DR 耦合过程也存在于气体传感实验中, 从而为气体扩散理论在气体传感研究中的应用奠定了坚实的基础。

$$S_A = BC_A^r \tag{6.36}$$

$$\frac{\partial S_A}{\partial t} = \frac{D_A}{B^{\frac{1}{r}}} \frac{\partial^2 S_A^{\frac{1}{r}}}{\partial x^2} \tag{6.37}$$

2. Yamazoe 的扩散模型

2001 年, Yamazoe 课题组[36]系统性地将 DR 耦合模型和气体扩散控制的响应

灵敏度机理引入 SnO_2 气体传感中。该模型主要基于两个假定：

(1) SnO_2 薄膜中发生 Knudsen 扩散；

(2) 扩散遵循一级动力学过程。

对于第一条假定，研究通过在溶胶悬浮液中进行旋涂合成了孔径大小为 10～15 nm 的 SnO_2 晶粒，处于 Knudsen 扩散范围内。基于以上两大假定与 DR 耦合模型，Knudsen 扩散方程可表示为式 (6.38)，其中 C_A、x 和 k 分别代表目标气体的浓度、扩散深度和反应速率常数。对于稳定状态，式 (6.38) 等于零。通过引入边界条件以及 Hatta 数 [Ha，式 (6.32)]，方程的特解可表示并简化为式 (6.39)，其中 $C_{A,s}$ 和 L 分别代表材料表面的气体浓度和 SnO_2 传感层的厚度[68]。如前所述，Ha 反映了 DR 耦合过程中扩散与反应之间的竞争关系[见图 6.25(a)]。Ha 值越大表明反应活性越强，从而降低了扩散深度；Ha 值越小则扩散行为越显著。此外，该研究又进行了额外的假定，即 SnO_2 传感层接触到测试气体后的电导变化与气体浓度呈线性关系[式 (6.40)]，其中 $\sigma(x)$ 和 σ_0 分别代表材料接触到测试气体和在空气中时的电导，a 是个常数。这里假定的线性关系是基于气体浓度较低的条件，因此其实际有效性可能在其他条件下存疑。通过对电导积分，薄膜的响应（气体灵敏度 S）可由式 (6.41) 定义。具体而言，Hatta 数包含气体传感研究中的三个主要因素：反应速率、Knudsen 扩散系数，以及传感层厚度。极低的 Hatta 数（接近零）表明内扩散速率远大于反应速率，导致 MOS 传感器传感层中气体分子均匀分布，几乎没有浓度梯度，因此整个体系中几乎不受气体扩散的影响。随着 Ha 值的增加，内扩散速率将逐步和反应速率相当，也因此起到决定性作用。在实际 MOS 气体传感体系的 DR 耦合过程中，如果要充分利用传感器，应该控制气体内扩散速率与反应速率相当，在此条件下传感器的响应值或灵敏度的变化趋势最显著。如图 6.25(b) 所示，这一现象发生在 Ha 介于 1～10 时。对于特定温度下的特定气体，k 和 D_K 值均为常数，因此可以通过精确调控传感器传感层厚度 L 来获得特定的 Ha 值，这也提供了一种可行的定量实验设计的思路。两个主要的参数，即 Knudsen 扩散系数与反应速率常数都与温度有关，并且温度升高会使扩散速率和反应速率同时提升。虽然 D_K 和 k 都与温度成正相关，但后者速率增加得比前者快得多。因此在高温下，反应过程将占主导，使得扩散成为决速步骤；低温下则相反。

$$\frac{\partial C_A}{\partial t} = D_K \frac{\partial^2 C_A}{\partial x^2} - kC_A \tag{6.38}$$

$$C_A = C_{A,s} \frac{\cosh\left[(L-x)\sqrt{\frac{k}{D_K}}\right]}{\cosh\left(L\sqrt{\frac{k}{D_K}}\right)} = C_{A,s} \frac{\cosh\left[\left(L - \frac{x}{L}\right)Ha\right]}{\cosh Ha} \tag{6.39}$$

$$\sigma(x) = \sigma_0(1 + aC_A) \tag{6.40}$$

$$S = \frac{R_a}{R_g} = 1 + \frac{aC_{A,s}}{L\sqrt{\frac{k}{D_K}}} \tanh\left(L\sqrt{\frac{k}{D_K}}\right) = 1 + \frac{aC_{A,s}}{Ha} \tanh Ha \tag{6.41}$$

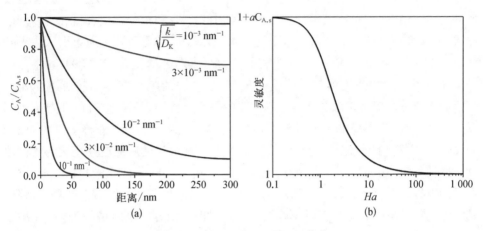

图 6.25 气体传感灵敏度定义[36]

(a) 在固定温度下,不同 $\sqrt{\frac{k}{D_K}}$ 值的传感层(厚度为 300 nm)内模拟的气体浓度曲线;(b) 固定温度下薄膜的气体灵敏度的通用表示

虽然该模型相对正式,但其局限性仍然限制了其应用,主要的局限性来源于两大假定:SnO_2 中发生的 Knudsen 扩散以及电导线性关系[式(6.40)]。除了 Knudsen 扩散,该研究还忽略了介孔材料中存在的表面扩散。当然,也有研究者对有序介孔 SiO_2 纤维进行了分析,并指出室温下表面扩散在总扩散流量中只占约 10%,证明有序介孔材料能够促进 Knudsen 扩散[179]。此外,该模型的主要短板在于假定的提出没有援引任何实验数据。虽然 Yamazoe 团队后来提出了 MOS 传感器幂律规则的理论基础(前文已讨论)[40,110],但该扩散模型并未通过进一步修正来适应幂律规则[180]。该模型的另一个局限性在于假定扩散为一级动力学过程,这在掺杂其他材料(如 PdO)或潮湿条件下无效[181-182]。

2002 年,Yamazoe 课题组[37]进一步将扩散模型扩展至非稳定状态[式(6.42)]。与式(6.38)相比,在非稳定状态下,气体浓度与扩散时间也有关。非稳定状态下的一个典型状况是过冲现象(见图 6.26),它表明了 DR 耦合过程中的竞争效应。在接近材料表面的地方,由于气体分子的内扩散比反应早发生,气体浓度值将过冲到超过平衡态,然后由于反应消耗而稳步下降至平衡态。过冲现象的发生是由于气体内扩散速率和反应速率不一致造成的局部浓度不均。在 Yamazoe 课题组的研究中[37],过冲现象的时间尺度约为 10^{-6} s,通过裸眼或实验设备无法检测。但在实际操作中,尤其是 D_A 和/或 L 取极端值时,在达到平衡态前可能观察到一个小但可检测的峰,产生一条非平坦的响应-恢复曲线[64],该峰可以通过调整实验操作或参数来消除。在实际传感研究中,气体分子首先在空气中发生外扩散,然后吸附到 MOS 材料表面并向更深处进行内扩散。尽管气体的外扩散对传感器的响应值几乎没有贡献,也因而常常被忽略,但当注入大量测试气体时,吸附速率的增加将同时促进其逆过程脱附的发生,使得气体分子重新进入空气中,进而发生另一个外扩散过程。在这种情况下,气体分子外扩散的影响不可忽略,并且整个体系将达到表面快速吸脱附、内扩散与反应的三方平衡,在此过程中局部的非平衡气体浓度快速变化,导致材料电阻出现略微甚至显著的变化,从而在最终到达平衡前显示出一个甚至多个过冲的峰。

$$\frac{\partial C_A(x,t)}{\partial t} = D_A \frac{\partial^2 C_A(x,t)}{\partial x^2} - kC_A(x,t) \tag{6.42}$$

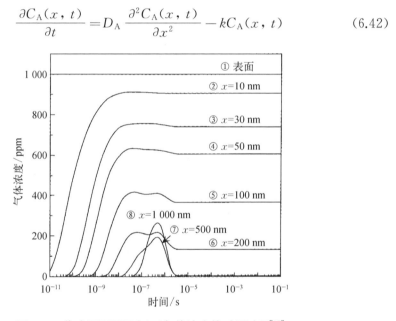

图 6.26 传感层不同深度(x)气体浓度的时间过程[37]

2003 年,Yamazoe 课题组[38]研究了响应和恢复过程中的气体扩散动力学,并提出了一个新的扩散模型。与之前的模型相比,这一模型在推导中没有包含任何假定,因此研究认为其在整个 DR 过程中均有效。对于扩散方程[式(6.42)],气体浓度被拆分成一个齐次函数和另一个非齐次函数的和。虽然该模型没有基于任何假定,因此不会限制其应用范围,但它只考虑了扩散常数与反应速率,没有考虑温度与传感层厚度的影响,而后两者对于瞬态过程也起到关键作用[183],阐明二元介孔-大孔结构中响应时间减少的实验已证明了这一观点[184]。

一般认为三个基本因素，即受体功能、换能器功能和效用因子会影响气体传感性能[109,185-186]。2003 年，Yamazoe 团队把效用因子[来源于式(6.33)中的液相效用因子]引入 MOS 气体传感器[见图 6.27(a)]，用以衡量材料内部氧化物晶粒对目标气体的可及性，从而评估传感层内部的 DR 耦合效应[68,187]。虽然在先前的研究中已观察到传感器响应值与工作温度关系的火山形曲线，但研究者直到对 SnO_2 薄膜的研究才获得了定量结果[36-37,187]。特别地，传感器的效用因子[U,式(6.43)]定义为实际响应值(S)与 $Ha=0$ 条件下理想内部响应值(S_i)的比值。图 6.27(b)展现了 U 和 Ha 的关系，出现了和图 6.25(b)相似的反 S 形曲线，并且两张图的自变量均为 Ha。为了提升效用因子，Ha 的值应该保持较低，因而降低传感层厚度 L、降低 k/D_K 的比值将有利于提升 U 的值，这已通过先前的实验得到证实[36]。此外，由于 D_K 与孔道半径 r 成正比[式(6.25)]，而后者被认为与晶粒大小(D)有大致的正相关关系，因此通过调控晶粒大小也能改变效用因子。

$$U = \frac{S}{S_i} = \frac{\tanh Ha}{Ha} \quad (6.43)$$

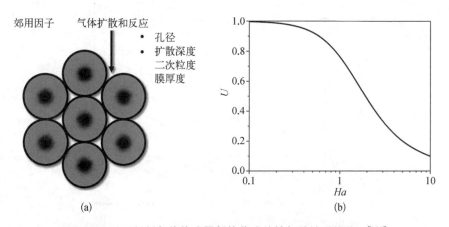

图 6.27　MOS 材料气体传感器气体传感特性与其性质的关系[187]

(a) MOS 材料气体传感器中效用因子及物理化学或材料特性；(b) 效用因子与 Ha 的关系

3. 进一步的气体扩散理论模型与修正

以上扩散模型已经在气体传感研究中有广泛应用，但仍然存在一些局限性，因此也有研究者尝试去修正。2009 年，Liu 等[104]对先前的扩散模型进行了修正，基于 SnO_2 薄膜的 H_2S 传感实验，发现材料电导和气体浓度之间的线性关系[式(6.40)]与实验结果契合度不高，而事实上得到了幂指数为 0.5 的幂律关系。如图 6.28(a)和图 6.28(b)所示，和先前的模型相比，新模型中响应值与 Ha 或气体浓度关联的曲线变得平缓，主要是因为幂指数的下降。如前所述，在先前的模型中，气体响应值变化最明显的区间是 Ha 介于 $1\sim10$,并且这一点可能在定量实验设计中可行。然而,在新模型中,这一变化趋势减弱,表明传感器响应值与气体浓度的相关性减弱。虽然新模型得到的拟合曲线与一些实验数据相符,但在 H_2S 浓度超过 10 ppm 时,实验数据依然产生了偏离[见图 6.28(c)]。此时材料的电导率足够高以至于影响幂指数,因而响应值达到了饱和。事实上,之前的研究已证明幂指数会因目标气体不同或气体浓度足够高而发生变化[103]。因此,新模型被进一步扩展到更大的范围,此时幂律规则的指数 n 被考虑在内。这样,电导率和气体响应可分别表示为式(6.44)和式(6.45)。类似地,对于 n 值从 0.25 到 1 变化时($n=1$ 即为先前的模型),传感器响应与 Ha 或传感层厚度[见图 6.28(d) 和图 6.28(e)]等参数间的变化趋势变得越来越显著,表明整个体系中气体扩散产生的影响增加。

$$\sigma(x) = \sigma_0 1 + a\left[C(x)\right]^n \tag{6.44}$$

$$S = 1 + \frac{a(6+nHa^2)}{6}\left(\frac{C_{A,s}}{\cosh Ha}\right) \tag{6.45}$$

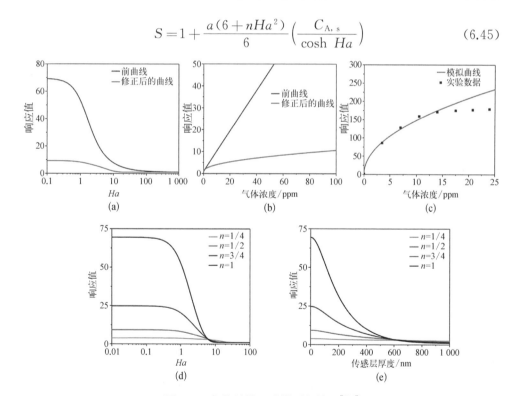

图 6.28 气体扩散理论模型与修正[104]

(a) 使用先前模型和修正的模型对响应值和 Ha 的函数关系进行模拟;(b) 使用先前模型和修正的模型对响应值和气体浓度的函数关系进行模拟;(c) 修正的模型与不同气体浓度下薄膜实验传感器的响应值的相关性;(d) 还原性气体中不同幂指数(n)值时传感器响应值与 Ha 的函数关系;(e) 固定温度下还原性气体中不同幂指数(n)值时传感器响应值与传感层厚度的关系

2017年,Ghosh 和 Majumder[108]提出了一个类似的模型,阐述了材料电导率与气体浓度之间的非线性关系。先前认为气体响应值以及最佳工作温度(temperature corresponding to maximum response,T_{opt})将随着传感层厚度的增加而降低[见图6.29(a)][36]。然而,Ghosh 和 Majumder 在一系列 CO 传感实验中发现气体响应值与 T_{opt} 先随传感层厚度的增加而增加至一最大值(320 nm 处),然后再下降至较低结果[见图6.29(b)]。此外,该研究也对不同 n 值下响应值与温度的关系进行了模拟,与 Liu 等的结果相似[104]。

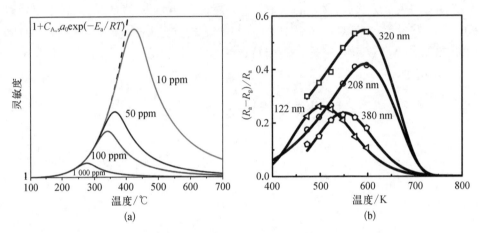

图 6.29　气体灵敏度与温度的关系

(a) 不同传感层厚度下气体灵敏度与温度的关系,在如下条件下模拟:$E_a = 50\ \text{kJ} \cdot \text{mol}^{-1}$,$E_k = 200\ \text{kJ} \cdot \text{mol}^{-1}$,$A = 1.7 \times 10^7\ \text{nm}^{-1} \cdot \text{K}^{1/4}$,$a_0 = 3\ 400\ \text{ppm}^{-1}$[36];(b) 实验获得的传感层厚度(122～380 nm)下对 500 ppm CO 的响应值随温度的变化(点)以及非线性拟合曲线[108]。

4. 孔道性质调控以应用 DR 耦合模型

除了基于数学和物理方程建立起的扩散模型,擅长材料设计的研究者也热衷于在 MOS 气体传感器中提升气体扩散能力。如前所述,在气体传感器中,气体分子在多孔传感材料的孔道内部发生约束性扩散是内扩散的一种典型过程[150]。因此,孔道性质(孔径大小、孔道形貌等)的调控是调节扩散的常见思路。孔径大小通常与比表面积和扩散能力本身有关。孔径越大,扩散速率越高,同时比表面积越低;孔径越小,比表面积越大,但同时会抑制气体扩散[69]。因此,在应用 DR 耦合模型时,合适的孔径大小十分关键。

Kida 等[188]合成了一系列可控晶粒尺寸的 SnO_2 纳米颗粒,并对 H_2、CO 和 H_2S 进行了气敏研究。研究通过控制 SnO_2 纳米颗粒的晶粒大小,对传感层上的孔径进行调控。相比于 CO 和 H_2S,H_2 浓度受离薄膜表面深度的影响较弱[见图6.30(a)],这是因为 H_2 的分子量最小,因而扩散速率最高,此时 SnO_2 薄膜的效用因子也得到提升。换言之,对于 H_2 分子,孔径大小对扩散几乎没有影响。

对于扩散能力较弱的 CO 和 H_2S 分子,应使用更大晶粒尺寸和更大孔径的材料来提升其扩散速率[见图 6.30(b)和图 6.30(c)]及效用因子。因此,当进行 H_2 传感研究时,较小孔径的材料可能提升选择性。此外,该研究进一步对比了 CO 和 H_2S,发现两者的 Knudsen 扩散系数 D_K 值接近[根据式(6.25),CO 的 D_K 值约为 H_2S 的 1.05 倍],但后者的响应值远比前者高,这主要是因为 CO 的表面活性较高[见图 6.30(d)和图 6.30(e)]。对于活性较高的 CO 分子,它们可能在 SnO_2 传感层的表面附近直接被消耗掉,难以扩散至材料深处,导致材料的效用因子较低;与之相比,H_2S 分子活性较弱,因此会扩散至材料深处而提升传感器响应值以及材料的效用因子。因此通过气体扩散和反应的组合来提升效用因子十分重要。

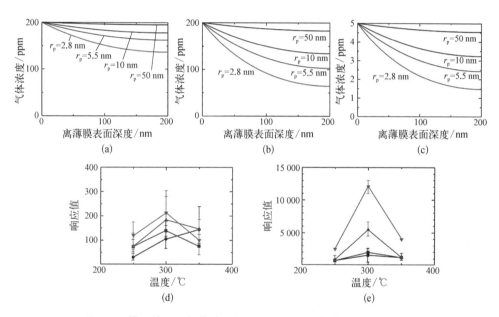

图 6.30 模拟的不同气体浓度与不同孔径下离薄膜表面深度的关系以及传感器响应和温度的关系[188]

(a)~(c) H_2、CO、H_2S 浓度与不同孔径下离薄膜表面深度的关系;(d)、(e) 传感器对 CO 和 H_2S 的响应和温度的关系,实心圆、方块、三角和倒三角分别对应晶粒大小为 7、11、14 和 18 nm 的样品

Wang 等[129]提出了一个能够评估气体传感过程中各因素,包括表面化学反应 (surface chemical reaction,SCR)、气体扩散、最佳测试温度(T_{opt})、气体种类等关系的模型。对于 T_{opt} 低于 300℃ 的情况,气体扩散与 SCR 在阐述气敏机理时起到关键作用。在低温下[见图 6.31(a)],SCR 速率很低,因此成为决速步骤,整个过程由表面控制,故有更大比表面积的材料响应值更高。在高温下[见图 6.31(c)和图 6.31(d)],扩散速率提高的程度不如 SCR 快,整个过程的扩散控制,此时孔径更

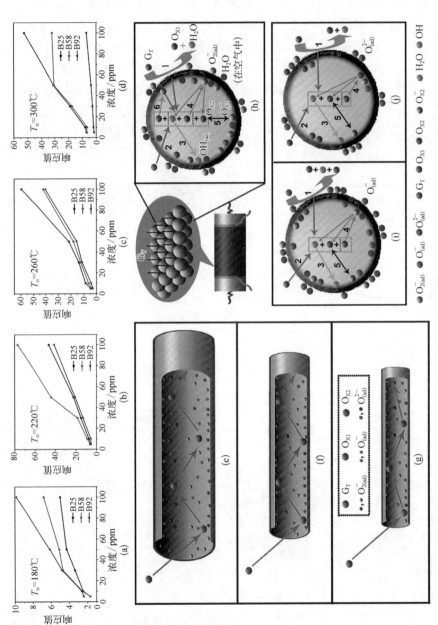

图 6.31 SnO_2 空心微球丙酮传感性能及机理研究（后附彩图）[129]

(a)~(d) 丙酮在不同温度下的浓度-响应曲线，三种材料 B25、B58、B92 以其比表面积来命名；(e)~(g) 目标气体在不同孔径下扩散的示意图，其中 (e) 表示孔径较大，(f) 表示孔径适中，(g) 表示孔径较小；(h)~(j) 不同温度下 SnO_2 空心微球下目标气体扩散的示意图，其中 (h) 表示 $<150℃$，(i) 表示 $150\sim200℃$，(j) 表示 $>200℃$

大的材料响应值更高。而在中间温度下[见图 6.31(b)]，气体扩散和 SCR 在相互竞争中旗鼓相当，整个体系受到双重控制，拥有适中结构的材料响应值最高。简言之，随着温度的提升，SCR 控制替换为扩散控制。随后研究提出了一个新的孔道模型[见图 6.31(e)～图 6.31(g)]来描述气体分子在不同孔径大小的材料中发生 Knudsen 扩散的过程。在低温下，气体分子的平均自由程大于孔径，从而阻碍气体分子扩散进入孔道，因此大部分响应过程发生在孔道外部。扩散速率和 SCR 速率都很低，但后者起决定作用。随着温度上升，气体分子的平均自由程逐步下降，当然此时仍然与孔径大小相当，这样更多的分子可以扩散至孔道内部并与其中的氧反应。这是一个过渡状态。在更高温度下，平均自由程小于孔径，使得更多分子能够扩散，然而，SCR 速率增加得远比扩散速率快，最终产生了扩散控制状态。之后该研究又建立了空心球模型[见图 6.31(h)～图 6.31(j)]来描述 SCR 和气体扩散的相互关系。在低温下[见图 6.31(h)]，SnO_2 表面吸附的氧主要是不活泼的 O_2^-，从而限制了 SCR 速率。气体分子有些在表面部分氧化，另一些扩散至材料内部，并经历一系列不同的氧化反应，包括部分氧化、完全氧化和离子化。在过渡状态[见图 6.31(i)]或更高温度下[见图 6.31(j)]，氧负离子主要以更活泼的 O^- 和 O^{2-} 形式存在。SCR 变得更活泼，使得空心球内部的分子得到充分氧化，因而气体扩散成为决速步骤。

此外，精确控制传感器材料的形貌可以进一步提高传感性能。对于 MOS 材料而言，合成有序介孔金属氧化物传感材料是一条有前途的道路。对于孔道结构（孔径大小、孔容、结晶度、孔壁厚度和孔道对称性）的精确控制对于得到高响应值和选择性十分关键[123]。笔者团队长期以来致力于基于模板法进行高度有序介孔金属氧化物传感材料的合成开发，使用了各种结构导向剂，包含软模板（两亲性嵌段共聚物）聚环氧乙烷-聚苯乙烯[poly(ethylene oxide)-*block*-polystyrene，PEO-*b*-PS]、聚苯乙烯-聚(4-乙烯基吡啶)[polystyrene-*block*-poly(4-vinylpyridine)，PS-*b*-P4VP][88-89,91-93,144,147-149,151,189-199]，以及硬模板脲醛树脂(urea-formaldehyde，UF)[90]。特别地，在软模板法中，基于溶剂挥发诱导自组装(evaporation-induced self-assembly，EISA)手段[200-209]，开展了介孔的造孔工程研究，实现了各种孔径大小以及形貌的材料的合成。由于有序介孔材料的孔径大小接近大多数测试气体的平均自由程，有利于气体分子在具有高比表面积和互相连通介孔的材料中进行传质和扩散[179,195]。Liu 等[93]合成了 Ce 掺杂介孔 WO_3 气体传感器(标记为 Ce-2/mWO_3)，并使用该传感器进行 H_2S 检测，在解释机理时应用了表面催化与 Knudsen 扩散模型(见图 6.32)。H_2S 分子在介孔中首先经历 Knudsen 扩散，然后在丰富的活性位点处发生表面催化反应。一方面，传感材料的介孔结构促进了 H_2S 分子向更深处扩散并加速了 EDL 的形成以及载流子的传输；另一方面，表面催化过程促进了 H_2S 分子同时氧化产生气相的 SO_2 和固相的 WS_2。产生

的 WS_2 相比原先的 WO_3 拥有更窄的带隙以及更低的电阻,进一步提升了传感器的响应。

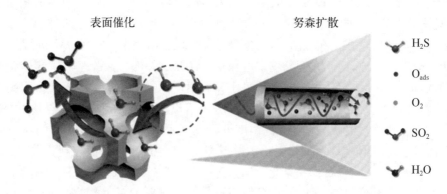

图 6.32　H_2S 和 Ce–2/mWO_3 之间的表面催化和 Knudsen 扩散模型[93]

5. 脉冲激发气体传感器中气体扩散的加强

虽然气体扩散理论仍然需要通过定量分析和建模来优化[64],但它在气体传感研究中有重要作用,并已被广泛用于解释 MOS 材料的不同形貌对响应值产生的影响[117,210-212]。近年来,研究者们在提升传感性能方面实现了一些新的突破[150],包括光学气体传感器[9,31,213-215]、表面等离子体共振(surface plasmon resonance,SPR)加强的气体传感器[216-218]、脉冲激发气体传感器[219-221]和 FET 气体传感器[101,222-224],其中脉冲激发气体传感器能提升气体扩散能力。Shimanoe 等将脉冲加热装置应用于挥发性有机物的检测。在加热关闭时,气体分子先扩散至材料的深处并沉积下来,最终在加热开启时经历燃烧反应[220,225-228]。值得指出的是脉冲激发气体传感器最初受到研究者关注是因为其耗能较低[150,220]。然而,通过脉冲模式也能提升气体扩散能力。在传统条件下,外部加热在整个传感过程中持续进行,高温下气体分子常常在传感层表面附近直接燃烧耗尽,使得响应值和效用因子较低。然而,脉冲加热装置使得即使在较高操作温度下,气体分子仍然能够扩散到材料的深处,从而提高传感器的效用因子[220]。此外,新定义的几种不同阶段的响应值[始态、终态和两者的比例,分别表示为 S_i、S_e 和 S_p,见图 6.33(a)]清晰地展现了气体传感过程中气体分子所经历的不同阶段(扩散、沉积与燃烧),说明了该传感器对 VOC 分子产生了明确反应并针对其他气体有良好的选择性[225-226]。在脉冲加热过程前引入预热过程[称为双重脉冲模式,见图 6.33(b)]可进一步通过表面改性来增加 O^{2-} 的吸附从而提高传感性能,其成功应用于 ppt 级别的 VOC 气体检测[227-228]。

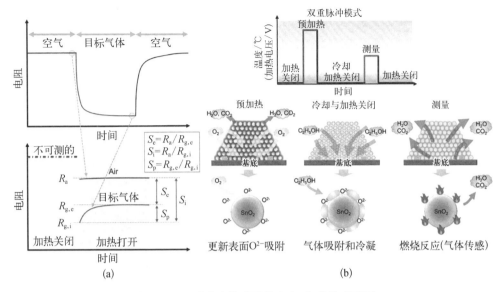

图 6.33 双重脉冲模式下的 SnO_2 气体传感机理

(a) 传感器响应的确定：S_e、S_i 和 S_p[226]；(b) 在 SnO_2 作为半导体气敏材料的使用双重脉冲模式进行气体检测的示意图[227]

6.3 结论与展望

在 MOS 材料气体传感器的领域中，研究者的大多数研究都主要集中于对气体分子和敏感材料之间化学反应的调控。然而，气体分子的动态过程，包括吸附、脱附和扩散，同样与传感过程有密切联系。对于吸附、脱附和扩散过程的调节有利于调控传感过程甚至是传感器的性能，包括灵敏度、选择性，以及响应/恢复动力学，这对于开发高性能气体传感器十分关键。

氧吸附是传感材料暴露到空气中时发生的最普遍的现象，但会产生 EDL 或 HAL 而导致材料电阻发生变化，测试气体与预吸附的氧之间将发生进一步的吸附，产生不同的响应。在无氧条件下，还原性气体的化学吸附以及与晶格氧的反应将产生更高的响应。如果测试气体能够直接与传感材料发生反应，这一化学吸附会进一步提升响应值或使得材料的响应发生倒转。在 H_2O 分子存在时，化学吸附、物理吸附甚至共吸附会同时发生，并且常常成为低温下提升响应值的阻碍。室温下 O_2 的物理吸附也可能占主导，并能导致材料电阻略微上升。通过将 MOS 材料表面耗尽层理论和吸附造成的材料电阻变化两者结合起来，研究者推导得出了材料电阻与测试气体分压间的幂律规则。此外，本章还详细讨论了响应/恢复过程的动力学，指出吸脱附过程在其中作为决速步骤。

虽然气体扩散在气体传感过程中只起辅助作用，但对扩散的有效操控也能提升传感器响应值和选择性。总体而言，气体分子与 MOS 传感材料的反应遵循 DR 耦合过程。随着传感材料孔径的增加，气体分子将逐步遵循表面扩散、Knudsen 扩散和分子扩散。在 Knudsen 扩散中，气体分子更可能与活性位点发生碰撞从而发生反应。在不同纳米结构的材料中，有序介孔材料有利于 Knudsen 扩散的发生。通过假定 Knudsen 扩散和一级动力学，研究者们建立了一系列气体扩散模型并进行了修正，并在各种实验中验证了其各自的有效性，但仍然需要进一步的定量分析。随着温度升高，气-固界面反应而非扩散成为主导步骤，从而使扩散成为决速步骤。高温使得气体分子的平均自由程下降，并能够同时活化气体分子与传感材料，但它会造成响应值下降。更大的孔径将提升扩散能力，但它们相应的低比表面积不利于提升反应速率。对于高扩散率的气体（H_2），可以通过减小材料孔径来提升选择性。作为一种新型测试模式，脉冲激发气体传感器将扩散与反应过程分开，从而可在高温下促进气体扩散并提升传感器的利用率。

众所周知，一个理想的 MOS 气体传感器应该展现出以下优秀的特性：高响应值和选择性、快速的响应/恢复时间和低成本。气体分子的动态过程，即吸脱附和扩散在用以制备更好传感器的材料领域起到重要作用。基于上述内容，我们提出一些展望，期待可以在高性能 MOS 气体传感器中进一步提升气体的吸脱附与扩散。① 气体扩散模型可以进一步改进。目前整个传感过程中只有内扩散的影响被考虑到，这与气体吸附标志着传感过程的"开始"的假定相一致。然而，气体注入和吸附之间发生的外扩散同样会对传感结果造成影响。例如，在实际测试过程中，过冲现象是常见的外扩散起关键作用的情况。不管是外扩散还是内扩散，最终都会导致实际气体浓度的损失，因此对气体响应值有负面影响。在将来的模型建立中，内扩散中效用因子的概念可以被进一步拓展到外扩散中。此外，扩散模型中的线性假定也应该进行修正以符合幂律规则，这些改进可被用来实现对实际传感过程更精准的描述。② 对于传感材料的合成，表面改性是加强气体分子吸脱附的可行手段。例如，可以在 MOS 框架中掺杂异质元素，如稀土，来调节表面碱度。这些材料的活性位点通常位于非最高价的阳离子处（如 Ce^{3+}），属于常见的 Lewis 碱，能够加强酸性气体分子的吸附。类似地，引入疏水组分可能消除湿度影响同时加强疏水性气体分子的吸附。此外，也可以合成具有不同形貌和暴露晶面的 MOS 材料用于加强气体吸附。③ 对传感材料中孔道结构（孔径大小、孔容、结晶度、孔壁厚度以及孔道对称性）的精确调控对于提升传感过程的气体扩散十分关键，也能提高传感器响应值和/或选择性。对每一种特定测试气体的扩散与反应速率的竞争关系进行比较，可以实现有目的性的孔道结构优化。以下提出一些潜在的假说。对于同时具有高扩散速率和高活性的气体分子（如 H_2），相比其他气体，孔道结构对其响应值的影响较小。通过缩小孔径或增加传感层厚度等手段可能有利于提升其选择性,尽管此时响应值会有一定程度的降低。对于扩散速率低而反应活性高的气体分子(如 H_2S),扩散的影响应该彻底消除。拥有大孔径和低传感层厚度的 MOS 材料将成为解决低传感器利用率的良方。此外,H_2S 和 MOS 材料之间的化学吸附和直接反应能进一步提升响应值,恢复时间则可通过表面改性在孔壁中产生更多氧空位而缩短。对于扩散速率高而反应活性低的气体分子(如 CH_4),能确保传感器本身的效用因子维持在高位,但需要提升反应速率。通常而言,高性能贵金属催化剂,如 Pd、Pt、Au、Ag、Rh 和 Ru,被广泛应用于低碳烷烃的传感和活化中。在这种情况下,通过多组分共组装(如 EISA)方法将贵金属原位负载在介孔内部将是理想的手段。气体分子能与多孔传感层中的巨大界面发生相互作用,同时促进效用因子和响应值的提升。再者,通过调控孔道结构,实现具有大孔口直径小窗口内径的喇叭口圆柱介孔被认为有利于加强气体分子扩散进入介孔内部而抑制向外流失,从而促进气体分子在传感层内部富集,实现反应的发生。对于扩散速率极低而反应活性一般的 VOC 分子,合成具有超高效用因子的 2D 材料将对消除扩散影响起到作用。引入脉冲激发气体传感器也可能是提升传感器响应值的重要补充手段。最后,根据测试气体分子的相应性质,可以通过调节传感层的厚度来对应合适的 Hatta 数。这样一来,传感材料的研究需要从传统的以材料为中心转变为新的以气体为中心。从结果来看,对于气体传感过程中 MOS 材料内部性质的研究,即吸附、脱附和气体扩散,可以通过理论数学与物理研究以及精确材料设计(如调控组分与晶面、调变孔径和孔结构)的组合来实现,这也为 MOS 气体传感器的未来指明了关键的方向。

参考文献

[1] Lewis A, Edwards P. Validate personal air-pollution sensors[J]. Nature, 2016, 535: 29-31.

[2] Lombardo L, Corbellini S, Parvis M, et al. Wireless sensor network for distributed environmental monitoring[J]. IEEE Trans. Instrum. Meas. 2018, 67: 1214-1222.

[3] Ardi S, Abdurrahman H. 2017 4th International Conference on Information Technology, Computer, and Electrical Engineering (ICITACEE)[J]. Semarang, Indonesia: IEEE Xplore, 2017, 192-196.

[4] Iyer B, Patil N. IoT enabled tracking and monitoring sensor for military applications[J]. Int. J. Syst. Assur. Eng. Manage. 2018, 9: 1294-1301.

[5] Sall M L, Fall B, Diédhiou I, et al. Toxicity and electrochemical detection of lead, cadmium and nitrite ions by organic conducting polymers: a review[J]. Chem. Africa, 2020, 3: 499-512.

[6] Eguchi K. Optical Gas Sensors[J]. Netherlands: Springer, 1992, 307-328.

[7] Bonifacio L D, Ozin G A, Arsenault A C. Photonic nose-sensor platform for water and food quality control[J]. Small, 2011, 7: 3153-3157.

[8] Guillén M G, Gámez F, Lopes-Costa T, et al. A fluorescence gas sensor based on Förster resonance energy transfer between polyfluorene and bromocresol green assembled in thin films[J]. Sens. Actuators B, 2016, 236: 136 – 143.

[9] da Silva L F, M'Peko J C, Catto A C, et al. UV-enhanced ozone gas sensing response of $ZnO-SnO_2$ heterojunctions at room temperature[J]. Sens. Actuators B, 2017, 240: 573 – 579.

[10] Lee Y, Cheng S, Cheng C, et al. 2017 19th International Conference on Solid-State Sensors, Actuators and Microsystems (TRANSDUCERS)[J]. Kaohsiung, Taiwan (ROC): IEEE, 2017, 672 – 675.

[11] Regulacio M, Wang Y, Seh Z, et al. Tailoring porosity in copper-based multinary sulfide nanostructures for energy, biomedical, catalytic, and sensing applications[J]. ACS Appl. Nano Mater. 2018, 1: 3042 – 3062.

[12] Ma Y, Cametti M, Džolić Z, et al. AIE-active bis-cyanostilbene-based organogels for quantitative fluorescence sensing of CO_2 based on molecular recognition principles[J]. J. Mater. Chem. C, 2018, 6: 9232 – 9237.

[13] Chong X, Zhang Y, Li E, et al. Surface-enhanced infrared absorption: pushing the frontier for on-chip gas sensing[J]. ACS Sens. 2018, 3: 230 – 238.

[14] Muthusamy S, Charles J, Renganathan B, et al. In situ growth of Prussian blue nanocubes on polypyrrole nanoparticles: facile synthesis, characterization and their application as fiber optic gas sensor[J]. J. Mater. Sci., 2018, 53: 15401 – 15417.

[15] Yao Q, Ren G, Xu K, et al. 2D plasmonic tungsten oxide enabled ultrasensitive fiber optics gas sensor[J]. Adv. Opt. Mater., 2019, 7: 1901383.

[16] Dai J, Ogbeide O, Macadam N, et al. Printed gas sensors[J]. Chem. Soc. Rev., 2020, 49: 1756 – 1789.

[17] Tabassum R, Kant R. Recent trends in surface plasmon resonance based fiber – optic gas sensors utilizing metal oxides and carbon nanomaterials as functional entities[J]. Sens. Actuators B, 2020, 310: 127813.

[18] Li G, She C, Zhang Y, et al. A "Turn-on" fluorescence perovskite sensor based on $MAPbBr_3$/mesoporous TiO_2 for NH_3 and amine vapor detections[J]. Sens. Actuators B, 2021, 327: 128918.

[19] Seiyama T, Kato A, Fujiishi K, et al. A new detector for gaseous components using semiconductive thin films[J]. Anal. Chem., 1962, 34: 1502 – 1503.

[20] Seiyama T, Kagawa S. Study on a detector for gaseous components using semiconductive thin films[J]. Anal. Chem., 1966, 38: 1069 – 1073.

[21] Keyu C, Wenhe X, Yu D, et al. Alkaloid precipitant reaction inspired controllable synthesis of mesoporous tungsten oxide spheres for biomarker sensing[J]. ACS Nano. 2023, 17: 15763 – 15775.

[22] Lee K C, Chiang Y J, Lin Y C, et al. Effects of PdO decoration on the sensing behavior of SnO_2 toward carbon monoxide[J]. Sens. Actuators B, 2016, 226: 457 – 464.

[23] Xue Z, Cheng Z, Xu J, et al. Controllable evolution of dual defect Zn_i and V_O associate-rich ZnO nanodishes with (0001) exposed facet and its multiple sensitization effect for ethanol

detection[J]. ACS Appl. Mater. Interfaces, 2017, 9: 41559–41567.
[24] Bulemo P M, Cho H J, Kim N H, et al. Mesoporous SnO_2 nanotubes via electrospinning-etching route: highly sensitive and selective detection of H_2S molecule[J]. ACS Appl. Mater. Interfaces, 2017, 9: 26304–26313.
[25] Kim H H, Pak Y S, Jeong Y G, et al. Amorphous Pd-assisted H_2 detection of ZnO nanorod gas sensor with enhanced sensitivity and stability[J]. Sens. Actuators B, 2018, 262: 460–468.
[26] Krivetskiy V, Zamanskiy K, Krotova A. Effect of AuPd bimetal sensitization on gas sensing performance of nanocrystalline SnO_2 obtained by single step flame spray pyrolysis[J]. Proceedings, 2019, 14: 46–64.
[27] Yin X, Zhou W, Li J, et al. A highly sensitivity and selectivity Pt–SnO_2 nanoparticles for sensing applications at extremely low level hydrogen gas detection[J]. J. Alloys Compd., 2019, 805: 229–236.
[28] Maziarz W. TiO_2/SnO_2 and TiO_2/CuO thin film nano-heterostructures as gas sensors[J]. Appl. Surf. Sci., 2019, 480: 361–370.
[29] Wei F, Wang T, Jiang X, et al. Controllably engineering mesoporous surface and dimensionality of SnO_2 toward high-performance CO_2 electroreduction[J]. Adv. Funct. Mater., 2020, 30: 2002092.
[30] Zhang W, Fan Y, Yuan T, et al. Ultrafine tungsten oxide nanowires: synthesis and highly selective acetone sensing and mechanism analysis[J]. ACS Appl. Mater. Interfaces, 2020, 12: 3755–3763.
[31] You F, Wan J, Qi J, et al. Lattice distortion in hollow multi-shelled structures for efficient visible-light CO_2 reduction with a SnS_2/SnO_2 junction[J]. Angew. Chem. Int. Ed. 2020, 59: 721–724.
[32] Jaaniso R, Tan O K. Semiconductor gas sensors [J]. Cambridge, UK: Woodhead Publishing Limited, 2013.
[33] Gardner J W. A diffusion-reaction model of electrical conduction in tin oxide gas sensors[J]. Semicond. Sci. Technol., 1989, 4: 345–350.
[34] Lundström I. Approaches and mechanisms to solid state based sensing[J]. Sens. Actuators B-Chem., 1996, 35–36: 11–19.
[35] Vilanova X, Llobet E, Alcubilla R, et al. Analysis of the conductance transient in thick-film tin oxide gas sensors[J]. Sens. Actuators B, 1996, 31: 175–180.
[36] Sakai G, Matsunaga N, Shimanoe K, et al. Theory of gas-diffusion controlled sensitivity for thin film semiconductor gas sensor[J]. Sens. Actuators B, 2001, 80: 125–131.
[37] Matsunaga N, Sakai G, Shimanoe K, et al. Diffusion equation-based study of thin film semiconductor gas sensor-response transient[J]. Sens. Actuators B, 2002, 83: 216–221.
[38] Matsunaga N, Sakai G, Shimanoe K, et al. Formulation of gas diffusion dynamics for thin film semiconductor gas sensor based on simple reaction–diffusion equation[J]. Sens. Actuators B, 2003, 96: 226–233.
[39] Korotcenkov G, Brinzari V, Golovanov V, et al. Kinetics of gas response to reducing gases of SnO_2 films, deposited by spray pyrolysis[J]. Sens. Actuators B, 2004, 98: 41–45.

[40] Yamazoe N, Shimanoe K. Theory of power laws for semiconductor gas sensors[J]. Sens. Actuators B, 2008, 128: 566-573.

[41] Walker J M, Akbar S A, Morris P A. Synergistic effects in gas sensing semiconducting oxide nano-heterostructures: a review[J]. Sens. Actuators B, 2019, 286: 624-640.

[42] Moradi H, Azizpour H, Bahmanyar H, et al. Molecular dynamics simulation of H_2S adsorption behavior on the surface of activated carbon[J]. Inorg. Chem. Commun., 2020, 118: 108048.

[43] Masel R I. Principles of adsorption and reaction on solid surfaces[J]. Hoboken: John Wiley & Sons, 1996.

[44] Burke G M, Wurster D E, Buraphacheep V, et al. Model selection for the adsorption of phenobarbital by activated charcoal[J]. Pharm. Res., 1991, 8: 228-231.

[45] Sing K. Adsorption methods for the characterization of porous materials[J]. Adv. Colloid Interface Sci., 1998, 76: 3-11.

[46] Fang Q, Chetwynd D G, Covington J A, et al. Micro-gas-sensor with conducting polymers [J]. Sens. Actuators B, 2002, 84: 66-71.

[47] Gomri S, Seguin J L, Guerin J, et al. Adsorption-desorption noise in gas sensors: modelling using Langmuir and Wolkenstein models for adsorption[J]. Sens. Actuators B, 2006, 114: 451-459.

[48] Zhang C, Kaluvan S, Zhang H, et al. A study on the Langmuir adsorption for quartz crystal resonator based low pressure CO_2 gas sensor[J]. Measurement, 2018, 124: 286-290.

[49] Thompson W A, Sanchez Fernandez E, Maroto-Valer M M. Probability Langmuir-Hinshelwood based CO_2 photoreduction kinetic models[J]. Chem. Eng. J., 2020, 384: 123356.

[50] Razdan N K, Bhan A. Catalytic site ensembles: a context to reexamine the Langmuir-Hinshelwood kinetic description[J]. J. Catal., 2021, 404: 726-744.

[51] Batebi D, Abedini R, Mosayebi A. Kinetic modeling of combined steam and CO_2 reforming of methane over the Ni-Pd/Al_2O_3 catalyst using Langmuir-Hinshelwood and Langmuir-Freundlich isotherms[J]. Ind. Eng. Chem. Res., 2021, 60: 851-863.

[52] Prins R. Eley-Rideal, the other mechanism[J]. Top. Catal., 2018, 61: 714-721.

[53] Quan J, Muttaqien F, Kondo T, et al. Vibration-driven reaction of CO_2 on Cu surfaces via Eley-Rideal-type mechanism[J]. Nat. Chem., 2019, 11: 722-729.

[54] Engelmann Y, van't Veer K, Gorbanev Y, et al. Plasma catalysis for ammonia synthesis: a microkinetic modeling study on the contributions of Eley-Rideal reactions[J]. ACS Sustainable Chem. Eng., 2021, 9: 13151-13163.

[55] Mars P, Van Krevelen D W. Oxidations carried out by means of vanadium oxide catalysts [J]. Chem. Eng. Sci., 1954, 3: 41-59.

[56] Doornkamp C, Ponec V. The universal character of the Mars and Van Krevelen mechanism [J]. J. Mol. Catal. A, 2000, 162: 19-32.

[57] Campbell K D, Lunsford J H. Contribution of gas-phase radical coupling in the catalytic oxidation of methane[J]. J. Phys. Chem., 1988, 92: 5792-5796.

[58] Wolkenstein T. Electron transitions in chemisorption[J]. New York: Springer, 1991.
[59] Rothschild A, Komem Y. The effect of grain size on the sensitivity of nanocrystalline metal-oxide gas sensors[J]. J. Appl. Phys., 2004, 95: 6374-6380.
[60] Kissine V V, Sysoev V V, Voroshilov S A. Individual and collective effects of oxygen and ethanol on the conductance of SnO_2 thin films[J]. Appl. Phys. Lett., 2000, 76: 2391-2393.
[61] Geistlinger H. Electron theory of thin-film gas sensors[J]. Sens. Actuators B, 1993, 17: 47-60.
[62] Medford A J, Vojvodic A, Hummelshøj J S, et al. From the Sabatier principle to a predictive theory of transition-metal heterogeneous catalysis[J]. J. Catal., 2015, 328: 36-42.
[63] Balandin A A. Modern state of the multiplet theor of heterogeneous catalysis[J]. Adv. Catal., 1969, 19: 1-210.
[64] Ji H, Zeng W, Li Y. Gas sensing mechanisms of metal oxide semiconductors: a focus review[J]. Nanoscale, 2019, 11: 22664-22684.
[65] Shankar P, Rayappan J. Gas sensing mechanism of metal oxides: the role of ambient atmosphere, type of semiconductor and gases—a review[J]. Sci. Lett. J., 2015, 4: 126-144.
[66] Hoa N D, An S Y, Dung N Q, et al. Synthesis of p-type semiconducting cupric oxide thin films and their application to hydrogen detection[J]. Sens. Actuators B, 2010, 146: 239-244.
[67] Barsan N, Weimar U. Conduction model of metal oxide gas sensors[J]. J. Electroceram., 2001, 7: 143-167.
[68] Yamazoe N, Sakai G, Shimanoe K. Oxide semiconductor gas sensors[J]. Catal. Surv. Asia, 2003, 7: 63-75.
[69] Zhou X, Cheng X, Zhu Y, et al. Ordered porous metal oxide semiconductors for gas sensing[J]. Chin. Chem. Lett., 2018, 29: 405-416.
[70] Kim H J, Lee J H. Highly sensitive and selective gas sensors using p-type oxide semiconductors: overview[J]. Sens. Actuators B, 2014, 192: 607-627.
[71] Li T, Zeng W, Long H, et al. Nanosheet-assembled hierarchical SnO_2 nanostructures for efficient gas-sensing applications[J]. Sens. Actuators B, 2016, 231: 120-128.
[72] Yamazoe N. New approaches for improving semiconductor gas sensors[J]. Sens. Actuators B, 1991, 5: 7-19.
[73] Sysoev V V, Strelcov E, Kar S, et al. The electrical characterization of a multi-electrode odor detection sensor array based on the single SnO_2 nanowire[J]. Thin Solid Films, 2011, 520: 898-903.
[74] Klier K, Nováková J, Jíru P. Exchange reactions of oxygen between oxygen molecules and solid oxides[J]. J. Catal., 1963, 2: 479-484.
[75] Lee E J, Yoon Y S, Kim D J. Two-dimensional transition metal dichalcogenides and metal oxide hybrids for gas sensing[J]. ACS Sens., 2018, 3: 2045-2060.
[76] Kortidis I, Swart H C, Ray S S, et al. Characteristics of point defects on the room

temperature ferromagnetic and highly NO_2 selectivity gas sensing of p-type Mn_3O_4 nanorods[J]. Sens. Actuators B, 2019, 285: 92-107.

[77] Iwamoto M, Yoda Y, Yamazoe N, et al. Study of metal oxide catalysts by temperature programmed desorption. 4. Oxygen adsorption on various metal oxides[J]. J. Phys. Chem., 1978, 82: 2564-2570.

[78] Hahn S H, Bârsan N, Weimar U, et al. CO sensing with SnO_2 thick film sensors: role of oxygen and water vapour[J]. Thin Solid Films, 2003, 436: 17-24.

[79] Hübner M, Pavelko R G, Barsan N, et al. Influence of oxygen backgrounds on hydrogen sensing with SnO_2 nanomaterials[J]. Sens. Actuators B, 2011, 154: 264-269.

[80] Bârsan N, Hübner M, Weimar U. Conduction mechanisms in SnO_2 based polycrystalline thick film gas sensors exposed to CO and H_2 in different oxygen backgrounds[J]. Sens. Actuators B, 2011, 157: 510-517.

[81] Henrich V E, Cox P A. The surface science of metal oxides[J]. Cambridge, UK: Cambridge University Press, 1996.

[82] Zhu L, Zeng W, Li Y. A non-oxygen adsorption mechanism for hydrogen detection of nanostructured SnO_2 based sensors[J]. Mater. Res. Bull., 2019, 109: 108-116.

[83] Lu Z, Ma D, Yang L, et al. Direct CO oxidation by lattice oxygen on the SnO_2 (110) surface: a DFT study[J]. Phys. Chem. Chem. Phys., 2014, 16: 12488-12494.

[84] Zakaryan H, Aroutiounian V M. CO gas adsorption on SnO_2 surfaces: density functional theory study[J]. Sens. Transducers, 2017: 212: 50-56.

[85] Xu Z, Luo Y, Duan G. Self-assembly of Cu_2O monolayer colloidal particle film allows the fabrication of CuO sensor with superselectivity for hydrogen sulfide[J]. ACS Appl. Mater. Interfaces, 2019, 11: 8164-8174.

[86] Miao J, Chen C, Meng L, et al. Self-assembled monolayer of metal oxide nanosheet and structure and gas-sensing property relationship[J]. ACS Sens., 2019, 4: 1279-1290.

[87] Speight J G. Lange's handbook of chemistry[J]. New York, NY, USA: McGraw-Hill Education, 2017.

[88] Li Y, Luo W, Qin N, et al. Highly ordered mesoporous tungsten oxides with a large pore size and crystalline framework for H_2S sensing[J]. Angew. Chem., 2014, 53: 9035-9040.

[89] Li Y, Zhou X, Luo W, et al. Pore engineering of mesoporous tungsten oxides for ultrasensitive gas sensing[J]. Adv. Mater. Interfaces, 2019, 6: 1801269.

[90] Wan L, Song H, Ma J, et al. Polymerization-induced colloid assembly route to iron oxide-based mesoporous microspheres for gas sensing and Fenton catalysis[J]. ACS Appl. Mater. Interfaces, 2018, 10: 13028-13039.

[91] Xiao X, Liu L, Ma J, et al. Ordered mesoporous tin oxide semiconductors with large pores and crystallized walls for high-performance gas sensing[J]. ACS Appl. Mater. Interfaces, 2018, 10: 1871-1880.

[92] Xiao X, Zhou X, Ma J, et al. Rational synthesis and gas sensing performance of ordered mesoporous semiconducting WO_3/NiO composites[J]. ACS Appl. Mater. Interfaces, 2019, 11: 26268-26276.

[93] Liu Y, Guo R, Yuan K, et al. Engineering pore walls of mesoporous tungsten oxides via

Ce doping for the development of high-performance smart gas sensors[J]. Chem. Mater. 2022, 34: 2321-2332.

[94] Heiland G, Kohl D, Seiyama T. Physical and chemical aspects of oxidic semiconductor gas sensors[J]. Chemical Sensor Technology, 1988: 15-38.

[95] Wang L, Zhao F, Han Q, et al. Spontaneous formation of $Cu_2O-g-C_3N_4$ core-shell nanowires for photocurrent and humidity responses[J]. Nanoscale, 2015, 7: 9694-9702.

[96] Morrison S R. The chemical physics of surfaces[J]. New York, NY, USA: Springer Science & Business Media, 2013.

[97] Agmon N. The Grotthuss mechanism[J]. Chem. Phys. Lett., 1995, 244: 456-462.

[98] Deng Z, Tong B, Meng G, et al. Insight into the humidity dependent pseudo-n-type response of $p-CuScO_2$ toward ammonia[J]. Inorg. Chem., 2019, 58: 9974-9981.

[99] Španěl P, Smith D. Reactions of hydrated hydronium ions and hydrated hydroxide ions with some hydrocarbons and oxygen-bearing organic molecules[J]. J. Phys. Chem., 1995, 99: 15551-15556.

[100] Kannan P K, Saraswathi R, Rayappan J B B. A highly sensitive humidity sensor based on DC reactive magnetron sputtered zinc oxide thin film[J]. Sens. Actuators A, 2010, 164: 8-14.

[101] Hong S B, Shin J M, Hong Y K, et al. Observation of physisorption in a high-performance FET-type oxygen gas sensor operating at room temperature[J]. Nanoscale, 2018, 10: 18019-18027.

[102] Pazniak H, Varezhnikov A S, Kolosov D A, et al. 2D molybdenum carbide MXenes for enhanced selective detection of humidity in air[J]. Adv. Mater., 2021, 33: 2104878.

[103] Morrison S R. Mechanism of semiconductor gas sensor operation[J]. Sens. Actuators, 1987, 11: 283-287.

[104] Liu J, Gong S, Xia J, et al. The sensor response of tin oxide thin films to different gas concentration and the modification of the gas diffusion theory[J]. Sens. Actuators B, 2009, 138: 289-295.

[105] Gardner J W. A non-linear diffusion-reaction model of electrical conduction in semiconductor gas sensors[J]. Sens. Actuators B, 1990, 1: 166-170.

[106] Duy L T, Kim D J, Trung T Q, et al. High performance three-dimensional chemical sensor platform using reduced graphene oxide formed on high aspect-ratio micro-pillars [J]. Adv. Funct. Mater., 2015, 25: 883-890.

[107] Wu J, Feng S, Wei X, et al. Facile synthesis of 3D graphene flowers for ultrasensitive and highly reversible gas sensing[J]. Adv. Funct. Mater., 2016, 26: 7462-7469.

[108] Ghosh A, Majumder S B. Modeling the sensing characteristics of chemi-resistive thin film semi-conducting gas sensors[J]. Phys. Chem. Chem. Phys., 2017, 19: 23431-23443.

[109] Yamazoe N, Shimanoe K, Sawada C. Contribution of electron tunneling transport in semiconductor gas sensor[J]. Thin Solid Films, 2007, 515: 8302-8309.

[110] Yamazoe N, Suematsu K, Shimanoe K. Extension of receptor function theory to include two types of adsorbed oxygen for oxide semiconductor gas sensors[J]. Sens. Actuators B, 2012, 163: 128-135.

[111] Windischmann H, Mark P. A model for the operation of a thin-film SnO_x conductance-modulation carbon monoxide sensor[J]. J. Electrochem. Soc., 1979, 126: 627-633.

[112] Scott R W J, Yang S M, Chabanis G, et al. Tin dioxide opals and inverted opals: near-ideal microstructures for gas sensors[J]. Adv. Mater., 2001, 13: 1468-1472.

[113] D'Arienzo M, Armelao L, Mari C M, et al. Macroporous WO_3 thin films active in NH_3 sensing: role of the hosted Cr isolated centers and Pt nanoclusters[J]. J. Am. Chem. Soc., 2011, 133: 5296-5304.

[114] Scott R W J, Yang S M, Coombs N, et al. Engineered sensitivity of structured tin dioxide chemical sensors: opaline architectures with controlled necking[J]. Adv. Funct. Mater. 2002, 13: 225-231.

[115] Bai J, Luo Y, Chen C, et al. Functionalization of 1D In_2O_3 nanotubes with abundant oxygen vacancies by rare earth dopant for ultra-high sensitive ethanol detection[J]. Sens. Actuators B, 2020, 324: 128755.

[116] Chen Y, Xu P, Li X, et al. High-performance H_2 sensors with selectively hydrophobic micro-plate for self-aligned upload of Pd nanodots modified mesoporous In_2O_3 sensing-material[J]. Sens. Actuators B, 2018, 267: 83-92.

[117] Li G, Wang X, Yan L, et al. PdPt bimetal-functionalized SnO_2 nanosheets: controllable synthesis and its dual selectivity for detection of carbon monoxide and methane[J]. ACS Appl. Mater. Interfaces, 2019, 11: 26116-26126.

[118] Committee A M. Recommendations for the definition, estimation and use of the detection limit[J]. Analyst, 1987, 112: 199-204.

[119] Shrivastava A, Gupta V B. Methods for the determination of limit of detection and limit of quantitation of the analytical methods[J]. Chron. Young Sci., 2011, 2: 21-25.

[120] Mukherjee K, Majumder S B. Analyses of response and recovery kinetics of zinc ferrite as hydrogen gas sensor[J]. J. Appl. Phys., 2009, 106: 064912.

[121] Lundström I, DiStefano T. Hydrogen induced interfacial polarization at Pd SiO_2 interfaces [J]. Surf. Sci., 1976, 59: 23-32.

[122] Korotcenkov G, Ivanov M, Blinov I, et al. Kinetics of indium oxide-based thin film gas sensor response: the role of "redox" and adsorption/desorption processes in gas sensing effects[J]. Thin Solid Films, 2007, 515: 3987-3996.

[123] Hu H, Trejo M, Nicho M E, et al. Adsorption kinetics of optochemical NH_3 gas sensing with semiconductor polyaniline films[J]. Sens. Actuators B, 2002, 82: 14-23.

[124] Aygün S, Cann D. Response kinetics of doped CuO/ZnO heterocontacts[J]. J. Phys. Chem. B, 2005, 109: 7878-7882.

[125] Cabot A, Arbiol J, Morante J R, et al. Analysis of the noble metal catalytic additives introduced by impregnation of as obtained SnO_2 sol-gel nanocrystals for gas sensors[J]. Sens. Actuators B, 2000, 70: 87-100.

[126] Kranendonk W G T, Frenkel D. Simulation of the adhesive-hard-sphere model[J]. Mol. Phys., 1988, 64: 403-424.

[127] Hassan H A, Hash D B. A generalized hard-sphere model for Monte Carlo simulation[J]. Phys. Fluids A, 1993, 5: 738-744.

[128] Sun C, Bai B. Diffusion of gas molecules on multilayer graphene surfaces: dependence on the number of graphene layers[J]. Appl. Therm. Engi., 2017, 116: 724-730.

[129] Wang X, Wang Y, Tian F, et al. From the surface reaction control to gas-diffusion control: the synthesis of hierarchical porous SnO_2 microspheres and their gas-sensing mechanism[J]. J. Phys. Chem. C, 2015, 119: 15963-15976.

[130] Sun C, Bai B. Gas diffusion on graphene surfaces[J]. Phys. Chem. Chem. Phys., 2017, 19: 3894-3902.

[131] Raccis R, Nikoubashman A, Retsch M, et al. Confined diffusion in periodic porous nanostructures[J]. ACS Nano, 2011, 5: 4607-4616.

[132] Mitzithras A, Strange J H. Diffusion of fluids in confined geometry[J]. Magn. Reson. Imaging, 1994, 12: 261-263.

[133] Bickel T. A note on confined diffusion[J]. Physica A, 2007, 377: 24-32.

[134] Karniadakis G, Beskok A, Aluru N. Microflows and nanoflows: fundamentals and simulation[J]. New York: Springer Science & Business Media, 2006.

[135] Li T, Wu Y, Huang J, et al. Gas sensors based on membrane diffusion for environmental monitoring[J]. Sens. Actuators B, 2017, 243: 566-578.

[136] Xiao J, Wei J. Diffusion mechanism of hydrocarbons in zeolites—I. Theory[J]. Chem. Eng. Sci., 1992, 47: 1123-1141.

[137] Hahn K, Kärger J, Kukla V. Single-file diffusion observation[J]. Phys. Rev. Lett., 1996, 76: 2762.

[138] Dubbeldam D, Calero S, Maesen T L M, et al. Incommensurate diffusion in confined systems[J]. Phys. Rev. Lett., 2003, 90: 245901.

[139] Dubbeldam D, Smit B. Computer simulation of incommensurate diffusion in zeolites: understanding window effects[J]. J. Phys. Chem. B, 2003, 107: 12138-12152.

[140] Ghorai P K, Yashonath S, Demontis P, et al. Diffusion anomaly as a function of molecular length of linear molecules: levitation effect[J]. J. Am. Chem. Soc., 2003, 125: 7116-7123.

[141] Nag S, Ananthakrishna G, Maiti P K, et al. Separating hydrocarbon mixtures by driving the components in opposite directions: high degree of separation factor and energy efficiency[J]. Phys. Rev. Lett., 2020, 124: 255901.

[142] Liu Z, Yuan J, van Baten J M, et al. Synergistically enhance confined diffusion by continuum intersecting channels in zeolites[J]. Sci. Adv., 2021, 7: eabf0775.

[143] Rauch W L, Liu M. Development of a selective gas sensor utilizing a perm-selective zeolite membrane[J]. J. Mater. Sci., 2003, 38: 4307-4317.

[144] Zou Y, Xi S, Bo T, et al. Mesoporous amorphous Al_2O_3/crystalline WO_3 heterophase hybrids for electrocatalysis and gas sensing applications[J]. J. Mater. Chem. A, 2019, 7: 21874-21883.

[145] Lei M, Gao M, Yang X, et al. Size-controlled Au nanoparticles incorporating mesoporous ZnO for sensitive ethanol sensing[J]. ACS Appl. Mater. Interfaces, 2021, 13: 51933-51944.

[146] Lei M, Zhou X, Zou Y, et al. A facile construction of heterostructured ZnO/Co_3O_4

mesoporous spheres and superior acetone sensing performance[J]. Chin. Chem. Lett., 2021, 32: 1998 - 2004.

[147] Ma J, Li Y, Li J, et al. Rationally designed sual-mesoporous transition metal oxides/noble metal nanocomposites for fabrication of gas sensors in real-time detection of 3-hydroxy-2-butanone biomarker[J]. Adv. Func. Mater., 2022, 32: 2107439.

[148] Ma J, Ren Y, Zhou X, et al. Pt nanoparticles sensitized ordered mesoporous WO_3 semiconductor: gas sensing performance and mechanism study[J]. Adv. Func. Mater., 2018, 28: 1705268.

[149] Zhu Y, Zhao Y, Ma J, et al. Mesoporous tungsten oxides with crystalline framework for highly sensitive and selective detection of foodborne pathogens[J]. J. Am. Chem. Soc., 2017, 139: 10365 - 10373.

[150] Deng Y. Semiconducting metal oxides for gas sensing[J]. Singapore: Springer, 2019.

[151] Ma J, Li Y, Zhou X, et al. Au nanoparticles decorated mesoporous SiO_2 - WO_3 hybrid materials with improved pore connectivity for ultratrace ethanol detection at low operating temperature[J]. Small, 2020, 16: 2004772.

[152] Malek K, Coppens M O. Knudsen self- and Fickian diffusion in rough nanoporous media [J]. J. Chem. Phys., 2003, 119: 2801 - 2811.

[153] Skaug M J, Mabry J, Schwartz D K. Intermittent molecular hopping at the solid-liquid interface[J]. Phys. Rev. Lett., 2013, 110: 256101.

[154] Chen Y D, Yang R T. Concentration dependence of surface diffusion and zeolitic diffusion [J]. AIChE J., 1991, 37: 1579 - 1582.

[155] Radovic L R, Suarez A, Vallejos-Burgos F, et al. Oxygen migration on the graphene surface. 2. Thermochemistry of basal-plane diffusion (hopping)[J]. Carbon, 2011, 49: 4226 - 4238.

[156] Nakajima H. The discovery and acceptance of the Kirkendall effect: the result of a short research career[J]. JoM, 1997, 49: 15 - 19.

[157] Paul A, van Dal M J H, Kodentsov A, et al. The Kirkendall effect in multiphase diffusion [J]. Acta Mater., 2004, 52: 623 - 630.

[158] Yin Y, Rioux R, Erdonmez C, et al. Formation of hollow nanocrystals through the nanoscale Kirkendall Effect[J]. Science, 2004, 304: 711 - 714.

[159] Fogler H S. Essentials of chemical reaction engineering: essenti chemica reactio engi[J]. London: Pearson Education, 2010.

[160] Vannice M A, Joyce W H. Kinetics of catalytic reactions[J]. New York: Springer, 2005.

[161] Mears D E. Diagnostic criteria for heat transport limitations in fixed bed reactors[J]. J. Catal., 1971, 20: 127 - 131.

[162] Mohagheghi M, Bakeri G, Saeedizad M. Study of the effects of external and internal diffusion on the propane dehydrogenation reaction over Pt - Sn/Al_2O_3 Catalyst[J]. Chem. Eng. Technol., 2007, 30: 1721 - 1725.

[163] Klaewkla R, Arend M, Hölderich W F. A review of mass transfer controlling the reaction rate in heterogeneous catalytic systems[J]. London: INTECH Open Access Publisher, 2011.

[164] Weisz P, Prater C. Interpretation of measurements in experimental catalysis[J]. Adv. Catal., 1954, 6: 143-196.
[165] Cohen D, Merchuk J, Zeiri Y, et al. Catalytic effectiveness of porous particles: a continuum analytic model including internal and external surfaces[J]. Chem. Eng. Sci., 2017, 166: 101-106.
[166] Thiele E W. Relation between catalytic activity and size of particle[J]. Ind. Eng. Chem., 1939, 31: 916-920.
[167] Aris R. Communication. Normalization for the Thiele modulus[J]. Ind. Eng. Chem. Fundam., 1965, 4: 227-229.
[168] Kashid M N, Renken A, Kiwi-Minsker L. Gas-liquid and liquid-liquid mass transfer in microstructured reactors[J]. Chem. Eng. Sci., 2011, 66: 3876-3897.
[169] Whitman W. Preliminary experimental confirmation of the two-film theory of gas absorption[J]. Chem. Metall. Eng., 1923, 29: 146-148.
[170] Wang J. Flow reactor models for fluid-fluid systems, based on the two-film theory[J]. Chem. Eng. J., 1995, 60: 105-110.
[171] Higbie R. The rate of absorption of a pure gas into a still liquid during short periods of exposure[J]. Trans. AIChE, 1935, 31: 365-389.
[172] Cussler E. Diffusion: mass transfer in fluid systems[J]. Cambridge: Cambridge University Press, 2009.
[173] Toor H L, Marchello J M. Film-penetration model for mass and heat transfer[J]. AIChE J., 1958, 4: 97-101.
[174] Huang H, Chatterjee S G. Transient physical gas absorption in a stirred liquid using the surface renewal model of mass transfer[J]. Chem. Eng. Sci., 2021, 234: 116449-11643.
[175] Wang J, Yuan Q, Dong M, et al. Experimental investigation of gas mass transport and diffusion coefficients in porous media with nanopores[J]. Int. J. Heat Mass Transf., 2017, 115: 566-579.
[176] Bird R, Stewart W, Lightfoot E. Transport phenomena[J]. Hoboken: John Wiley & Sons, 2006.
[177] Crank J. The mathematics of diffusion[J]. Oxford: Oxford University Press, 1979.
[178] Lu H, Ma W, Gao J, et al. Diffusion-reaction theory for conductance response in metal oxide gas sensing thin films[J]. Sens. Actuators B, 2000, 66: 228-231.
[179] Alsyouri H M, Lin J Y S. Gas diffusion and microstructural properties of ordered mesoporous silica fibers[J]. J. Phys. Chem. B, 2005, 109: 13623-13629.
[180] Yamazoe N, Shimanoe K. Theoretical approach to the gas response of oxide semiconductor film devices under control of gas diffusion and reaction effects[J]. Sens. Actuators B, 2011, 154: 277-282.
[181] Tyagi P, Sharma A, Tomar M, et al. Metal oxide catalyst assisted SnO_2 thin film based SO_2 gas sensor[J]. Sens. Actuators B, 2016, 224: 282-289.
[182] Hsu C L, Tsai J Y, Hsueh T J. Ethanol gas and humidity sensors of CuO/Cu_2O composite nanowires based on a Cu through-silicon via approach[J]. Sens. Actuators B, 2016, 224: 95-102.

[183] Wagner T, Haffer S, Weinberger C, et al. Mesoporous materials as gas sensors[J]. Chem. Soc. Rev., 2013, 42: 4036-4053.

[184] Li H, Meng F, Liu J, et al. Synthesis and gas sensing properties of hierarchical meso-macroporous SnO_2 for detection of indoor air pollutants[J]. Sens. Actuators B-Chem., 2012, 166-167: 519-525.

[185] Yamazoe N, Shimanoe K. Roles of shape and size of component crystals in semiconductor gas sensors: I. Response to oxygen[J]. J. Electrochem. Soc., 2008, 155: J85-J92.

[186] Yamazoe N, Shimanoe K. New perspectives of gas sensor technology[J]. Sens. Actuators B, 2009, 138: 100-107.

[187] Yamazoe N. Toward innovations of gas sensor technology[J]. Sens. Actuators B, 2005, 108: 2-14.

[188] Kida T, Fujiyama S, Suematsu K, et al. Pore and particle size control of gas sensing films using SnO_2 nanoparticles synthesized by seed-mediated growth: design of highly sensitive gas sensors[J]. J. Phys. Chem. C, 2013, 117: 17574-17582.

[189] Zou Y, Zhou X, Ma J, et al. Recent advances in amphiphilic block copolymer templated mesoporous metal-based materials: assembly engineering and applications[J]. Chem. Soc. Rev., 2020, 49: 1173-1208.

[190] Luo W, Zhao T, Li Y, et al. A micelle fusion-aggregation assembly approach to mesoporous carbon materials with rich active sites for ultrasensitive ammonia sensing[J]. J. Am. Chem. Soc., 2016, 138: 12586-12595.

[191] Ren Y, Zhou X, Luo W, et al. Amphiphilic block copolymer templated synthesis of mesoporous indium oxides with nanosheet-assembled pore walls[J]. Chem. Mater., 2016, 28: 7997-8005.

[192] Wang Z, Zhu Y, Luo W, et al. Controlled synthesis of ordered mesoporous carbon-cobalt oxide nanocomposites with large mesopores and graphitic walls[J]. Chem. Mater., 2016, 28: 7773-7780.

[193] Zhou X, Zhu Y, Luo W, et al. Chelation-assisted soft-template synthesis of ordered mesoporous zinc oxides for low concentration gas sensing[J]. J. Mater. Chem. A, 2016, 4: 15064-15071.

[194] Ma J, Xiao X, Zou Y, et al. A general and straightforward route to noble metal-decorated mesoporous transition-metal oxides with enhanced gas sensing performance[J]. Small, 2019, 15: 1904240.

[195] Zhou X, Zou Y, Ma J, et al. Cementing mesoporous ZnO with silica for controllable and switchable gas sensing selectivity[J]. Chem. Mater., 2019, 31: 8112-8120.

[196] Ren Y, Yang X, Zhou X, et al. Amphiphilic block copolymers directed synthesis of mesoporous nickel-based oxides with bimodal mesopores and nanocrystal-assembled walls[J]. Chin. Chem. Lett., 2019, 30: 2003-2008.

[197] Ren Y, Zou Y, Liu Y, et al. Synthesis of orthogonally assembled 3D cross-stacked metal oxide semiconducting nanowires[J]. Nat. Mater., 2020, 19: 203-211.

[198] Wang C, Li Y, Qiu P, et al. Controllable synthesis of highly crystallized mesoporous TiO_2/WO_3 heterojunctions for acetone gas sensing[J]. Chin. Chem. Lett., 2020, 31:

1119-1123.

[199] Ren Y, Xie W, Li Y, et al. Noble metal nanoparticles decorated metal oxide semiconducting nanowire arrays interwoven into 3D mesoporous superstructures for low-temperature gas sensing[J]. ACS Cent. Sci., 2021, 7: 1885-1897.

[200] Deng Y, Yu T, Wan Y, et al. Ordered mesoporous silicas and carbons with large accessible pores templated from amphiphilic diblock copolymer poly(ethylene oxide)-b-polystyrene[J]. J. Am. Chem. Soc., 2007, 129: 1690-1697.

[201] Deng Y, Liu C, Yu T, et al. Facile synthesis of hierarchically porous carbons from dual colloidal crystal/block copolymer template approach[J]. Chem. Mater., 2007, 19: 3271-3277.

[202] Deng Y, Liu C, Gu D, et al. Thick wall mesoporous carbons with a large pore structure templated from a weakly hydrophobic PEO-PMMA diblock copolymer[J]. J. Mater. Chem., 2008, 18: 91-97.

[203] Deng Y, Liu J, Liu C, et al. Ultra-large-pore mesoporous carbons templated from poly(ethylene oxide)-b-polystyrene diblock copolymer by adding polystyrene homopolymer as a pore expander[J]. Chem. Mater., 2008, 20: 7281-7286.

[204] Deng Y, Cai Y, Sun Z, et al. Controlled synthesis and functionalization of ordered large-pore mesoporous carbons[J]. Adv. Func. Mater., 2010, 20: 3658-3665.

[205] Wei J, Deng Y, Zhang J, et al. Large-pore ordered mesoporous carbons with tunable structures and pore sizes templated from poly(ethylene oxide)-b-poly(methyl methacrylate)[J]. Solid State Sci., 2011, 13: 784-792.

[206] Deng Y, Wei J, Sun Z, et al. Large-pore ordered mesoporous materials templated from non-Pluronic amphiphilic block copolymers[J]. Chem. Soc. Rev., 2013, 42: 4054-4070.

[207] Wei J, Zhou D, Sun Z, et al. A controllable synthesis of rich nitrogen-doped ordered mesoporous carbon for CO_2 capture and supercapacitors[J]. Adv. Func. Mater., 2013, 23: 2322-2328.

[208] Wei J, Li Y, Wang M, et al. A systematic investigation of the formation of ordered mesoporous silicas using poly(ethylene oxide)-b-poly(methyl methacrylate) as the template[J]. J. Mater. Chem. A, 2013, 1: 8819-8827.

[209] Wei J, Sun Z, Luo W, et al. New insight into the synthesis of large-pore ordered mesoporous materials[J]. J. Am. Chem. Soc., 2017, 139: 1706-1713.

[210] Sun X, Hao H, Ji H, et al. Nanocasting synthesis of In_2O_3 with appropriate mesostructured ordering and enhanced gas-sensing property[J]. ACS Appl. Mater. Interfaces, 2014, 6: 401-409.

[211] Yang S, Wang Z, Zou Y, et al. Remarkably accelerated room-temperature hydrogen sensing of MoO_3 nanoribbon/graphene composites by suppressing the nanojunction effects [J]. Sens. Actuators B, 2017, 248: 160-168.

[212] Mohammad-Yousefi S, Rahbarpour S, Ghafoorifard H. Describing the effect of Ag/Au modification on operating temperature and gas sensing properties of thick film SnO_2 gas sensors by gas diffusion theory[J]. Mater. Chem. Phys., 2019, 227: 148-156.

[213] Ab Kadir R, Rani R A, Alsaif M M, et al. Optical gas sensing properties of nanoporous

Nb$_2$O$_5$ films[J]. ACS Appl. Mater. Interfaces, 2015, 7: 4751-4758.

[214] Lochbaum A, Fedoryshyn Y, Dorodnyy A, et al. On-chip narrowband thermal emitter for mid-IR optical gas sensing[J]. ACS Photonics, 2017, 4: 1371-1380.

[215] Hodgkinson J, Tatam R P. Optical gas sensing: a review[J]. Meas. Sci. Technol., 2013, 24: 012004.

[216] Manera M G, Montagna G, Ferreiro-Vila E, et al. Enhanced gas sensing performance of TiO$_2$ functionalized magneto-optical SPR sensors[J]. J. Mater. Chem., 2011, 21: 16049-16056.

[217] Manera M G, Rella R. Improved gas sensing performances in SPR sensors by transducers activation[J]. Sens. Actuators B, 2013, 179: 175-186.

[218] Gahlot A P S, Paliwal A, Kapoor A. Theoretical and experimental investigation on SPR gas sensor based on ZnO/polypyrrole interface for ammonia sensing applications[J]. Plasmonics, 2022, 17: 1619-1632.

[219] Ruiz A M, Illa X, Díaz R, et al. Analyses of the ammonia response of integrated gas sensors working in pulsed mode[J]. Sens. Actuators B, 2006, 118: 318-322.

[220] Suematsu K, Shin Y, Ma N, et al. Pulse-driven micro gas sensor fitted with clustered Pd/SnO$_2$ nanoparticles[J]. Anal. Chem., 2015, 87: 8407-8415.

[221] Yang T, Yang Q, Xiao Y, et al. A pulse-driven sensor based on ordered mesoporous Ag$_2$O/SnO$_2$ with improved H$_2$S-sensing performance[J]. Sens. Actuators B, 2016, 228: 529-538.

[222] Sharma B, Kim J S. MEMS based highly sensitive dual FET gas sensor using graphene decorated Pd-Ag alloy nanoparticles for H$_2$ detection[J]. Sci. Rep., 2018, 8: 5902.

[223] Tabata H, Matsuyama H, Goto T, et al. Visible-light-activated response originating from carrier-mobility modulation of NO$_2$ gas sensors based on MoS$_2$ monolayers[J]. ACS Nano, 2021, 15: 2542-2553.

[224] Hong S B, Wu M, Hong Y K, et al. FET-type gas sensors: a review[J]. Sens. Actuators B, 2021, 330: 129240.

[225] Suematsu K, Harano W, Oyama T, et al. Pulse-driven semiconductor gas sensors toward ppt level toluene detection[J]. Anal. Chem., 2018, 90: 11219-11223.

[226] Suematsu K, Oyama T, Mizukami W, et al. Selective detection of toluene using pulse-driven SnO$_2$ micro gas sensors[J]. ACS Appl. Electron. Mater., 2020, 2: 2913-2920.

[227] Suematsu K, Hiroyama Y, Harano W, et al. Double-step modulation of the pulse-driven mode for a high-performance SnO$_2$ micro gas sensor: designing the particle surface via a rapid preheating process[J]. ACS Sens., 2020, 5: 3449-3456.

[228] Suematsu K, Harano W, Yamasaki S, et al. One-trillionth level toluene detection using a dual-designed semiconductor gas sensor: material and sensor-driven designs[J]. ACS Appl. Electron. Mater., 2020, 2: 4122-4126.

第7章
提高气敏性能的新方法

为了使基于金属氧化物的半导体气体传感器具有更好的传感性能,研究者们采用了许多新的技术,包括光激活(如紫外光)气体传感、表面等离子体共振(SPR)增强气体传感、脉冲驱动气体传感和场效应晶体管(FET)气体传感器,以开发具有更高灵敏度、更快响应-恢复速度和更低检出限(LoD)的传感器。

7.1 光学气体传感

在紫外光照射下工作的金属氧化物半导体气体传感器已经被证实可以在室温下检测各种浓度的化学物质。研究结果还表明,对于基于紫外光 LED 的气体传感器,通过优化传感器平台设计和波长、功率强度等紫外光源参数,可以提高其性能。此外,可以通过改变半导体层结构或将波长调整至最佳值来调整气体传感检测选择性。

图 7.1 说明了光激活气体传感的机理,它受各种电化学传感反应的控制,这些反应涉及光子与金属氧化物半导体表面在室温下相互作用产生的电子/空穴[1],显示了光激活化学电阻气体传感器中发生的表面反应。利用表面吸附理论和能带理论可以解释传感过程。当金属氧化物半导体暴露在空气中时,氧在低温或高温下以阴离子形式(O^{2-}、O^- 和 O_2^-)被分子吸附[见图 7.1(a)]。然而,只有一小部分表面在黑暗条件下与空气分子相互作用,导致可忽略的电导变化。光照通过增加导带中载流子的数量来增强表面的化学活性,从而在表面上提供更多的活性位点[见图 7.1(b)],因此,它通过提供更多的电子来提高表面吸附浓度。吸附是传感过程的重要组成部分,因为它激活了反应物的决定性化学键。目标气体分子与激发的电子/空穴和吸附的氧离子反应并转化为产物[见图 7.1(c)],这会导致由于电子数的变化而出现的层导电性的显著变化,通过监测电阻随时间的变化可实现气体浓度检测。最后系统在没有目标气体的情况下恢复到初始状态[见图 7.1(d)]。

Fan 等[2]报道了紫外光照射对氧化锌电子性能和气体传感性能的影响,发现紫外光可以提高传感器的灵敏度和响应/恢复速度。对氧化锌光响应行为的研究

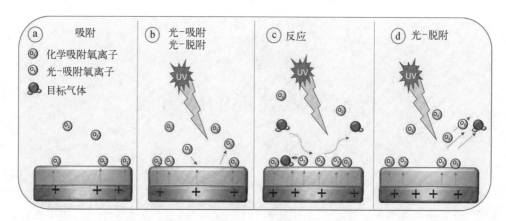

图7.1 紫外光照射下的光激活传感机制[1]

发现,紫外光产生的电子促进了氧的吸附,形成了光诱导的氧离子 $O_2^-(h\nu)$,这些 $O_2^-(h\nu)$ 使该氧化锌材料具有室温气体传感性能,可通过进一步优化提高传感器性能。

如图 7.2 所示,使用能量为 3.4 eV 的紫外光大大提高了一维和二维氧化锌纳米结构的灵敏度和响应/恢复速度,使其可以检测 ppm 水平的 H_2。紫外光产生的电子促进了氧的吸附,这些光致氧离子 $O_2^-(h\nu)$ 具有很高的反应活性,并导致了室温下的气体敏感性。氧吸附不仅受紫外光照射的影响,而且与氧化锌粒径有关。粒径越小,表面/体积比越大,产生的光致氧离子越多。由于氧化锌纳米线的粒径小于薄膜,因此纳米线具有更好的传感器性能。

当氧化锌在紫外光下时,光生空穴与吸附的氧离子(O_2^-)相互作用,使氧从氧化锌表面解吸,反应如下:

$$h\nu \rightarrow h^+ + e^- \tag{7.1}$$

$$h^+ + O_{2(ad)}^- \rightarrow O_{2(g)} \tag{7.2}$$

同时,由于环境氧分子与光电子发生反应,产生了额外的光诱导氧离子,如以下反应方程式所示:

$$O_2 + e^-(h\nu) \rightarrow O_2^-(h\nu) \tag{7.3}$$

这导致氧化锌电阻略有增加,如图 7.2(a)所示。由于紫外光促进的氧吸附,电阻变化为 50%。当氧吸附速率等于解吸速率时,氧化锌的电性能达到稳定状态。尽管部分光电子与周围的氧气结合,但由于大量光诱导载流子的存在,氧化锌的电阻仍然比在黑暗中时低得多。与强烈附着在氧化锌表面的化学吸附氧离子不同,光诱导氧离子 $O_2^-(h\nu)$ 与氧化锌的结合较弱,只需关闭紫外光即可轻易去除。暴露于 H_2 后,室温下参与氧化还原反应的是光诱导氧离子 $O_2^-(h\nu)$,

$$O_2^-(h\nu) + 2H_{2(g)} \rightarrow +2H_2O_{(g)} + e^- \tag{7.4}$$

这个过程中释放的电子导致了氧化锌电性能的变化,图 7.2(b)显示了在传感过程中的湿度变化,证实了这一论点。当 H_2 注入室内时,水分子的数量大大增加,这些过量的水分子是紫外光活化的室温气敏反应的结果。此外,这里使用的紫外光具有足够大的能量,可引起 H_2 的解吸,H_2 的吸附能为 0.9 eV,有利于解吸过程,从而提高回收率。

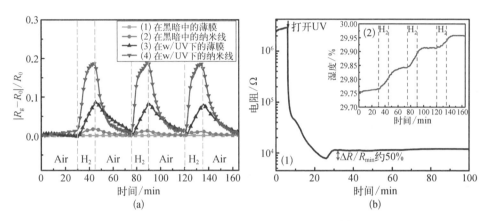

图 7.2 氧化锌薄膜的 H_2 传感性能研究[2]

(a) (1)~(4)分别表示氧化锌薄膜在黑暗中、400 nm 宽氧化锌线在黑暗中、氧化锌薄膜在紫外光下和 400 nm 宽氧化锌线在紫外光下对 100 ppm H_2 的动态响应灵敏度;(b) (1)、(2)分别为打开 UV-LED 后,氧化锌薄膜电阻的变化和气体感应过程中的湿度变化

Costello 等[3]描述了一项利用紫外光发光二极管(LED)提高氧化锌厚膜传感器室温下气体灵敏度的工作。基于氧化锌纳米粒子的传感器通过波长为 400 nm、入射光强度为 2.2 mW/cm² 的紫外光 LED 激活,能够检测非常低浓度(1 ppb)的丙酮和乙醛。相同的传感器在相同条件下工作,还能够检测低 ppm 范围内的其他挥发性有机物,包括碳氢化合物,如己烷、丁烷、丙烷和甲烷。在高湿度(相对湿度为 100%)条件下操作时,传感器对低 ppm 水平的挥发性有机物也很敏感。研究发现,传感器的最佳灵敏度取决于应用的光强度,给出最大响应的最佳光强度取决于分析物,这表明通过改变应用的光强度来调整传感器的选择性是可能的。GC-MS 研究发现,紫外光照射的氧化锌传感器能够在室温下催化一系列挥发性有机物的分解。催化分解的大致类型与在金属氧化物上加热传感的机理一致。

小型化高性能氢气传感器在基线漂移和周围空气流量方面经常会遇到可靠性问题。一般来说,传感器是在微加热器上实现的,以促进氧化还原反应(高温下产生的水分子的解吸)和避免基线漂移导致的结构不稳定。Lakshmanan 等[4]报道了一种紫外光（λ 为 400 nm，P_o 为 400 μW，P_i 为 18 mW）激发的基于氧化锌纳米球的 H_2 传感器，该传感器几乎不受周围气流的影响，且基线漂移很小。该研究采用微波辅助法合成了氧化锌纳米球，并在不同温度下紫外光存在和不存在的条件下进行了表征，以优化操作条件。可以观察到，紫外光照射的传感器获得了足够的活化能，可使电流在不同温度（60℃和100℃）下调制，包括室温（27℃）时不同浓度 H_2（1%～4%）下调制。利用紫外光照射，证明了在室温（27℃）下，具有可忽略基线漂移的 H_2 传感器与流量无关（250～1 000 sccm），且具有快速响应恢复时间。通过扫描电子显微镜、X 射线衍射、紫外-可见近红外光谱和光致发光光谱的表征，解释了传感特性的理论推理，还进行了流量依赖性、重复性、交叉选择性和湿度测试，以研究可靠性问题。

Wagner 等[5]报道说，In_2O_3 显示了一种有趣的光催化行为，可用于气体传感应用，暴露于蓝光（460 nm）下的介孔 In_2O_3 对臭氧（O_3）和 NO_2 产生了更快和更强的传感器响应，使得在室温下对氧化气体的传感成为可能（见图 7.3）。研究还发现，湿度对臭氧反应的影响主要是由于活性表面的中毒，以及利用光的正影响可进行低温 NO_2 检测。研究观察到的特性也导致了一种新的纳米结构 In_2O_3 传感模型的产生，并通过基于氧进出扩散而非氧表面基团的结构-性质关系解释了光的影响（见图 7.4）。

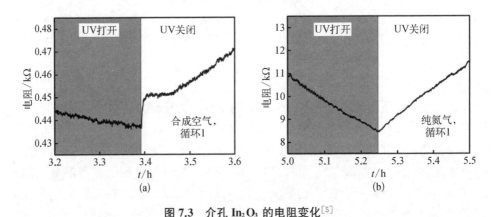

图 7.3　介孔 In_2O_3 的电阻变化[5]

(a)、(b) 合成空气（20.5%氧气，79.5%氮气）和纯氮气（氮气）中介孔 In_2O_3 在紫外光源关闭后的电子电阻变化

可见光照射是一种提高室温半导体气体传感器灵敏度、缩短响应/恢复时间的有效策略。然而，半导体在光照下的宽禁带和严重的载流子复合限制了室温气体传感器的发展。Sun 等[6]合成了多金属氧酸盐（POM）和有机染料分子修饰的 TiO_2 薄膜，并在可见光照明下实现了室温（25℃）NO_2 气体传感性能的提升。POM 分子作为染料/TiO_2 薄膜中的电子受体，可以实现光生载流子的快速分离和

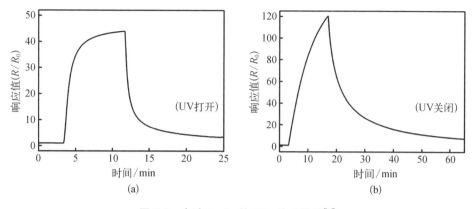

图 7.4 介孔 In_2O_3 的 NO_2 传感性能[5]

(a)、(b) 在紫外光(350 nm)照射下和没有紫外光时,介孔 In_2O_3 对 5 ppm NO_2(100℃)的气体响应(R 和 R_0 是存在和不存在 NO_2 时的电子电阻)

传输。与有机染料结合,POM 修饰染料/TiO_2 薄膜在广泛的 NO_2 浓度范围内 (50 ppb~5 ppm)表现出优异的传感特性,如高灵敏度(R_a/R_g=233.1 vs 1 ppm)、相对较低的检测浓度和高选择性。此外,在无须任何加热的条件下,该敏感膜表面 NO_2 的响应和恢复时间可分别持续控制在 48 s 和 66 s(见图 7.5)。

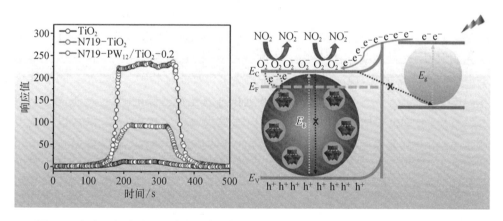

图 7.5 多金属氧酸盐作为染料/TiO_2 薄膜中的电子受体,以加速室温 NO_2 气体传感[6]

7.2 表面等离子体共振增强气体传感

表面等离子体(SP)被认为是非定域电子的集体相干振荡,在金属(如铜、金和银)与电介质的界面受到入射光的激发。由表面等离子体产生的表面等离子体共振(SPR)可以极大地增强局域电磁场。此外,SPR 对金属薄膜表面的折射率非常敏感。当介质条件发生变化时,SPR 的共振谱响应会发生变化,这可以反映系统的某些特性[7]。

基于 SPR 的光学气体传感器具有结构简单、可靠性高、可室温工作、响应速度快等优点。一些研究者已经探索了利用 Kretschmann 配置的金属膜上的敏感膜进行气体传感应用的 SPR 技术。利用钯作为传感金属的 SPR 传感器已经被证明能够检测出钯中 H_2 分子的强烈吸附作用产生的氢。使用 SPR 技术检测气体需要在贵金属表面涂上合适的敏感层,以便表面等离子体波(SPW)在金属/介电层的界面处传播。由于气体的存在,敏感层折射率的变化将导致 SP 色散关系的变化。传感层在暴露于目标气体时折射率的变化,直接关系到与之相互作用的气体分子的浓度。因此,为了得到高灵敏度、高选择性的传感器,对传感膜进行优化是非常重要的[8]。

过渡金属氧化物薄膜在传感应用中具有重要的技术意义。在不同的材料中,氧化锌(ZnO)作为一种具有宽频带隙(3.37 eV)的Ⅱ-Ⅵ族半导体材料被广泛研究,并被发现在其他应用中对气体传感具有重要性。

Do 等[9]报道了金修饰氧化锌结构的表面等离子体增强紫外发射,以增强气体传感性能。研究采用简单的化学镀液沉积和光还原方法,在玻璃基片上成功地生长了纯的和金修饰的亚微米氧化锌球。扫描电子显微镜(SEM)和透射电子显微镜(TEM)图像分析、能量分散 X 射线谱(EDS)、紫外-可见吸收和光致发光(PL)光谱结果均验证了等离子体金纳米粒子(NP)在氧化锌薄膜上的结合。时间分辨光致发光(TRPL)光谱表明,表面等离子体效应存在,从金纳米粒子到亚微米氧化锌球的电荷转移速率很快,说明该材料可用于设计高效催化装置。在 120℃ 的最佳传感温度下,Do 等研究了不同气体浓度下沉积态氧化锌薄膜的 NO_2 传感能力。研究结果显示,等离子体金纳米粒子的表面修饰提高了灵敏度(141 倍),以及响应($\tau_{Res}=9$ s)和恢复速度($\tau_{Rec}=39$ s)。增强的气体传感性能和光催化降解过程不仅归因于表面等离子体共振效应,还归因于等离子体金纳米粒子和氧化锌结构之间的肖特基势垒(见图 7.6 和图 7.7)。

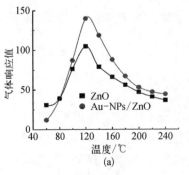
(a)

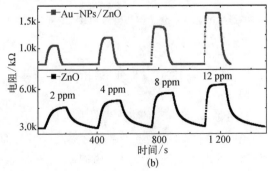
(b)

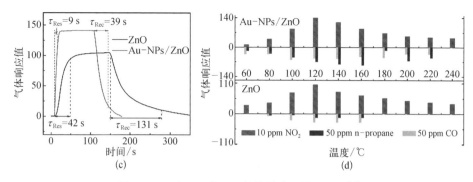

图 7.6 Au/ZnO 对 NO_2 气体的传感性能研究[9]

(a) 传感器在不同温度下暴露于 10 ppm NO_2 时的响应；(b) 在 120℃ 下响应 NO_2 的动态电阻瞬态；
(c) 响应-恢复特性；(d) 传感器对不同目标气体的响应

Thepudom 等[10] 利用氧化锌传感层制备了一种用于 CO 气体检测的高灵敏度、高效率的 SPR 气体传感器 (Prism/Au/ZnO)。在 250℃ 基板温度下生长的 200 nm 厚的优化氧化锌薄膜在宽浓度范围 (0.5~100 ppm) 内对 CO 气体表现出增强和稳定的传感响应。开发的传感器显示出对 CO 气体的快速响应 (1 s) 和高灵敏度 (0.091/ppm)。选择性研究还表明，研制的 SPR 传感器对 CO 高度敏感，对其他气体 (NH_3、CO_2、NO_x、LPG、H_2) 的干扰可忽略不计 (见图 7.8 和图 7.9)。

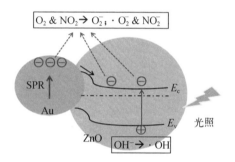

图 7.7 金纳米粒子/氧化锌结构增强气体传感和光活性机制示意图[9]

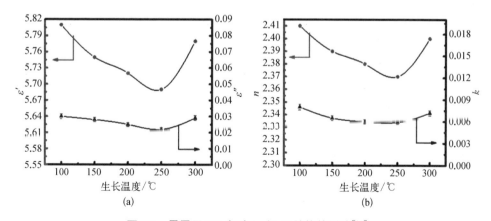

图 7.8 暴露于 CO 中对 Au/ZnO 结构的影响[10]

(a) 不同浓度 CO 气体暴露的 Prism/Au/ZnO 结构的 SPR 反射数据；(b) 不同浓度 CO 气体的 SPR 传感系统的 SPR 倾角和 R_{min} 变化 (校准曲线)

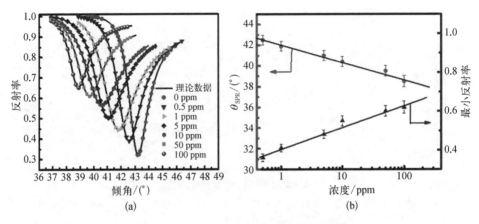

图 7.9　Au/ZnO 对 CO 的气体传感性能[10]

(a) SPR 传感器(Prism/Au/ZnO 系统,其 ZnO 薄膜在 250℃下生长)对 CO 气体的瞬态响应;(b) SPR 传感器对 CO 气体的校准曲线

Thepudom 等[10]开发了一种 PEC CPF 传感器,其光电极由镀金光栅基板/TiO_2/P_3HT/AuNP 组成。金光栅上传播的 SPR 与 P_3HT 薄膜的吸收峰波长吻合良好,增强了对 CPF 的高灵敏度检测的短路光电流。此外,由于局部表面等离子体效应,AuNP 的负载进一步增强了短路光电流。与无表面等离子体激元结构的光阳极相比,改进的混合等离子体光阳极得到了增强。利用这两种等离子体的协同效应,可将短路光电流放大到 μA 标度,明显高于以前报道的方法所获得的信号,通过该方法在所有 CPF 浓度下获得了数十 nA 水平的信号。此外,使用该系统可在 7.5 nm 的极低浓度下检测出 CPF。所证明的新方法提供了一种多功能光伏效应,可用于多种基于 PEC 的传感应用(见图 7.10 和图 7.11)。

Sharma 等[12]基于表面等离子体共振(SPR)设计了一种新型光纤六氯苯(HCB)传感器,该传感器采用混合石墨烯纳米板(GNP)、氧化锡纳米颗粒(SnO_2 NP)为载体。在纳米杂化反应中,将预先合成的 SnO_2 纳米颗粒修饰在 GNP 上,并利用两者的综合作用增强 HCB 检测的化学反应性。GNP/SnO_2 纳米复合物由于具有高导电性、大表面积、高反应性和催化行为等协同特性,对 HCB 具有良好的灵敏度和选择性。通过优化参数,该探针在 $0\sim10^{-2}$ g/L 范围内实现了良好的检测 HCB 浓度的传感性能,在该浓度范围内观察到峰值吸收波长红移了 74 nm。传感器具有 8.7×10^{-13} g/L 的检测限,小于文献中使用其他各种技术的传感器的检测限值。实际样品中的 HCB 浓度使用拟议探针进行测试,结果与文献中报告的值相近。此外,该研究还报道了其他分析物对传感器性能的干扰。综上,该传感器可用于在线监测和遥感。

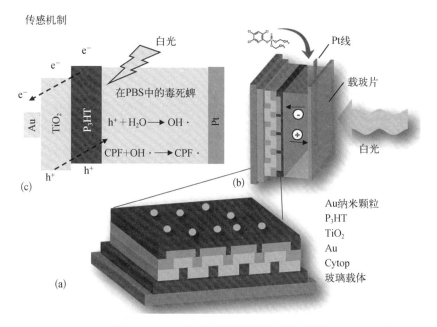

图 7.10 CPF 传感系统结构及传感机制[11]

(a) 具有光栅结构的制造光电极;(b) 用于 CPF 测量的 PEC 单元;(c) CPF 传感机制的示意图

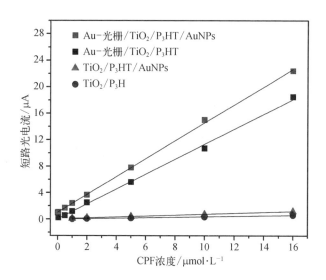

图 7.11 不同成分的预制 PEC 电池的短路光电流与 CPF 浓度的关系图(在 50°的最佳入射角下测量短路电流)[11]

贵金属-金属氧化物纳米复合物在光催化应用中发挥着越来越重要的作用。Zhang 等[13]用改进的聚合物网络凝胶法制备了 ZnO-Ag 纳米颗粒,用于室温光助 NO_2 气敏。由于两种材料之间形成了异质结,表面氧空位增加,因此与纯氧化锌传感器相比，该传感器在不同光照条件下($\lambda=520\sim365$ nm)对 NO_2 气体($0.5\sim5$ ppm)的敏感性增强。结果表明，表面等离子体共振(SPR)可以使这种氧化锌-银纳米结构具有良好的可见光性能。更重要的是，通过使用不同的 LED 光源调整工作波长，可以获得最佳的灵敏度。当使用蓝绿色 LED(470 nm, 75 mW/cm^2)时，3%[①] ZnO-Ag 传感器显示出最高的灵敏度以及优越的稳定性和选择性。该研究还详细讨论了湿度对传感器性能的影响。

7.3 脉冲驱动气体传感

近年来，新发展的脉冲驱动的微型气体传感器具有结构紧凑、功耗低等优点。基于金属氧化物的脉冲驱动气体传感器使用微加热器和电极，通过微机电系统(MEMS)技术安装，该技术每秒重复加热微加热器。换言之，即加热该传感器一秒钟，然后在加热后冷却一秒钟，不断重复上述步骤。在这种驱动模式下，颗粒表面的催化活化只在加热开启阶段进行，而易燃气体在加热关闭阶段渗入传感层。图 7.12 显示了使用 SnO_2 纳米粒子(NPs)描述传感层中脉冲驱动气体扩散行为的示意图。换句话说，该脉冲驱动气体传感器通过驱动目标气体进入传感层来工作。在先前的研究中，有研究者报道了脉冲加热模式下催化燃烧传感器对 VOC 气体的响应在达到稳态前出现较大的峰值。脉冲驱动气体传感器结合了两个关键因素，即气体氧化过程的激活和气体渗透进入传感层，有助于提高对可燃气体的敏感性，并降低浓度检测限值。

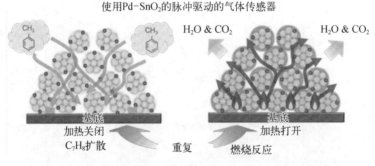

图 7.12 在加热关闭和加热打开阶段气体扩散行为的示意图[14]

Suematsu 等[14]介绍了一种以 Pd-SnO_2 簇状纳米颗粒为传感材料，由微加热器和传感器电极组成的开关式脉冲驱动气体传感器。通过向微加热器施加电压，反复加热传感器并使其冷却，VOC 气体在未加热状态下渗入传感层内部。因此，

---

① 3% 指摩尔分数。

脉冲驱动气体传感器的效用因子大于传统的连续加热传感器,进而传感器对甲苯的响应增强,实际上,传感器对甲苯的检测浓度为 1 ppb。此外,根据响应与甲苯浓度的关系,该研究中的脉冲驱动气体传感器可以检测浓度为 200 ppt 甚至更低的甲苯。综上,脉冲驱动的微加热器和用于检测甲苯的材料的结合提高了传感器的响应,并促进了 ppt 级别的甲苯检测,该传感器对基于人体呼吸的医学诊断的发展具有重要作用。

Yang 等[15]制备了一种基于介孔 Ag_2O/SnO_2 复合材料的微球型直接加热气体传感器,并对其气体传感性能进行了研究。结果表明,介孔 Ag_2O/SnO_2 基气体传感器在 100℃时具有良好的选择性、高响应性和对 H_2S 的稳定性,采用脉冲驱动方法可提高对 H_2S 的敏感性能。在脉冲驱动下,传感器对 300 ppb H_2S 的响应为 5.7,约为恒流下的 2 倍,检测限(LoD)提高到 50 ppb(见图 7.13)。由于介孔 Ag_2O/SnO_2 的组成、结构以及脉冲驱动方式,该传感器具有较高的传感性能。

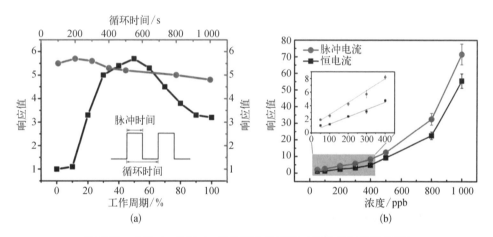

图 7.13　介孔 Ag_2O/SnO_2 脉冲驱动传感器对 H_2S 的气敏性能[15]

(a) 不同占空比和循环时间的介孔 Ag_2O/SnO_2 脉冲驱动传感器对 300 ppb H_2S 的响应;
(b) 脉冲恒流介孔 Ag_2O/SnO_2 传感器对 H_2S 浓度的响应

Suematsu 等[16]开发了一种脉冲驱动的微型传感器,该传感器安装了聚集的 Pd/SnO_2 纳米粒子,用于检测挥发性有机物气体。将少量纳米颗粒悬浮液滴沉积在集成有微电极的微型加热器上,形成了微传感膜。即使加热器以脉冲加热模式驱动,该微传感器对氢气和甲苯的响应也很好。特别是当加热器打开时,传感器在 0.1 s 内即对甲苯迅速作出响应,表明其燃烧反应和扩散在微膜中有效地发生,并且微观结构可控。研究结果表明,脉冲驱动微传感器作为一种电池可操作的便携式气体传感器具有广阔的应用前景。

Triatafyllopoulou 和 Tsamis[17]报道了基于纳米多孔硅微热板的低功率 SnO_2

气体传感器的制备和表征,以及用于检测包括 CO 和 NO 在内的有毒气体,同时测量了 SnO₂ 气体传感器对各种气体浓度(100～500 ppm)的响应。在恒温模式下,保持微热板温度恒定和脉冲温度模式,通过向加热器施加电压脉冲进行分析。在这两种情况下,传感器的响应均随着温度和检测气体浓度的增加而增加(见图 7.14)。通过对两种不同工作模式的比较可知,在脉冲温度模式下,传感器具有较高的灵敏度,且利用这种技术可以显著降低功耗。此外,传感器在检测 NO 时显示出明显的选择性,因此可以实现两种气体之间的区分。

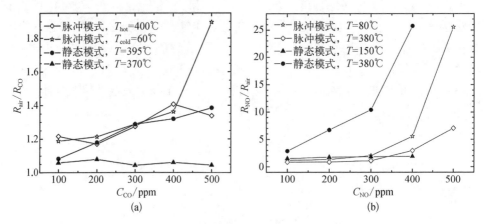

图 7.14　脉冲温度模式下 CO 传感性能研究[17]

(a) 与恒温模式相比,脉冲温度模式下的传感器对 CO 的敏感度;
(b) 与静态模式相比,脉冲温度模式下的传感器的敏感度

Ruiz 等[18]采用溶胶-凝胶法制备了 Cr-WO₃、Cr-TiO₂ 和 Pd-SnO₂ 纳米颗粒,并采用微滴技术在热板平台上沉积了添加剂改性的金属氧化物纳米材料。对氨气敏感层的功能性气体试验是在静态模式和脉冲驱动模式下进行的,脉冲驱动模式固定了表面的再生,从而增强了传感器的响应并缩短了瞬态时间。基于 Cr-WO₃ 的气体传感器对氨气的传感响应有明显改善。

7.4　场效应晶体管气体传感器

基于金属氧化物半导体(MO_x)和场效应晶体管的传感器由于在气体检测中的广泛应用而日益受到重视。各种研究证实,气体传感特性取决于金属氧化物和催化材料的灵敏度。MO_x 传感器作为商业化传感器使用得最为广泛。此外,FET 型气体传感器与传统的气体传感器相比,具有许多优点,这归因于它们的形状、尺寸和较低的生产成本。然而,加工参数和再现性需要加强,才能扩大其应用范围。

Zhou 等[19]开发了一种锰掺杂氧化锌纳米粒子(MZO)的新型互锁 p+n 场效

应晶体管电路,用于检测低浓度丙酮气体,特别是接近 1.8 ppm 的丙酮气体(见图 7.15)。值得注意的是,该联锁放大电路中的 MZO 对丙酮(<2 ppm)显示出低电压信号(<0.3 V),而对>2 ppm 丙酮的瞬态响应的电压信号>4.0 V。换句话说,对丙酮的浓度从 1 ppm 增加到 2 ppm,其响应增加了约 1 233%,这可以将糖尿病患者从健康人中分离出来。另外,85% 的高相对湿度对 2 ppm 丙酮的响应几乎没有影响。同时,与甲醛、乙醛、甲苯和乙醇相比,该连锁电路中的 MZO 具有较高的丙酮选择性,为糖尿病的广泛定性筛选提供了一种有前景的可能性。重要的是,这种联锁电路也适用于其他类型的金属氧化物半导体气体传感器。研究认为,p 场效应晶体管和 n 场效应晶体管因栅极电压的变化而引起电阻跳变,使联锁电路产生瞬态响应。

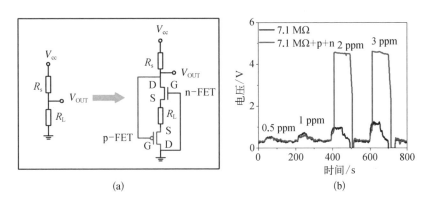

图 7.15　场效应晶体管的丙酮传感研究[19]

(a) 金属氧化物丙酮传感器传统电路和互锁 p+n 场效应晶体管(FET)电路的设计方案;
(b) 传统电路和互锁 p+n 场效应晶体管电路(R_L 为 7.1 mW)在 25% 湿度下的对比,其中锰掺杂氧化锌(MZO)对丙酮的输出电压为 0.5~3 ppm

MOS-FET H_2 传感器可在室温或高温(约 150℃)下工作,改善传感器响应并降低变化,包括温度和相对黏性的影响。但这种高温加热增加了传感器操作的功耗,限制了电池驱动传感器的应用[20]。有研究报道了一些降低 MOS-FET H_2 传感器功耗的创新结构。Yokosawa 等[21]报道了在 H_2 暴露期间催化金属层的比热。该研究在栅极金属和保护层(悬浮栅极)之间形成了一个具有空气开口的 MOS-FET,使得气体分子能够自由地迁移到两个表面,即使在室温下也可以测量 H_2。Ahn 等[22]用掺杂钯的 Si 纳米线(SiNW 传感器)制作了双面栅极,成形场效应晶体管的长度为 1 μm,通道宽度为 100 nm[见图 7.16(a)]。如图 7.16(b)所示,为了提高氢气的敏感性和选择性,在硅纳米线的顶部表面沉积了钯(约 1 nm)[22-23]。传输特性显示通道电流减小,阈值电压增加,这表明钯修饰后电子从通道中的撤回情况有所改善。

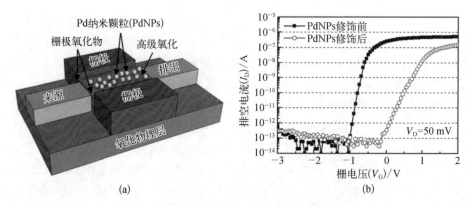

图 7.16 钯修饰的硅纳米线 FET 传感器[22-23]

(a) 用于 H_2 传感的硅纳米线 FET 的示意图；(b) 钯修饰的硅纳米线和硅纳米线的传输特性的变化

Wang 等[24]报道了一种独特设计的 DPPTT/苯六硫醇铜(Cu-BHT)-nT 异质结有机场效应晶体管(OFET)传感器,并用于 NO 的高效传感。据报道,Cu-BHT 由于具有高电导率和晶体缺陷 Cu_{2c} 的优势引起了人们的广泛关注。然而,具有二维 kagome 晶格的 Cu-BHT 无孔结构和小的比表面积总是限制了其在传感和催化方面的实际应用。该研究通过简单的均相反应设计并制备了 Cu-BHT 纳米管(Cu-BHT-NT)来解决上述问题。与传统的纳米棒状结构相比,Cu-BHT-NT 不仅具有更大的比表面积,而且具有更高比例的晶体缺陷(66.6%)。因此,成功配置的 DPPTT/Cu-BHT-nT 异质结有机 OFET 传感器对 NO 具有高达 13 610% 的灵敏度,最低检测限为 5 ppb,并表现出对 NO 的优异选择性。理论分析系统地表明,Cu-BHT-NT 中的 Cu_{2c} 位点增加了从异质结构转移到 NO 分子的电子数,证实了 Cu-BHT-nT 与 NO 分子之间强烈的键合作用提高了 NO 传感的灵敏度和选择性。此外,该研究还研制了一种基于异质结有机 OFET 传感器的全柔性装置,以确保佩戴和携带气体传感器的便利性,为下一代可穿戴智能电子设备的制造开辟了新的途径(见图 7.17)。

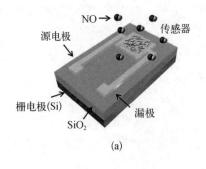
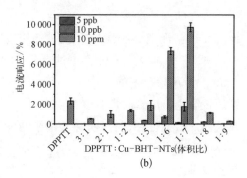

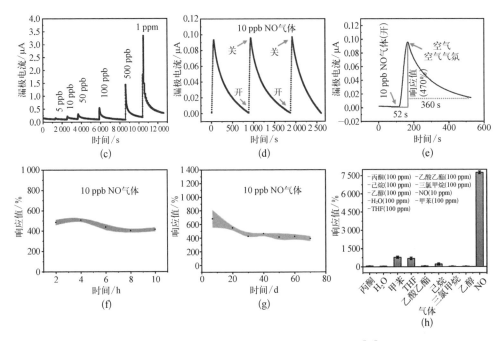

图 7.17 基于异质结有机 OFET 的传感器性能[24]

(a) OFET 气体传感器示意图；(b) 具有不同体积比的异质结器件对不同浓度 NO(5 ppb、10 ppb 和 10 ppm) 的响应性；(c) 响应-不同浓度 NO 的回收曲线(范围为 5 ppb～1 ppm)；(d) 对 10 ppb NO 响应的重复性；(e) 在 10 ppb NO 下的响应/恢复时间曲线；(f)～(h) 存在 10 ppb NO 时的环境稳定性研究和选择性行为

参考文献

［1］Espid E, Taghipour F. UV-LED photo-activated chemical gas sensors: a review[J]. Crit. Rev. Solid State, 2017, 42: 416-432.

［2］Fan S, Srivastava A, Dravid V. UV-activated room-temperature gas sensing mechanism of polycrystalline ZnO[J]. Appl. Phys. Lett., 2009, 95: 142106.

［3］Costello B, Ewen R, Ratcliffe N, et al. Highly sensitive room temperature sensors based on the UV-LED activation of zinc oxide nanoparticles[J]. Sens. Actuators B, 2008, 134: 945-952.

［4］Lakshmanan K, Vijayakumari A M, Basu P K. Reliable and flow independent hydrogen sensor based on microwave-assisted ZnO nanospheres: improved sensing performance under UV light at room temperature[J]. IEEE Sens. J., 2018, 18: 1810-1819.

［5］Wagner T, Kohl C, Morandi S, et al. Photoreduction of mesoporous In_2O_3: mechanistic model and utility in gas sensing[J]. Chem. Eur. J., 2012, 18: 8216-8223.

［6］Sun X, Lan Q, Geng J, et al. Polyoxometalate as electron acceptor in dye/TiO_2 films to accelerate room-temperature NO_2 gas sensing[J]. Sens. Actuators B, 2023, 374: 132795-132803.

［7］Homola J, Yee S. Surface plasmon resonance sensors: review[J]. Sens. Actuators B, 1999, 54: 3-15.

[8] Wang J, Lin W, Cao E, et al. Surface plasmon resonance sensors on Raman and fluorescence spectroscopy[J]. Sensors, 2017, 17: 2719-2738.

[9] Do T, Ho T, Bui T, et al. Surface-plasmon-enhanced ultraviolet emission of Au-decorated ZnO structures for gas sensing and photocatalytic devices[J]. Beilstein J. Nanotechnol., 2018, 9: 771-779.

[10] Thepudom T, Lertvachirapaiboon C, Shinbo K, et al. Surface plasmon resonance-enhanced photoelectrochemical sensor for detection of an organophosphate pesticide chlorpyrifos[J]. MRS Commun., 2018, 8: 107-112.

[11] Paliwal A, Sharma A, Tomar M, et al. Carbon monoxide (CO) optical gas sensor based on ZnO thin films[J]. Sens. Actuators B, 2017, 250: 679-685.

[12] Sharma S, Usha S, Shrivastav A, et al. A novel method of SPR based SnO_2: GNP nanohybrid decorated optical fiber platform for hexachlorobenzene sensing[J]. Sens. Actuators B, 2017, 246: 927-936.

[13] Zhang Q, Xie G, Xu M, et al. Visible light-assisted room temperature gas sensing with ZnO-Ag heterostructure nanoparticles[J]. Sens. Actuators B, 2018, 259: 269-281.

[14] Suematsu K, Harano W, Oyama T, et al. Pulse-driven semiconductor gas sensors toward ppt level toluene detection[J]. Anal. Chem., 2018, 90: 11219-11223.

[15] Yang T, Yang Q, Xiao Y, et al. A pulse-driven sensor based on ordered mesoporous Ag_2O/SnO_2 with improved H_2S-sensing performance[J]. Sens. Actuators B, 2016, 228: 529-538.

[16] Suematsu K, Shin Y, Ma N, et al. Pulse-driven micro gas sensor fitted with clustered Pd/SnO_2 nanoparticles[J]. Anal. Chem. 2015, 87: 8407-8415.

[17] Triantafyllopoulou R, Tsamis C. Detection of CO and NO using low power metal oxide sensors[J]. Phys. Status Solidi A, 2008, 205: 2643-2646.

[18] Ruiz A, Illa X, Diaz R, et al. Analyses of the ammonia response of integrated gas sensors working in pulsed mode[J]. Sens. Actuators B, 2006, 118: 318-322.

[19] Zhou X, Wang J, Wang Z, et al. Transilient response to acetone gas using the interlocking p+n field-effect transistor circuit[J]. Sensors, 2018, 18: 1914-1925.

[20] Scharnagl K, Karthigeyan A, Burgmair M, et al. Low temperature hydrogen detection at high concentrations: comparison of platinum and iridium[J]. Sens. Actuators B, 2001, 80: 163-168.

[21] Yokosawa K, Saitoh K, Nakano S, et al. FET hydrogen-gas sensor with direct heating of catalytic metal[J]. Sens. Actuators B, 2008, 130: 90-99.

[22] Ahn J, Yun J, Choi Y, et al. Palladium nanoparticle decorated silicon nanowire field-effect transistor with side-gates for hydrogen gas detection[J]. Appl. Phys. Lett. 2014, 104: 013508.

[23] Sharma B, Sharma A, Kim J. Recent advances on H_2 sensor technologies based on MOX and FET devices: a review[J]. Sens. Actuators B-Chem, 2018, 262: 758-770.

[24] Wang L, Chen X, Yi Z, et al. Facile synthesis of conductive metal-organic frameworks nanotubes for ultrahigh-performance flexible NO sensors[J]. Small Methods, 2022, 6: 2200581.

第 8 章
不同类型的常用半导体金属氧化物气体传感器

在过去的几十年中,研究者们已经开发了各种半导体气体传感器,有的甚至已经出现在了市场上。然而,由于半导体金属氧化物气体传感器的工作原理不同,很难对其进行分类[1-2]。本章将介绍迄今已开发或已提出的各种半导体气体传感器的基本情况。一般而言,半导体气体传感器由一个接收器和一个换能器组成。接收器提供材料或材料系统,其在与目标气体相互作用时,要么引起自身性质(功函数、介电常数、电极电势、质量等)的改变,要么发热或发光。换能器是将这种效应转换成电信号(即传感器响应)的装置。传感器的构造是由植入了受体的换能器确定的。从这个角度看,半导体气体传感器可以定义为半导体材料被用作受体或换能器的一种传感器。气体传感器可以根据传感方法来分类[3-5],即基于电特性变化的方法和基于其他性质变化的方法,如半导体金属氧化物(SMO)、碳纳米管和聚合物的材料能够基于电特性的变化而表现出对目标气体的响应。气体传感器也可以根据所采用的换能器的类型进行分类,包括电阻器[6-7]、二极管[8-9]、MIS(金属-绝缘体-半导体)电容器[10-11]、金属-绝缘体-半导体场效应晶体管(MISFET)[12-13]和氧浓度电池[14-15]的气体传感器。

在上述不同的传感器类型中,最常用的是化学电阻型传感器,也就是所谓的电阻式气敏元件。在化学电阻型传感器中,半导体金属氧化物通常被用作气体传感材料,在外界施加氧化或还原气体时能改变电阻,这些传感器通常称为半导体金属氧化物气体传感器,其根据敏感机理的不同可以分为两类,即表面敏感型传感器和体敏感型传感器。本章专门讨论表面敏感型传感器。高性能传感器的发展一直追求比传统传感器更快、更便宜、更灵敏、更稳定和选择性更高,极大地推动了气体传感领域向高性能纳米传感器的发展[16]。微加工技术是一种基于集成电路(IC)技术的微制造技术,它可用于制造小型三维器件,能够以低成本大规模地制造可靠的小尺寸新型化学和气体传感器。微制造技术采用了最新发展的超大规模集成电路(VLSI)技术,它对于制造进一步减小尺寸和提高效率的金属氧化物半导体(MOS)气体传感器是非常有利的。为了实现小型化,研究者们还对基于 Si 或 SiC 通道的金属氧化物半导体场效应晶体管（MOSFET）气体传感器进行了大量的研究。自 Janata 等提出悬浮栅场效应晶体管的概念，业界对基于氧化锌或氧化锡纳米线等传感材料的场效应晶体管型传感器的传感方案进行了广泛的研究[17]，并积极地将其投入实际应用。本章重点讨论和阐述传统 MOS、MOSMEMS 和 MOSFET 气体传感器的器件结构、传感材料和工作原理。

8.1 电阻型气体传感器

基于半导体金属氧化物的电导气体传感器或化学电阻器实际上是研究得最多的气体传感器系列之一。它们具有成本低、生产灵活性强、操作简单、可检测气体量大等突出特点，在基础研究和工业领域引起了广泛关注。气体传感器发展的动力来源于 Brattein 等[18]对锗和 Heiland[19]对氧化锌金属氧化物与气体相互反应的发现。Taguchi[20]将基于金属氧化物的半导体传感器商业化，创造了重要的工业产品（Taguchi 型传感器），这是半导体气体传感器发展的里程碑。目前，世界上有许多上述类型传感器的商业化公司，例如 Figaro（日本）、NissaFIS（日本）、UST（中国）、CityTech（英国）、Alphaseense（英国）和 Honeywell（美国）公司，它们的应用范围涵盖从简单的对爆炸或有毒气体的警报到汽车工业中的进气控制[21]。

气体传感特性可由三个基本因素解释，分别是受体功能、传感器功能和利用系数[21-23]，如图 8.1 所示。第一个因素与每个金属氧化物晶体如何响应激发气体有关，而激发气体又与特定半导体金属氧化物的固有性质直接相关，与表面和目标分析物的理想特定相互作用相关。第二个因素是关于如何将每个晶体的传感响应转换成器件电阻，即将该分子信息有效地转换成宏观上可获得的信号，也就是电阻的变化。第三个因素描述了由于刺激气体在其内部扩散期间的消耗，装置响应（即电阻变化）如何在实际多孔传感体中衰减[24]。因此，对于给定类型的基材，传感器性质取决于结构特征、催化活性表面掺杂剂的存在状态以及工作温度。

图 8.1 控制半导体气体传感器的三个基本因素[21]

8.1.1 电阻型气体传感器的器件结构与制作

传感器装置被制造成电阻型气体传感器的过程中,传感材料的多孔叠层附着有加热器和电阻测量元件(通常是一对金属电极)。首先制造一个带有 Au 电极和 Pt 丝的烧结陶瓷管;然后,将 Ni-Cr 合金线圈插入陶瓷管中以控制传感器的操作温度,最后将敏感材料涂层涂敷到氧化铝陶瓷管表面。高质量气体传感器的制造开始于"湿法"工艺制备氧化物半导体的细粉末(直径约 10 nm 的微晶尺寸),如图 8.2 所示,为了获得基于 ZnO 的电阻型传感器,需先在玛瑙研钵中用松油醇黏合剂轻微研磨 ZnO 材料以形成浆料悬浮液;然后将浆料涂覆到陶瓷管的表面上以形成厚膜[25],并将 Ni-Cr 合金线圈放置在管内作为加热器以提供气体传感器的工作温度;最后将上述制备好的传感器在最佳工作温度下老化一周,以提高测试前的稳定性。负载电阻器(R_L)与气体传感器连接,以便将传感器上的电压调节在最佳范围内。将电路电压(V_C)设置为 5 V,输出电压(V_{out})是负载电阻器的两端负载电压,在典型的传感过程中,将测试气体注入测试室中并用空气稀释,测试气体可与敏感层上吸附的氧物质发生反应并释放自由电子,导致传感器的电阻和电压降低,而 V_{out} 增加。

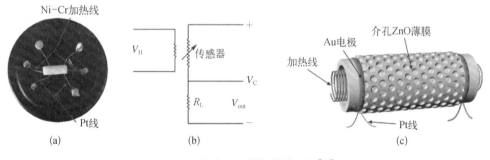

图 8.2 介孔 ZnO 基气体传感器[25]

(a) 组装装置照片;(b) 气体传感测量电路;(c) 旁热式介孔 ZnO 基气体传感器的结构示意图

在实际应用中,为了避免触发气体爆炸,每个器件都与 2 个连接器引脚相结合并放置在顶部具有孔的金属盖内。此外,将吸附剂如活性炭(通常称为过滤器)放置在孔正后方,以除去干扰气体。与传统传感器相比,对更便宜、更快、更灵敏、更具有选择性和稳定性的高性能传感器的持续需求已经推动气体传感领域朝先进的纳米传感器方向发展。作为这种结构的微版本,微机电系统(MEMS)传感器目前正被广泛地研究和开发[26-29],我们将在 8.2 节对此进行阐述。

8.1.2 敏感材料

在不同类型的传感器中,电导气体传感器或化学电阻器是研究得最多的传感器系列之一。表面敏感型电阻传感器的工作原理非常简单，当在高温下暴露于空气中的目标气体时，其电阻作为气体分压的函数会降低或增加。在众多金属氧化物中，n 型氧化物（如 SnO_2、In_2O_3、WO_3、ZnO 和 $\gamma-Fe_2O_3$）和 p 型氧化物（如 NiO、CuO、Co_3O_4、Cr_2O_3 和 Mn_3O_4）表现出显著的气敏性质[30-32]。当使用 n 型氧化物时，在正常大气压条件和 200～400℃的典型工作温度下，它具有表面电子耗尽层。表面电子耗尽层的产生是由于像 O_2^- 或 O^- 种类的吸附氧的吸附束缚了电子载体。表面电子耗尽层对气体高度敏感，还原性气体（如 CO 或 H_2）可与表面反应并除去化学吸附的氧，因此耗尽层减小，而氧化气体（如 NO_2）可导致耗尽层增大。除了氧化还原活性气体之外，已知 CO_2 和水蒸气也或多或少会影响电阻，其原理是上述气体分子在金属氧化物表面解离影响吸附氧的含量。于是，研究人员利用 CO_2 的上述性质开发了一种半导体 CO_2 传感器[33]。

在实际应用中，氧化锡半导体传感器广泛用于检测各种污染物和可燃气体。这种传感器具有灵敏度高、制作简单、重量轻、成本低等优点，然而，它的主要问题是选择性太低。氧化锡半导体传感器灵敏度和选择性的提高可通过引入合适的添加剂，如贵金属（如 Pd 和 Pt）、过渡金属氧化物（如 La_2O_3、Nd_2O_3 和 SrO）和非金属元素（如 Si、N、P）形成复合敏感材料。例如，已有研究使用二氧化铈来增加对 H_2S 的敏感性并提高 CO 选择性。使用贵金属和其他类型的掺杂剂是提高化学气体传感器选择性的一种典型方法，它是通过改变传感材料的原子和电子结构或目标气体在传感层上的表面反应途径来实现的[34-35]。用于改进对特定气体传感的典型复合金属氧化物可列出如下：SnO_2-PdO（CO，丙烷）、SnO_2-Pt 或 PdO（甲烷）、$SnO_2-Co_3O_4$（CO）、SnO_2-CuO（H_2S）、SnO_2-Ag_2O（H_2）、In_2O_3-PdO（CO，恶臭气体）、WO_3-Pt（CO）、WO_3-Au（NH_3）、WO_3-Si（丙酮）、WO_3-Ce（H_2S）、$SnO_2-La_2O_3-Pt$（乙醇）、SnO_2-CaO（乙醇）、$In_2O_3-Fe_2O_3$（臭氧）、$SnO_2-Fe_2O_3$（NO_2）、$TiO_2-Cr_2O_3$（NO）等[36-45]。

8.2 MEMS 平台气体传感器

微机电系统（MEMS）技术允许将标准 Si 晶片上的气体传感器、加热元件和温度传感器集成到标准互补金属氧化物半导体（CMOS）的电路中，目前在用于实现各种类型的物理传感器和执行器的研究中已经得到了很大的发展。CMOS 或 CMOS-MEMS 技术目前已用于传感器生产中，主要用于器件的小型化、降低功耗、更快的传感器响应、按工业标准批量制造、降低成本以及灵敏度的提高。此外，由于比表面积大大增加，用于制造传感装置的纳米材料得到了越来越多的关注，这为降低金属氧化物半导体气体传感器的工作温度提供了机会[46-53]。

在湿度传感器和有毒气体传感器的气体传感系统中，MEMS 微加热装置是关键部件，它的应用范围也正在迅速扩大。本节将讨论两种微机械加工的基底，即封闭膜和悬浮膜。由于微机械加工基底的机械脆性，传感层的沉积过程变得复杂。本节还将讨论用于形成敏感层的不同方法，如薄膜和厚膜沉积技术。最后，本节将对敏感层的气体传感功能进行分析，并针对这种新方法带来的传感器性能的提高，提出各种从中提取信息的方法。

8.2.1 MEMS 气体传感器的器件结构与制作

传统半导体金属氧化物气体传感器的缺点之一是其有限的比表面积会导致气体灵敏度太低。此外，陶瓷和薄膜气体传感器通常需要在超过 500℃ 的温度下操作来提高灵敏度。高工作温度(≥300℃)是大多数气体传感器系统(田口型)的缺点，会最终导致高功耗。低功耗是对具有理想电池寿命的传感器系统的基本要求，特别是在现场应用中，要求在整个传感层中高温分布必须均匀，因为这样的操作才能提高传感器的灵敏度。为此，MEMS 微加热器成为化学气体传感器提高温度均匀性的关键组件之一[54-59]，而温度均匀性又主要取决于膜材料和微加热器的几何形状。金属氧化物气体传感器应用的最新趋势是要求具有良好的热性能和机械性能，如高温下良好的机械稳定性[60-68]。为了实现这些目标，必须熟悉和优化微热板的热特性，并通过控制热损失、介电材料和加热器的配置来提升温度的均匀分布。研究人员使用不同金属和金属合金，结合不同加热器结构开发了微热板结构[69-70]。无论问题是什么，关键在于加热器材料应当在低热膨胀的情况下维持高温而不损坏，并且膜应当是具有低热导率的良好电介质。在此基础上，开发出的膜有两种类型，即封闭膜和悬浮膜。

在封闭膜类型中(见图 8.3)，加热器区域下方有一个硅膜，该硅膜不仅用作膜之间的热分布器，还能使其机械稳定[71]。通过从背面各向异性地蚀刻硅来形成封闭膜型传感器，通常使用如 KOH 或 EDP2 的湿蚀刻剂来进行，而合适的蚀刻材料包括氮化硅、氧化硅或掺硼硅。对于膜的形成，已知有两种不同的方法。第一种方法更流行，即使用氧化硅或氮化硅作为膜和绝缘材料以获得典型厚度为 1～2 mm 的膜。近来得到证明的第二种方法，是通过硅阳极氧化和之后的氮化获得厚度为 25～30 mm 的氮化多孔硅。

在悬浮膜类型中，蜘蛛状支承梁承载着带有传感层的热板[71](见图 8.4)。在热板下方，通过产生微腔蚀刻掉衬底，以提供良好的热隔离。在这种情况下，使用标准蚀刻剂或通过牺牲蚀刻从正面执行蚀刻，不需要从背面进行光刻，这就是其流行的原因[72]。尽管有上述几个优点，但由于这种类型的膜只有四根梁支撑，因此不能很好地证明其机械稳定性。

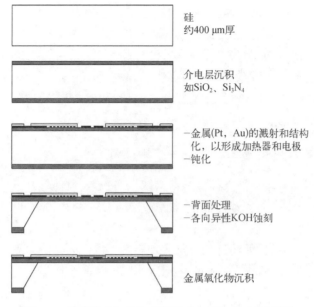

图 8.3 封闭膜式气体传感器形成工艺流程示意图[71]

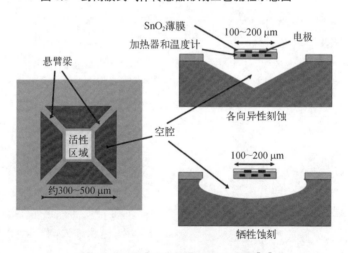

图 8.4 悬浮膜式气体传感器示意图[71]

8.2.2 传感材料

微机电系统(MEMS)技术已逐渐成为传感器小型化的流行技术。基于MEMS技术制备的不同类型半导体金属氧化物的气体传感器具有体积小、重量轻、性能好、易于批量生产、成本低等优点。许多研究在制造各种气体传感器时已采用了MEMS技术[71]，如Mitzner等[72]报道了ZnO材料可作为甲烷传感器，但其响应恢复时间较长。

如前所述，制造化学电阻型微传感器的关键技术之一在于需要将传感材料精确地加载到指定的微感测区域上,特别是对于大规模生产,仍然是一大难题。这里要提到一项特别的研究[73-74],在这项研究中,使用预先生长到微芯片非传感区域上的具有超疏水特性的图案化自组装单层(SAM)来引导传感材料自发地聚集到亲水感测区域上。显然,只是暴露出传感电极,其他所有导线都覆盖着SiO_2层。因此,加热电路和传感电路在平面内是隔离的。如图 8.5(a)和(b)所示,由于对各层

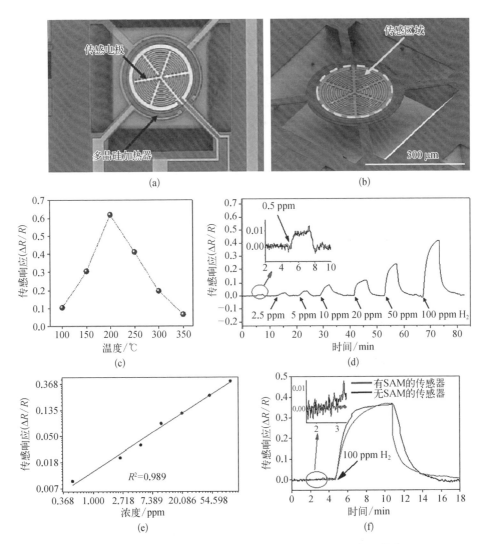

图 8.5 带有悬浮加热板的 MEMS 传感器芯片(后附彩图)[74]

(a)光学图像;(b)制造的微孔板传感器芯片的 SEM 图,其中感应区域被环形加热器包围,导线被一层SiO_2隔离,电极暴露在外以进行化学电阻连接;(c)同一传感器对 200 ppm H_2 的温度相关传感响应(200℃是最佳工作温度);(d)传感器在 200℃下对 H_2 的响应,浓度范围为 0.5~100 ppm,插图显示传感器可溶解于 0.5 ppm H_2;(e)传感响应与气体浓度之间的对数线性关系;(f)有 SAM(红色)和无 SAM(黑色)的传感器对 100 ppm H_2 的响应,插图为传感器基线,显示了传感器的噪声基底

的应力进行了优化设计和控制,在高温实验中,复合层引入的热失配应力较小,且板保持平坦,同时该研究成功把纳米点修饰的介孔 In_2O_3 成功装载到了微加热板传感器所需的微感测区域上[74]。

该研究还评估了传感器对 200 ppm 氢气的灵敏度,其中通过以 50℃ 的 I 增量加热板来实现 100～350℃ 的阶跃温度。在相对较低的温度范围(100～200℃)内,响应随测量温度的增加而增加,并在 200℃ 时达到最大响应信号。当温度进一步升高到 200℃ 以上时,响应迅速降低。传感器依次暴露于浓度为 0.5～100 ppm 的 H_2 中,传感器响应随着气体浓度的增加而增加,对 0.5 ppm H_2 显示出超高响应,响应约 0.01。对于 100 ppm H_2 的响应信号高达 0.36,并且本底噪声约 0.33%。可以看出,传感器响应随着氢气浓度的增加而增加。气敏响应与气体浓度呈对数线性关系,对 100 ppm H_2 的响应信号为 0.36,本底噪声为 1%。结果表明,该方法涂敷的传感材料可以有效地降低噪声,有利于提高传感器的分辨率[74]。

目前,制作 MEMS 衬底的高性能传感器的另一个难题是如何在保持器件稳定性、一致性的基础上,提高器件对待测气体的响应灵敏度。与传统的烧结陶瓷管式气体传感器不同,MEMS 敏感区域小、集成难度高,对敏感材料的制作工艺提出了更严苛的要求。Zhu 等[75]通过聚苯乙烯(PS)小球与锡源界面组装的方法,在 MEMS 基底上制备得到了有序单层 SnO_2 纳米碗,并通过原子层沉积(ALD)和水热生长工艺,在纳米碗壁上生长了一维线性 ZnO 纳米棒(见图 8.6)。由于具有高比表面积和独特的异质结结构,所获得的 SnO_2@ZnO NW 对 H_2S 具有超高的响应(R_a/R_g=6.24@1ppm)和良好的稳定性,这种工艺在保证灵敏度的前提下,具有

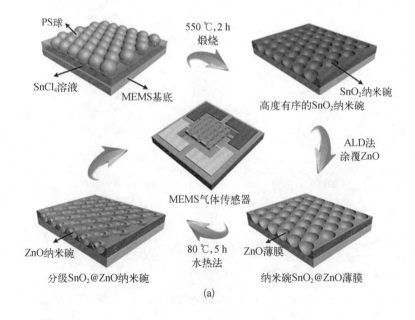

(a)

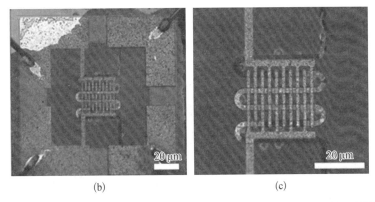

图 8.6 基于 $SnO_2@ZnO$ NW 的高灵敏度和稳定的 MEMS 硫化氢传感器[75]

晶圆级制备的可能,显示出巨大的应用前景。该研究还通过能带结构对敏化机理进行了深入分析,由于二氧化锡的功函数(4.9 eV)低于氧化锌(5.2 eV),电子从二氧化锡流向氧化锌并产生接触势垒,导致材料在空气中具有更高的电阻以及在接触到 H_2S 气体时具有更大的电阻变化。

Wu 等[76]通过原子层沉积、水热法和磁控溅射在微机电系统(MEMS)器件上原位生长了 Pt 纳米粒子修饰的 SnO_2-ZnO 核壳纳米片(SnO_2-ZnO-Pt NSs),并用于硫化氢(H_2S)气体传感,该传感器具有优越的灵敏度、高选择性、良好的再现性和低功耗。如图 8.7 所示,具体而言,当在 375℃ 的工作温度下检测 5 ppm H_2S 时,SnO_2-ZnO-Pt NSs 显示出高的灵敏度($R_a/R_g=30.43$)和良好的选择性,其抗性变化率为 29.43,分别为原始 SnO_2 NS(约 1.25)和 SnO_2-ZnO 芯-壳 NS(约 3.43)的 24 倍和 9 倍。如此优异的传感性能主要归因于异质结的形成、Pt 纳米粒子的催化敏化效应和 Pt 修饰导致的比表面积增加。因此,这项研究中所提出的 SnO_2-ZnO-Pt NS 气体传感器作为高性能 H_2S 气敏材料具有巨大潜力。

此外,Zhu 等[77]也基于 MEMS 衬底合成了碳纳米颗粒修饰的介孔 α-Fe_2O_3(C-d-mFe_2O_3)纳米棒(NRs),其可作为优异的丙酮气体传感器,表现出高灵敏度和稳定性。图 8.8 以 MEMS 作为主体器件,能够确保传感器的低功耗、小尺寸和高集成度。所获得的 C-d-mFe_2O_3 NRs 具有良好的热稳定性和优异的丙酮传感性能,在 225℃ 下表现出良好的响应($R_a/R_g=5.2$@2.5 ppm)和选择性、快速的响应/恢复速度(10/27 s),以及较低的检测限(500 ppb)。此外,该丙酮传感器即使在空气中储存超过 10 个月,也具有显著的长期稳定性和重复性。对其增强的丙酮传感机制进行探究,结果表明,其出色的传感性能可归因于介孔 α-Fe_2O_3 NRs 的大比表面积、界面处的高导电碳纳米颗粒与 α-Fe_2O_3 间形成的异质结。此外,密度泛函理论(DFT)计算有助于进一步验证优异的丙酮传感性能。C-d-mFe_2O_3-NRs 气体传感器的优异性能使其在环境中监测有害丙酮气体方面具有巨大的实际应用潜力。

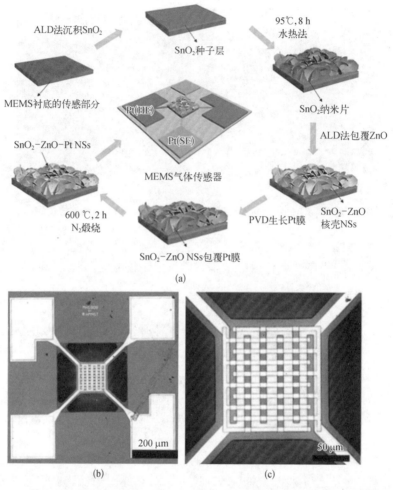

图 8.7 基于 Pt 纳米粒子改性的 SnO_2-ZnO 核壳纳米片的 MEMS 硫化氢传感器[76]

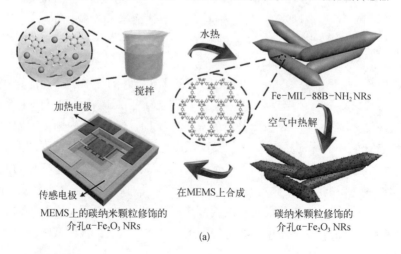

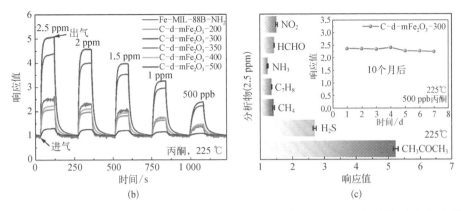

图 8.8 基于 α‑Fe₂O₃/C 介孔纳米棒的高灵敏度和稳定的 MEMS 丙酮传感器（后附彩图）[77]

8.3 场效应晶体管型气体传感器

小型化是半导体纳米线场效应晶体管(NW‑FET)型气体传感器发展的最终目标之一[78-79]。在本节中，我们以半导体纳米线为典型例子来讨论场效应晶体管型气体传感器[80-81]，通常使用单根的纳米线或纳米线网络膜作为活性检测元件。场效应晶体管的参数可以通过目标气体存在与否来调节，它们的变化与气体分子的类型和浓度密切相关。对于金属氧化物，主要气敏机制是通过气体分子在其表面的吸附和脱附影响载流子的浓度，从而引起其电阻的变化。如图 8.9 所示，在氧

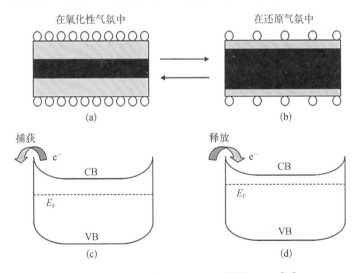

图 8.9 基于纳米线的场效应晶体管原理图[82]

(a) 在氧化性气氛中，部分自由电子被固定在纳米线表面；(b) 在还原性气氛中，部分捕获的电子释放到导带中；(c)、(d) (a)和(b)情况下相应的能带示意图

化性气氛中,金属氧化物的部分自由电子被俘获在表面,载流子浓度降低导致电阻升高,在还原性气氛中则与之相反。与传统电导式气体传感器不同,场效应晶体管型气体传感器可以通过调节栅极电压来改变灵敏度,并可以通过调节阈值电压和亚阈值摆幅来实现模式识别[82]。此外,借助于微制造技术,可以将纳米线场效应晶体管型气体传感器非常小地集成到智能系统中。本节简要讨论了基于半导体纳米线场效应晶体管型的气体传感器的制作和气敏机理。值得注意的是,术语"场效应晶体管型气体传感器"也可以指基于Si或SiC通道[83-85]的金属氧化物半导体场效应晶体管(MOSFET)气体传感器。这些传感器在金属栅极和气体分子之间的催化作用通过不同的感应机制工作。

8.4 传感装置和材料

8.4.1 纳米线场效应晶体管型气体传感器的结构与制备

根据工作机理和器件结构,半导体纳米线气体传感器可主要分为电导型和场效应晶体管(FET)型。栅极通过在纳米线的背面、顶部或周围的电介质层与纳米线分离,分别称为背栅、顶栅或环绕栅FET[86],如图8.10所示。电导型传感器是基于电阻变化的,电阻变化又是由传感元件暴露在目标气体中引起的,通常需要局部加热器来升高传感材料的温度以改善其反应性。通过感测部件将金属氧化物颗粒膜替换为纳米线膜,可完成田口型金属氧化物气体传感器的自然演变。相反,由于传感通道暴露于目标气体中,场效应晶体管型气体传感器基于场效应晶体管参数变化。除了电流变化之外,其他场效应晶体管参数(如阈值电压和亚阈值摆幅)的变化也可用于识别传感过程。这种原理和机制与电导型气体传感器不同,后者只使用电阻变化来解释气体传感。场效应晶体管型气体传感器具有如下有趣的特征。首先,它们能够在室温下工作,相比之下,电导型气体传感器通常在200~400℃下工作。其次,它们能够应用于各种传感材料,从而为产生用于敏感和选择性传感的传感表面带来更多的可能性。再次,由于场效应晶体管型气体传感器具有非常小的尺寸与微制造技术的兼容性,其适合于制造具有更强传感能力的传感器阵列[87]。关于基于半导体纳米线的气体传感器,特别是电导型气体传感器,已经有相对全面的综述[88-95]。

基于金属氧化物纳米线的FET气体传感器的制造取决于获得纳米线的方式。在使用自上向下方法(如光刻定义的Si纳米线)的情况下,纳米线的制造与后续器件制造工艺相兼容。相反,如果通过自下向上方法生产纳米线,则必须将纳米线转移到最终器件衬底上,并且制造工艺不是那么可控,还需要进一步改进。最简单的方法是先将纳米线分散到液体介质中,然后将悬浮溶液滴到基底上,这种方法可以使纳米线平铺在基底上,不需要控制它们的分散和排列。

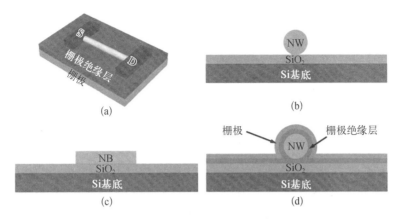

图 8.10 NW-FET 示意图[81]

(a) 背栅 NW-FET 的三维视图；(b) 圆柱形纳米线背栅 FET 的横截面；(c) 带状纳米线背栅 FET 的横截面；(d) 顶栅 NW-FET 的横截面

8.4.2 传感材料

基于金属氧化物纳米线的 FET 气体传感器的灵敏度与纳米线的半径有直接关系，并可以基于经典的 Drude 模型进行推导。在这个模型中，假设纳米线中电流密度是均匀的，于是纳米线的电导 G 与半径 r 和长度 l 的关系可以用式(8.1)表示，其中 n_e 和 μ_e 是电子密度和电子迁移率。电子密度 n_e 可以进一步表示为式(8.2)，其中 n_0 是暴露在待测气体之前的电导率，N_S 和 α 分别是气体分子的表面吸附密度和电荷转移系数，因此电阻变化率也就可以表示为式(8.3)，可以明显看出电阻变化率和 $1/r$ 成线性关系，这一结果表明随着纳米线半径的缩小，对气体检测的灵敏度将会提高。

$$G = \frac{\pi r^2}{l} n_e e \mu_e \tag{8.1}$$

$$n_e = n_0 - \frac{2\alpha N_S}{\gamma} \tag{8.2}$$

$$\frac{\Delta G}{G_0} = \frac{2}{\gamma} \times \frac{\alpha N_S}{n_0} \tag{8.3}$$

各种半导体金属氧化物纳米线，特别是 SnO_2、ZnO 和 V_2O_5 的金属氧化物纳米线，已经在 FET 气体传感器中得到了广泛研究。在本节中，根据基于 SnO_2、ZnO、In_2O_3 和其他氧化物纳米线的 FET 气体传感器中使用的材料类型[96-97]，将讨论分为四个部分。SnO_2 基气体传感器是目前主要使用的固态传感器[98-99]。1962 年，Taguchi 申请了关于多孔 SnO_2 陶瓷材料基气体传感器的日本专利，被认

为是 SnO_2 基气体传感器的开端[100]。后来,经过几年的努力,基于 SnO_2 的气体传感器于 1968 年投放市场。据报道,通过减小 SnO_2 纳米晶体的直径,当在测量期间将测试气体切换为空气时,可以实现更大的电阻变化[101]。单晶 SnO_2 纳米线的成功合成,使得利用微制造技术研究这种 FET 气体传感器成为可能,迄今为止,研究者们对使用 SnO_2 纳米线的 FET 气体传感器已经开展了许多研究工作[102]。

Moskovits 课题组[103]报道了在典型的 SnO_2 纳米线 FET 气体传感器中,通过调节栅极电压改变纳米线中的电子浓度,可以改变在纳米线表面发生的氧化和还原反应的速率和程度,这是纳米线 FET 气体传感器的一个特征。该项工作系统研究了该装置对三种不同气氛(N_2、N_2+O_2、N_2+O_2+CO)的响应。如图 8.11 所示,当气体仅包含 N_2 时,氧从纳米线的表面热脱附,形成表面氧空位。吸附氧捕获的电子被热激发到导带中,得到电流。当向 100 mL/min 的 N_2 流中加入 10 mL/min 的氧时,源极-漏极电流急剧减小并最终达到新的稳定状态。此后的某个时间 t_2,将 CO 引入气体混合物中(100 mL/min N_2、10 mL/min O_2 和 5 mL/min CO)。CO 与 SnO_2 反应产生 CO_2,留下氧空位,这会产生新的施主态,从而导致电导增加。同时,氧吸附-解吸也影响了空位的数量。整个反应是在 SnO_2 表面将 CO 催化氧化成 CO_2。通过改变可用于氧表面电子的数量,可以调节氧化速率和氧吸附的程度。当适当地调谐和选择栅极电压时,器件的灵敏度可以非常高。结果表明,通过调节纳米线内载流子浓度可影响纳米线表面的化学反应性,这对纳米线 FET 气体传感器具有重要意义。此外,原始 SnO_2 纳米线的传感能力可通过用金属纳米颗粒进行表面改性而得到进一步增强,因为在氧化物上修饰催化金属纳米颗粒可有效地改善氧化物表面的活性[104-106]。

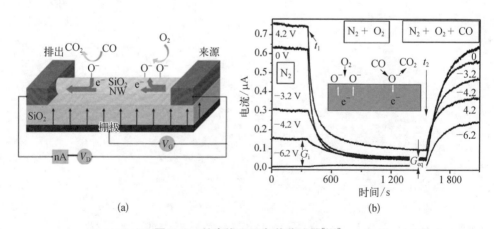

图 8.11 纳米线 FET 气体传感器[103]

(a) 结构示意图;(b) 纳米线电导在三种不同气氛条件下的演变,t_1 时 N_2 变为 N_2+O_2 的混合物,t_2 时变为 N_2+O_2+CO 的混合物,器件的响应曲线被栅极电势大幅调整,尤其是在衰减和上升区域

作为最突出的半导体金属氧化物之一,ZnO 也因其多重物理性质、化学性质,以及在电子、光电子和压电电子学中的潜在应用而受到广泛的研究。ZnO 是直接带隙半导体(室温下为 3.37 eV),具有稳定的纤锌矿晶体结构和极性表面。基于这些独特的性质,研究人员已经利用 ZnO 纳米线制造了各种纳米结构器件,如紫外激光器、发光二极管、光电检测器和化学传感器等。类似于 SnO_2 纳米线,在气体传感器领域中,ZnO 纳米线也被用于制造 FET 气体传感器[90],如图 8.12 所示。众所周知,ZnO 纳米线具有大量的表面缺陷,主要是氧空位,其可以吸附气体物质并充当散射和捕获中心。这些缺陷和化学物质对 ZnO NW-FET 的性能有重要影响。Chen 等[107]制作了基于 ZnO 的纳米线化学传感器,该传感器在室温下对 TNT 分子的检测极限为 60 ppb。随着 TNT 浓度的增加,器件电导从 3 mS(在空气中)单调抑制到 0.5 mS(在 1.36 ppm TNT 中),可观察到更明显的电导调制。灵敏度符合 $S=1/(A+B/C)$ 方程,其中 $A=0.086$、$B=16.74$、C 为浓度。在较低浓度下,化学传感器响应和 TNT 浓度之间呈现线性相关性,其表面覆盖趋于饱和,因此导致图 8.12(d)中所示的饱和响应,这是金属氧化物 NW-FET 气体传感器中的

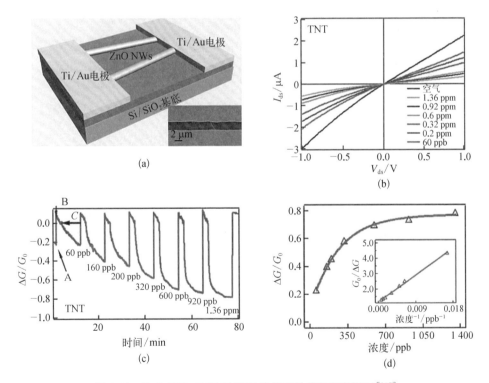

图 8.12　ZnO NW-FET 对 TNT 的传感性能(后附彩图)[107]

(a) ZnO NW-FET 结构示意图;(b) 在空气中和不同 TNT 浓度下的 I_{ds}-V_{ds} 曲线;(c) ZnO 纳米线化学传感器对 TNT 的传感响应;(d) 归一化电导变化相对于 TNT 浓度(C),使用 $S=1/(A+B/C)$ 进行拟合

常见现象。Fan 和 Lu[108]发现 ZnO NW-FET 在不同温度下对 NH₃ 的响应呈现出不同的规律,在室温状态下,NH₃ 分子将导致 ZnO 纳米线的电阻升高,而在高温时(500 K),NH₃ 分子将导致 ZnO 纳米线电阻降低。这是由于 ZnO 的费米能级和 NH₃ 分子的化学势发生了变化。在室温状态下,ZnO 的费米能级高于 NH₃ 分子的化学势,电子从 ZnO 流向 NH₃ 分子,导致 ZnO 纳米线电阻升高;高温时 ZnO 的费米能级低于 NH₃ 分子的化学势,导致 ZnO 纳米线电阻降低。值得注意的是,Cha 等[109]通过自对准平面栅电极和定义明确的纳米气隙电介质证明了 ZnO NW-FET 的高性能。这些独特的 ZnO NW-FET 表现出优异的性能:跨导为 3.06 mS、开关比为 106、场效应迁移率为 928 cm² · V⁻¹ · s⁻¹,这是 ZnO NW-FET 在没有任何特定处理(如钝化)的情况下的最高值。

In_2O_3 是另一种流行的气体传感器材料,关于 In_2O_3 纳米线 FET 气体传感器的研究也很多。Zeng 等[110]和 Shen[111]等均报道了自组装 In_2O_3 纳米线的合成,制备了单结 In_2O_3 纳米结构基场效应晶体管(见图 8.13),并且获得了高于 200 cm² · V⁻¹ · s⁻¹ 的迁移率。

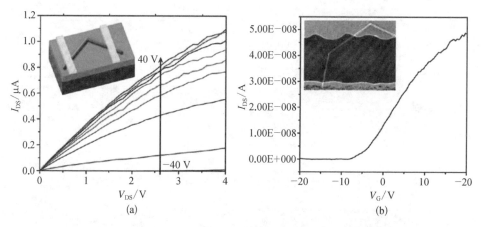

图 8.13 在单结 **In_2O_3 纳米线上构建的 FET 电流-电压数据**[111]

(a) 在不同栅极电压下以 5 V 步长测量的 I_{DS}-V_{DS} 曲线;
(b) I_{DS}-V_G 转移曲线,插图是所制造器件的 SEM 图

参考文献

[1] Wang C, Yin L, Zhang L, et al. Metal oxide gas sensors: sensitivity and influencing factors[J]. Sensors, 2010, 10: 2088-2106.

[2] Shimizu Y, Egashira M. Basic aspects and challenges of semiconductor gas sensors[J]. MRS Bull., 1999, 24: 18.

[3] Kentoro I. Hydrogen-sensitive Schottky barrier diodes[J]. Surf. Sci., 1979, 86: 345-352.

[4] Joo S, Muto I, Hara N. Hydrogen gas sensor using Pt- and Pd-Added anodic TiO_2

nanotube films[J]. J. Electrochem. Soc., 2010, 157: 221 - 226.

[5] Liu Y, Yu J, Lai P, et al. Investigation of WO_3/ZnO thin-film heterojunction-based Schottky diodes for H_2 gas sensing[J]. Hydrogen Energ., 2014, 39: 10313 - 10319.

[6] Sberveglieri G. Recent developments in semiconducting thin-film gas sensors[J]. Sens. Actuators B, 1995, 23: 103 - 109.

[7] Franke M E, Koplin T J, Simon U. Metal and metal oxide nanoparticles in chemiresistors: does the nanoscale matter? [J]. Small, 2006, 2: 36 - 50.

[8] Hyodo T, Shibata H, Shimizu Y, et al. H_2 sensing properties of diode-type gas sensors fabricated with Ti- and/or Nb-based materials[J]. Sens. Actuators B, 2009, 142: 97 - 104.

[9] Hyodo T, Yamashita T, Shimizu Y. Effects of surface modification of noble-metal sensing electrodes with Au on the hydrogen-sensing properties of diode-type gas sensors employing an anodized titania film[J]. Sens. Actuators B, 2015, 207: 105 - 116.

[10] Daugherty M, Janousek B. Surface potential relaxation in a biased $Hg_{1-x}Cd_x$Te metal-insulator-semiconductor capacitor[J]. Appl. Phys. Lett., 1983, 42: 290 - 292.

[11] Dhakal R, Kim E, Jo Y, et al. Characterization of micro-resonator based on enhanced metal insulator semiconductor capacitor for glucose recognition[J]. Med. Eng. Phys., 2017, 41: 55 - 62.

[12] Lim T, Bong J, Mills EM, et al. Highly stable operation of metal oxide nanowire transistors in ambient humidity, water, blood, and oxygen[J]. ACS Appl. Mater. Interfaces, 2015, 7: 16296 - 16302.

[13] Lundström I, Armgarth M, Spetz A, et al. Gas sensors based on catalytic metal-gate field-effect devices[J]. Sens. Actuators B, 1986, 10: 399 - 421.

[14] Wang C, Xu X, Li B. Ionic and electronic conduction of oxygen ion conductors in the Bi_2O_3 - Y_2O_3 system[J]. Solid State Ionics, 1983, 13: 135 - 140.

[15] Näfe H, Aldinger F. CO_2 sensor based on a solid state oxygen concentration cell[J]. Sens. Actuators B, 2000, 69: 46 - 50.

[16] Lupan O, Postica V, Wolff N, et al. Localized synthesis of iron oxide nanowires and fabrication of high performance nanosensors based on a single Fe_2O_3 nanowire[J]. Small, 2017, 13: 1602868.

[17] Feng P, Shao F, Shi Y, et al. Gas sensors based on semiconducting nanowire field-effect transistors[J]. Sensors, 2014, 14: 17406 - 17429.

[18] Brattain W, Bardeen J. Surface properties of germanium[J]. Tech. J., 1953, 32: 1.

[19] Heiland G. Zum Einfluß von adsorbiertem Sauerstoff auf die elektrische Leitfähigkeit von Zinkoxydkristallen[J]. Phys., 1954, 138: 459 - 464.

[20] Taguchi N, U.S. Patent 3, 631, 436[P]. 1971 - 12 - 28.

[21] Yamazoe N, Shimanoe K. New perspectives of gas sensor technology[J]. Sens. Actuators B, 2009, 138: 100 - 107.

[22] Yamazoe N, Shimanoe K. Roles of shape and size of component crystals in semiconductor gas sensors: I. Response to oxygen[J]. J. Electrochem. Soc., 2008, 155: 85.

[23] Yamazoe N, Shimanoe K, Sawada C. Contribution of electron tunneling transport in semiconductor gas sensor[J]. Thin Solid Films, 2007, 515: 8302 - 8309.

[24] Yamazoe N, Shimanoe K. Theory of power laws for semiconductor gas sensors[J]. Sens. Actuators B, 2008, 128: 566-573.

[25] Zhou X, Zhu Y, Luo W, et al. Chelation-assisted soft-template synthesis of ordered mesoporous zinc oxides for low concentration gas sensing[J]. J. Mater. Chem. A, 2016, 4: 15064-15071.

[26] Puigcorbe J, Vogel D, Michel B, et al. Thermal and mechanical analysis of micromachined gas sensors[J]. J. Micromech. Microeng., 2003, 13: 548-556.

[27] Rossi C, Temple-Boyer P, Esteve D. Realization and performance of thin SiO_2/SiN_x membrane for microheater applications[J]. Sens. Actuators A, 1998, 64: 241-245.

[28] Rossi C, Scheid E, Esteve D. Theoretical and experimental study of silicon micromachined microheater with dielectric stacked membranes[J]. Sens. Actuators A, 1997, 63: 183-189.

[29] Judy J W. Microelectromechanical systems (MEMS): fabrication, design and applications [J]. Smart Mater. Struct., 2001, 10: 1115-1134.

[30] Wang L, Kang Y, Liu X, et al. ZnO nanorod gas sensor for ethanol detection[J]. Sens. Actuators B, 2012, 162: 237-243.

[31] Heidari E K, Zamani C, Marzbanrad E, et al. WO_3-based NO_2 sensors fabricated through low frequency AC electrophoretic deposition[J]. Sens. Actuators B, 2010, 146: 165-170.

[32] Wetchakun K, Samerjai T, Tamaekong N, et al. Semiconducting metal oxides as sensors for environmentally hazardous gases[J]. Sens. Actuators B, 2011, 160: 580-591.

[33] Sun Y, Chen L, Wang Y, et al. Synthesis of MoO_3/WO_3 composite nanostructures for highly sensitive ethanol and acetone detection[J]. J. Mater. Sci., 2017, 52: 1561-1572.

[34] Ma J, Ren Y, Zhou X, et al. Pt nanoparticles sensitized ordered mesoporous WO_3 semiconductor: gas sensing performance and mechanism study[J]. Adv. Funct. Mater., 2018, 28: 1705268.

[35] Liu X, Chang Z, Luo L, et al. Sea urchin-like $Ag-\alpha-Fe_2O_3$ nanocomposite microspheres: synthesis and gas sensing applications[J]. J. Mater. Chem., 2012, 22: 7232.

[36] Gao J, Wang L, Kan K, et al. One-step synthesis of mesoporous $Al_2O_3-In_2O_3$ nanofibres with remarkable gas-sensing performance to NO_x at room temperature[J]. J. Mater. Chem. A, 2014, 2: 949-956.

[37] Nguyen H, El-Safty S. Meso- and macroporous Co_3O_4 nanorods for effective VOC gas sensors[J]. J. Phys. Chem. C, 2011, 115: 8466-8474.

[38] Liu Y, Guo R, Yuan K, et al. Engineering pore walls of mesoporous tungsten oxides via Ce doping for the development of high-performance smart gas sensors[J]. Chem. Mater., 2022, 34: 2321-2322.

[39] Ren Y, Zou Y, Liu Y, et al. Synthesis of orthogonally assembled 3D cross-stacked metal oxide semiconducting nanowires[J]. Nature Mater., 2020, 19: 203-211.

[40] Ma J, Ren Y, Zhou X, et al. Pt nanoparticles sensitized ordered mesoporous WO_3 semiconductor: gas sensing performance and mechanism study[J]. Adv. Funct. Mater., 2017, 28: 1705268.

[41] Kim H, Lee J. Highly sensitive and selective gas sensors using p-type oxide semiconductors: overview[J]. Sens. Actuators B Chem., 2014, 192: 607-627.

[42] Lou Z, Deng J, Wang L, et al. Toluene and ethanol sensing performances of pristine and PdO-decorated flower-like ZnO structures[J]. Sens. Actuators B Chem., 2013, 176: 323-329.

[43] Vallejos S, Stoycheva T, Umek P, et al. Au nanoparticle-functionalised WO_3 nanoneedles and their application in high sensitivity gas sensor devices[J]. Chem. Commun., 2011, 47: 565-567.

[44] Sun P, Zhou X, Wang C, et al. Hollow $SnO_2/\alpha-Fe_2O_3$ spheres with a double-shell structure for gas sensors[J]. J. Mater. Chem. A, 2014, 2: 1302-1308.

[45] Hong Y J, Yoon J, Lee J, et al. One-pot synthesis of Pd-loaded SnO_2 yolk-shell nanostructures for ultraselective methyl benzene sensors[J]. Chem. Eur. J., 2014, 20: 2737-2741.

[46] Kaneko H, Okamura T, Taimatsu H, et al. Performance of a miniature zirconia oxygen sensor with a Pd-PdO internal reference[J]. Sens. Actuators B Chem., 2005, 108: 331-334.

[47] Park J, Shen X, Wang G. Solvothermal synthesis and gas-sensing performance of Co_3O_4 hollow nanospheres[J]. Sens. Actuators B, 2009, 136: 494-498.

[48] Jinesh K B, Dam V A T, Swerts J, et al. Room-temperature CO_2 sensing using metal-insulator-semiconductor capacitors comprising atomic-layer-deposited La_2O_3 thin films [J]. Sens. Actuators B, 2011, 156: 276-282.

[49] Herrán J, Ga Mandayo G, Castaño E. Semiconducting $BaTiO_3$-CuO mixed oxide thin films for CO_2 detection[J]. Thin Solid Films, 2009, 517: 6192-6197.

[50] Puigcorbe J, Vila A, Cerda J, et al. Thermo-mechanical analysis of micro-drop coated gas sensors[J]. Sens. Actuators A, 2002, 97: 379-385.

[51] Gotz A, Gracia I, Cane C, et al. Thermal and mechanical aspects for designing micromachined low-power gas sensors[J]. J. Micromech. Microeng., 1997, 7: 247.

[52] Cavicchi R, Suehle J, Kreider K, et al. Growth of SnO_2 films on micromachined hotplates [J]. Appl. Phys. Lett., 1995, 66: 812-814.

[53] Kato Y, Myers R, Gossard A, et al. Observation of the spin hall effect in semiconductors [J]. Science, 2004, 306: 1910-1913.

[54] Moldovan C, Nedelcu O, Johander P, et al. Ceramic micro heater technology for gas sensors[J]. J. Inf. Sci. Technol., 2006, 10: 197-200.

[55] Gardner J, Pike A, de Rooij N, et al. Integrated array sensor for detecting organic solvents [J]. Sens. Actuators B Chem., 1995, 26: 135-139.

[56] Sberveglieri G, Hellmich W, MuÈller G. Silicon hotplates for metal oxide gas sensor elements[J]. Microsyst. Technol, 1997, 3: 183-190.

[57] Astie S, Gue A M, Scheid E, et al. Design of a low power SnO_2 gas sensor integrated on silicon oxynitride membrane[J]. Sens. Actuators B, 2000, 67: 84-88.

[58] Mele L, Santagata F, Iervolino E, et al. A molybdenum MEMS microhotplate for high-temperature operation[J]. Sens. Actuators A, 2012, 188: 173-180.

[59] Lee J, King W. Microcantilever hotplates: Design, fabrication, and characterization[J]. Sens. Actuators A, 2007, 136: 291-298.

[60] Spannhake J, Helwig A, Muller G, et al. SnO_2: Sb — A new material for high-temperature MEMS heater applications: Performance and limitations[J]. Sens. Actuators B, 2007, 124: 421-428.

[61] Kim H, Sigmund W. ZnO nanocrystals synthesized by physical vapor deposition[J]. Nanotechnology, 2004, 4: 275-278.

[62] Paul R, Das S N, Dalui S, et al. Synthesis of DLC films with different sp^2/sp^3 ratios and their hydrophobic behaviour[J]. J. Phys. D: Appl. Phys., 2008, 41: 055309.

[63] Kong X, Wang Z. Spontaneous polarization-induced nanohelixes, nanosprings, and nanorings of piezoelectric nanobelts[J]. Nano Lett., 2003, 3: 1625-1631.

[64] Wang X, Summers C, Wang Z. Mesoporous single-crystal ZnO nanowires epitaxially sheathed with Zn_2SiO_4[J]. Adv. Mater., 2004, 16: 1215-1218.

[65] Shi L, Hao Q, Yu C, et al. Thermal conductivities of individual tin dioxide nanobelts[J]. Appl. Phys. Lett., 2004, 84: 2638-2640.

[66] Wang Z. Zinc oxide nanostructures: growth, properties and applications[J]. J. Phys. Condens. Matter., 2004, 16: R829-R858.

[67] Bhattacharyya P, Basu P, Saha H, et al. Fast response methane sensor based on Pd(Ag)/ZnO/Zn MIM structure[J]. Sensors Lett., 2006, 4: 371.

[68] Basu P K, Bhattacharyya P, Saha N, et al. Methane sensing properties of platinum catalysed nano porous zinc oxide thin films derived by electrochemical anodization[J]. Sensors Lett., 2008, 6: 219.

[69] Fonash S, Roger J, Dupuy C. Ac equivalent circuits for MIM structures[J]. J. Appl. Phys., 1974, 45: 2907-2910.

[70] Suehle J, Cavicchi R, Gaitan M, et al. Tin oxide gas sensor fabricated using CMOS micro-hotplates and in-situ processing[J]. IEEE Electron Dev. Lett., 1993, 14: 118-120.

[71] Simon I, Bârsan N, Bauer M, et al. Micromachined metal oxide gas sensors: opportunities to improve sensor performance[J]. Sens. Actuators B Chem., 2001, 73: 1-26.

[72] Mitzner K, Sternhagen J, Galipeau D. Development of a micromachined hazardous gas sensor array[J]. Sens. Actuators B Chem., 2003, 93: 92-99.

[73] Puigcorbe J, Vila A, Cerda J, et al. Thermo-mechanical analysis of micro-drop coated gas sensors[J]. Sens. Actuators A. 2002, 97: 379-385.

[74] Chen Y, Xu P, Li X, et al. High-performance H_2 sensors with selectively hydrophobic micro-plate for self-aligned upload of Pd nanodots modified mesoporous In_2O_3 sensing-material[J]. Sens. Actuators B Chem., 2018, 267: 83-92.

[75] Zhu L, Yuan K, Yang J, et al. Hierarchical highly ordered SnO_2 nanobowl branched ZnO nanowires for ultrasensitive and selective hydrogen sulfide gas sensing[J]. Microsystems & Nanoengineering, 2020, 6: 30.

[76] Wu X, Zhu L, Sun J, et al. Pt nanoparticle-modified SnO_2-ZnO core-shell nanosheets on microelectromechanical systems for enhanced H_2S detection[J]. ACS Appl. Nano Mater., 2022, 5: 6627-6636.

[77] Zhu L, Yuan K, Li Z, et al. Highly sensitive and stable MEMS acetone sensors based on well-designed α-Fe_2O_3/C mesoporous nanorods[J]. J. Colloid Interface Sci., 2022, 622: 156.

[78] Karthigeyan A, Gupta R P, Scharnagl K, et al. A room temperature HSGFET ammonia sensor based on iridium oxide thin film[J]. Sens. Actuators B Chem., 2002, 85: 145-153.

[79] Das N, Kar J, Choi J, et al. Fabrication and characterization of ZnO single nanowire-based hydrogen sensor[J]. J. Phys. Chem. C, 2010, 114: 1689-1693.

[80] Li H, Yin Z, He Q, et al. Fabrication of single- and multilayer MoS_2 film-based field-effect transistors for sensing NO at room temperature[J]. Small, 2012, 8: 63-67.

[81] Kong J, Chapline M, Dai H. Functionalized carbon nanotubes for molecular hydrogen sensors[J]. Adv. Mater., 2001, 13: 1384-1386.

[82] Helmut G. Electron theory of thin-film gas sensors[J]. Sens. Actuators B Chem., 1993, 17: 47-60.

[83] Lundström I, Armgarth M, Spetz A, et al. Gas sensors based on catalytic metal-gate field-effect devices[J]. Sens. Actuators., 1986, 10: 399-421.

[84] Hong C, Kim J, Shin K, et al. Highly selective ZnO gas sensor based on MOSFET having a horizontal floating-gate[J]. Sens. Actuators B Chem., 2016, 232: 653-659.

[85] Sysoev V V, Goschnick J, Schneider T, et al. A gradient microarray electronic nose based on percolating SnO_2 nanowire sensing elements[J]. Nano Lett., 2007, 7: 3182-3188.

[86] Huang H, Liang B, Liu Z, et al. Metal oxidenanowire transistors[J]. J. Mater. Chem., 2012, 22: 13428.

[87] Kandasamy S, Wlodarski W, Holland A, et al. Electrical characterization and hydrogen gas sensing properties of a n-ZnO/p-SiC Pt-gate metal semiconductor field effect transistor[J]. Appl. Phys. Lett., 2007, 90: 064103.

[88] Huang J, Wan Q. Gas sensors based on semiconducting metal oxide one-dimensional nanostructures[J]. Sensors, 2009, 9: 9903-9924.

[89] Chen X, Wong C, Yuan C, et al. Nanowire-based gas sensors[J]. Sens. Actuators B Chem., 2013, 177: 178-195.

[90] Lao C, Liu J, Gao P, et al. ZnO nanobelt/nanowire Schottky diodes formed by dielectrophoresis alignment across Au electrodes[J]. Nano Lett., 2006, 6: 263-266.

[91] Li C, Zhang D, Liu X, et al. In_2O_3 nanowires as chemical sensors[J]. Appl. Phys. Lett., 2003, 82: 1613.

[92] Mubeen S, Moskovits M. Gate-tunable surface processes on a single-nanowire field-effect transistor[J]. Adv. Mater., 2011, 23: 2306.

[93] Dattoli E, Davydov A, Benkstein K. Tin oxidenanowire sensor with integrated temperature and gate control for multi-gas recognition[J]. Nanoscale, 2012, 4: 1760-1769.

[94] Yu C, Fu H, Yang S, et al. Controlled synthesis and enhanced gas sensing performance of zinc-doped indium oxide nanowires[J]. Nanomaterials, 2023, 13: 1170-1183.

[95] Fan Z, Lu J. Gate-refreshable nanowire chemical sensors[J]. Appl. Phys. Lett., 2005, 86: 123510.

[96] Yu H, Kang B, Pi U, et al. V_2O_5 nanowire-based nanoelectronic devices for helium

detection[J]. Appl. Phys. Lett., 2005, 86: 253102.

[97] Park J, Kim Y, Kim G, et al. Facile fabrication of SWCNT/SnO_2 nanowire heterojunction devices on flexible polyimide substrate[J]. Adv. Funct. Mater., 2011, 21: 4159.

[98] Gopel W, Schierbaum K. SnO_2 sensors: current status and future prospects[J]. Sens. Actuators B., 1995, 26: 1-12.

[99] Barsan N, Schweizer-Berberich M, Gopel W. Fundamental and practical aspects in the design of nanoscaled SnO_2 gas sensors: a status report[J]. Fresenius J. Anal. Chem., 1999, 365: 287.

[100] Chen P, Shen G, Zhou C. Chemical sensors and electronic noses based on 1-D metal oxide nanostructures[J]. IEEE Trans. Nanotechnol., 2008, 7: 668-682.

[101] Xu C, Tamaki J, Miura N, et al. Grain size effects on gas sensitivity of porous SnO_2-based elements[J]. Sens. Actuators B Chem., 1991, 3: 147-155.

[102] Freer E, Grachev O, Duan X, et al. High-yield self-limiting single-nanowire assembly with dielectrophoresis[J]. Nat. Nanotechnol., 2010, 5: 525-530.

[103] Zhang Y, Kolmakov A, Chretien S, et al. Control of catalytic reactions at the surface of a metal oxide nanowire by manipulating electron density inside it[J]. Nano Lett., 2004, 4: 403-407.

[104] Zhang J, Liu X, Wu S, et al. Au nanoparticle-decorated porous SnO_2 hollow spheres: a new model for a chemical sensor[J]. J. Mater. Chem. 2010, 20: 6453-6459.

[105] Kolmakov A, Chen X, Moskovits M, Functionalizing nanowires with catalytic nanoparticles for gas sensing application[J]. J. Nanosci. Nanotechnol. 2008, 8: 111-121.

[106] Moshfegh A. Nanoparticle catalysts[J]. J. Phys. D: Appl. Phys., 2009, 42: 233001.

[107] Chen P, Sukcharoenchoke S, Ryu K, et al. 2, 4, 6-Trinitrotoluene (TNT) chemical sensing based on aligned single-walled carbon nanotubes and ZnO nanowires[J]. Adv. Mater., 2010, 22: 1900.

[108] Fan Z, Lu G. Gate-refreshable nanowire chemical sensors[J]. Appl. Phys. Lett., 2005, 86: 123510.

[109] Cha S, Jang J, Choi Y, et al. High performance ZnO nanowire field effect transistor using self-aligned nanogap gate electrodes[J]. Appl. Phys. Lett., 2006, 89: 263102.

[110] Zeng Z, Wang K, Zhang Z, et al. The detection of H_2S at room temperature by using individual indium oxide nanowire transistors[J]. Nanotechnology, 2009, 20: 045503.

[111] Shen G, Liang B, Wang X, et al. Indium oxide nanospirals made of kinked nanowires[J]. ACS Nano, 2011, 5: 2155-2161.

第9章
气体传感器集成技术：电子鼻

金属氧化物半导体(MOS)是一类广泛使用的气体传感材料,对许多气体都可以产生较好的响应。然而MOS传感器大多选择性较差,即对多种气体均可产生响应,使得其在面对实际的复杂气氛环境时,无法单将所需检测气体的信号从干扰气体中分离出来,从而导致误报。这种交叉敏感性问题已成为MOS传感器实际应用中所面临的主要问题之一[1-2]。鉴于MOS传感器具有灵敏度高、成本低的特点,在此基础上若能进一步提高MOS传感器的选择性,将会对传感器工业产生重大实际帮助。研究表明,提高MOS传感器选择性的方法可归类为两种：一种是通过掺杂或改变微结构等措施来降低MOS敏感材料本身的交叉敏感性[3-6];另一种方法即构建具有不同交叉灵敏度分布的多个MOS传感器阵列,我们将在本章对这一方法进行详细介绍。与单一传感器不同的是,传感器阵列给出的是一系列响应数据,而非单个响应值,并通过引入模式算法对数据组进行整体分析,最后输出总体结果,类似于动物的嗅觉系统。电子鼻技术涉及材料、传感、微器件工艺、计算机、应用数学等多个领域,具有重要的理论意义和广阔的应用前景。

9.1 电子鼻：传感器阵列

1964年,Wilkens和Hartman[7]通过使用各种电极组合构建了一种多传感器装置,首先提出了一些不同的电极涂层物质能够对不同混合物产生不同响应的可能性,奠定了传感器阵列用于混合物检测的基础。首例气体分类用的智能化学传感器阵列系统于1982年由Persaud和Dodd研制成功,这一简单的系统可以分辨桉树脑、玫瑰油、丁香油等挥发性化学物质的气味[8]。1987年,Gardner重点提出了模式识别的概念,主张使用模式识别技术来区分传感器阵列的输出信号。1994年,Gardner和Bartlett提出了"电子鼻"这一术语,即一种包含一系列电子化学传感器的仪器,具有部分特异性和适当的模式识别系统,能够识别简单或复杂的气味[9]。自此之后,传感器阵列的设计和工艺都取得了突飞猛进的发展,1993年出现了第一批商用设备(Alpha MOS)。另外,传感器本身也不再仅限于电阻型的

MOS 传感器阵列,而是引入了其他多种类型的传感器,包括石英晶体微量天平(QCM)传感器、表面声波(SAW)传感器、导电聚合物(CP)传感器等[2,10-11]。表 9.1 对一些商品化的电子鼻产品及其传感技术和制造商进行了总结。

表 9.1 一些商品化的电子鼻产品及其传感技术和制造商[10]

制 造 商	产品名称	传 感 技 术
Airsense Analytics	i-Pen, PEN2, PEN3	MOS
	GDA 2	MOS, EC, IMS, PID
Alpha MOS	FOX 2000, 3000, 4000	MOS
	RQ Box, Prometheus	MOS, EC, PID, MS
Applied Sensor	Air quality module	MOS
Chemsensing	ChemSensing Sensor array	比色光学传感
CogniScent Inc.	ScenTrak	染料聚合物传感器
CSIRO	Cybernose	
Dr. Födisch AG	OMD 98, 1.10	MOS
Forschungszentrum	SAGAS	SAW
Gerstel GmbH Co.	QSC	MOS
GSG Mess- und	MOSES II	
Illumina Inc.	oNose	荧光光学传感
Microsensor Systems Inc	Hazmatcad, Fuel Sniffer, SAW MiniCAD mk II	SAW
Osmetech Plc	Aromascan A32S	导电高分子传感器
Sacmi	EOS 835, Ambiente	
Scensive Technol.	Bloodhound ST214	导电高分子传感器
Smiths Group Plc	Cyranose 320	碳黑-高分子传感器
Sysca AG	Artinose	MOS
Technobiochip	LibraNose 2.1	QMB
RST Rostock	FF2, GFD1	MOS, QMB, SAW

除电子鼻外,其他分析仪器如火焰离子化检测器(FID)或气相色谱-质谱(GC-MS)也可用于检测气体混合物的组成成分,这些分析手段大多根据混合气体样品中各个组分的化学特征进行分子级别的逐一分析和识别,因此具有较高准确度,但同时成本较高、检测时间长。与此不同的是,电子鼻在检测过程中不会单独识别每一组分,而是将样品作为一个整体,得到一系列平行的响应数据,并根据算法进行特征归类,从而得到样品的整体信息。相较于其他传统分析技术,电子鼻

的检测具有响应快、操作简便、成本低、检测范围广等优势,但定量分析的精确度低于 GC-MS。

Gardner 和 Bartlett 所提出的电子鼻是一个基于传感器阵列的复杂样品分析系统,包括以下基本单元:① 样品收集系统,用于将气体分子从被检测物处传送到传感器阵列;② 传感器阵列,用于对样品气体产生化学响应信号,须放在恒温恒湿的腔室中;③ 电子晶体管,用于将化学信号转换成放大和调节的电信号;④ 数字转换器,用于将电信号转换为数字信号;⑤ 计算机微处理器,用于对数字信号进行统计分析并显示输出。电子鼻的检测流程可以用图 9.1 简要概括。

图 9.1　电子鼻检测流程示意图

为达到最佳检测效果,电子鼻系统的设计需要考虑许多因素,其中最关键的一点是阵列中传感器的选择。一般来说,阵列中的每个传感器应具有不同且差异较大的气体选择性,根据应用需求为目标分析物提供不同的响应特征曲线。此外,为了有效识别或分类气体混合物,传感器还应满足其他要求,包括:① 适当的灵敏度,可以最大限度地提高仪器的整体灵敏度;② 快速响应和恢复;③ 良好的重复性;④ 对温度和湿度的高稳定性。缺乏上述任何条件均可能导致传感器误报,即未能检测到目标气体(假阴性结果)或过高估计分析物浓度(假阳性结果)。因此,应根据具体情况选择适当的传感器组成阵列,并定期校准传感器,才能保证电子鼻设备输出数据的有效性和准确性。

传感器集成一方面可以通过采用具有不同选择性特征的多个传感器来进行混合物分析,另一方面还可以扩展检测浓度范围。MOS 传感器大多仅在一定被测物浓度范围内才具有较好的灵敏度,气体浓度低于检测下限时无法给出响应信号,高于检测上限时响应区别度低、易达到饱和。因此,将具有相似选择性但测量浓度范围不同的传感器组成阵列,基于其互补的测量范围,可以扩展检测浓度上下限,根据实际需要,检测范围甚至可以达到 ppb 级到百分之百,同时确保高精度。

商用 TGS 系列气体传感器(Figaro Engineering,JP)是最广泛用于电子鼻的传感器之一,多种传感器组合适用于不同应用场所。Cyranose 320(Cyrano Science,Pasadena,CA,USA)是一种便携式电子鼻系统,其核心阵列由 32 个独立的聚合物传感器与碳黑复合材料混合而成。Airsense PEN2 和 PEN3(Airsense Analytics GmbH,Schwerin,Germany)电子鼻包含非常小巧便携的 10 个金属氧化物半导体(MOS)气体传感器阵列,并带有一个小体积测量室,可以与吸附剂捕集装置或顶空自动进样器连接,用于实验室分析。

9.2 信号-数据分析技术

电子鼻检测输出的信号来自阵列中每个传感器的响应信号,通常以一组数据的形式输出,或排列成图样,并对这个整体进行分析,无法从每个传感器的单个信号中分析出有效信息。因此,电子鼻的输出信号必须使用适当的模式识别(PARC)技术进行处理。数据分析方法有许多种,其主要目的是通过减少维数来归纳并给出数据集组件之间的共性和差异,这个分析过程主要包括:数据采集和预处理、特征提取、分类和决策(见图9.2)[12]。首先,通过采集器收集传感数据并将其转换成电信号,输出为图样空间中的 n 维向量,这一过程涉及电信号与数字信号的转化。接着进一步从图样空间中提取特征,并从图样空间转换为特征空间,以减少数据的维度。然后根据相似程度(通常通过图样中定义的距离函数来测量)对这些特征进行选择和分类特征,最后输出结果。

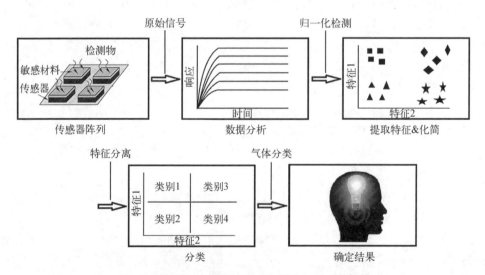

图9.2 电子鼻中的信号处理和模式识别过程[12]

在气敏检测过程中,由于待测气体分子性质不同,气体分子的吸附能、催化反应能垒以及脱附能往往不尽相同。因此,在宏观表现中,同一气敏器件对不同气体分子的响应值、响应时间、恢复时间、响应曲线的斜率、恢复曲线的斜率是不同的,换言之,气敏器件的响应信号中包含了气敏过程的指纹信息。如何对这些指纹信息进行提取,直接关系到模式识别的准确率。对此,Ogbeide 等[13]提出了基于动态响应曲线的特征提取办法,实现了 NO_2、NH_3 和水蒸气的分类及浓度检测,如图9.3所示。除采集气敏信号的基本信息外,还包括响应值(S_{max})和响应时间(T_{res}),

Ogbeide 运用拟合式(9.1)对响应和恢复曲线进行了拟合。其中 $\gamma_{res(rec)}$ 为响应/恢复过程的偏置参数，$C_{res(rec)}$ 为响应/恢复过程的常量拟合参数，$\tau_{res(rec)}$ 为响应/恢复过程的时间拟合常数。除此之外，研究还提取了响应和恢复时间的积分面积 $A_{res(rec)}$，通过共计 10 个特征参数实现了精确识别。

$$拟合曲线：\gamma_{res(rec)} + C_{res(rec)} \exp^{-t/\tau_{res(rec)}} \tag{9.1}$$

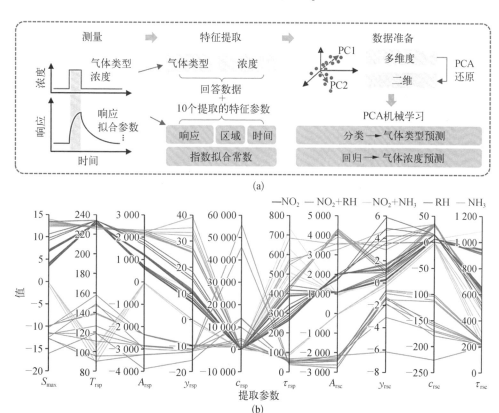

图 9.3　气敏过程的指纹信息[13]
(a) 特征提取示意图；(b) 提取特征的平行坐标图

商业上可用的分析技术分为三大类：图形分析、多维数据分析(MDA)和网络分析，如图 9.4 所示。根据在学习阶段是否提供参照，可以进一步分为监督技术和无监督技术。无监督技术通常用于探索性数据分析，无须先前已知样品的参考，旨在区分不同的样品而非对样品组分进行逐一识别。监督技术旨在根据预先开发的数据库进行气体识别，该数据库包含已知样品(组)的性质或特征。在实际应用中应根据对象及其变量的数量、问题的复杂性以及软件的计算能力选择适当的统计分析方法。

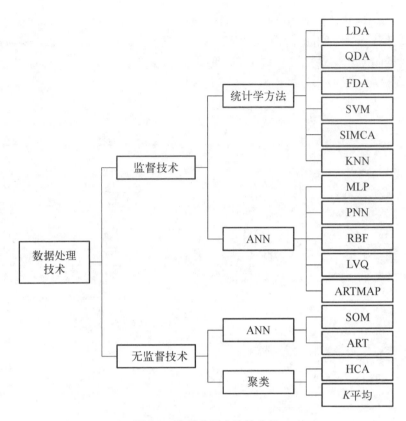

图 9.4 数据分析方法的分组

ANN：人工神经网络；LDA：线性判别分析；QDA：二次判别分析；FDA：判别因子分析；SVM：支持向量机；SIMCA：类比的软独立建模；KNN：K 近邻算法；MLP：多层感知器；PNN：概率神经网络；RBF：径向基函数；LVQ：学习矢量量化；ARTMAP：自适应共振理论图；SOM：自组织地图；ART：自适应共振理论；HCA：分层聚类分析

以树状图、极坐标图和层次聚类分析(HCA)为代表的图形分析方法是一类最为直观的可视化分析技术。图 9.5 给出了典型的极坐标图和树状图分析示意图。该方法简明易懂，可将未知分析物与单个指定参考物进行直观比较，适用于简单的测试体系。当需要使用多个参照样本数据或具有多个特征指标时，则可能需要其他更为复杂的分析方法。

在多维数据分析(MDA)技术中，最常使用的方法为主成分分析(PCA)、判别函数分析(DFA)和聚类分析(CA)[13]。MDA 的实质就是在处理多变量问题时降低变量维度，其是解决传感器重叠灵敏度最有效的数据分析方法。PCA 是一种应用最广泛的无监督化学计量学方法，它可以通过降低多变量问题中数据组的维数来显示模式中的所有信息[14-16]。PCA 有助于观察不同样本之间的差异，从而找出对差异贡献最大的变量，以初步评估类别之间的相似性。

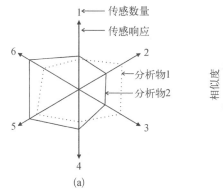
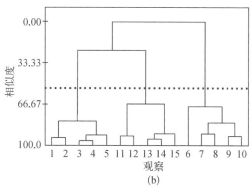

图 9.5 极坐标图和树状图分析示意图[13]

(a) 传感数据的典型极坐标图；(b) 说明 HCA 的树状图

DFA 是一种概率相关的监督算法,已被广泛使用在多种应用场所中,并已被证明有效性。DFA 包括两种类型,分别为线性判别分析(LDA)和二次判别分析(QDA)。根据 LDA 归纳出的类别呈均匀的正态分布,为了提高类别之间的分辨率,LDA 通常将类别之间的差异最大化,同时缩小同类别内样品之间的差异。QDA 通常比 LDA 具有更高的数据匹配度,但同时也涉及更多估算参数[13]。

CA 旨在根据数据集内样本点之间的相似性或距离将数据分成特定的组[17-18]。随着分类阈值逐渐降低,更多样本被合并归为同一类别,聚合成包含更多差异的更大的簇。分层聚类方法的结果通常通过树状图显示,使用合并的组内协方差矩阵计算从每个单独数据点到每个聚类的中心(质心)的距离,因此可以通过观察簇与簇之间的距离来确定它们之间的相似程度。

在网络分析方法中,最为常用的是人工神经网络(ANN)分析,这种算法是对人脑组织结构和运行机制的某种抽象、简化和模拟,由一组并行运行的互连处理算法组成,是商业化电子鼻统计软件中最为先进的一种算法[19]。该技术具有很大的潜力,可以处理基于提供非线性响应传感器的电子鼻产生的信号,分析结果通常以样本与参照数据库中已知气体图样匹配程度(百分比)的形式呈现。ANN 可分为多层感知和单层感知,每一层都包含若干神经元,各神经元之间用带可变权重的有向弧连接,网络通过对已知信息的反复学习训练,逐步调整改变神经元连接权重的方法,达到处理信息、模拟输入输出之间关系的目的。在所有人工神经网络(ANN)结构中,多层前向神经元网络(也称多层感知器,MLP)是目前最为常用的网络结构。

9.3 温度梯度法

MOS 气体传感器的一个共同特性是传感性能极易受工作温度影响,因为温度变化会导致 MOS 敏感材料的物理和化学性质、分析物吸附和解吸行为，甚至表面反应过程产生巨大差异。一方面，对一般的传感器来说，温度带来的影响是需要尽量避免的；但另一方面，可以将这种独特的性质加以利用，即通过产生工作温度梯度，使单一传感器也可以产生正交的数据组，从而将分析正交性引入传感系统。

Sysoev 等[20]报道了一种使用 KAMINA 平台的 SnO_2 纳米线的梯度微阵列电子鼻。为了使传感器的响应有所区别，研究通过在基板背面引入四个曲形 Pt 加热器使得薄膜表面温度产生横向变化区别（见图 9.6）。结果表明，尽管传感器阵列的等温（无热梯度）模式已经足以区分各种气体的信号，但当另外施加温度梯度时，识别显著增加，辨识度（平均马哈拉诺比斯距离）甚至超过等温时的两倍。

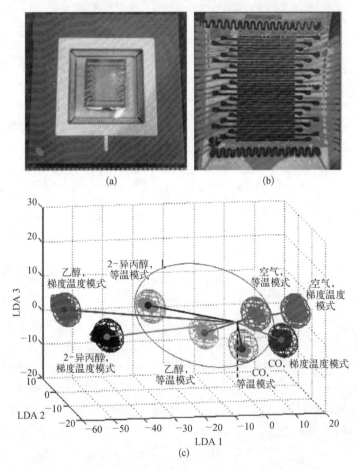

图 9.6 使用 KAMINA 平台的 SnO_2 纳米线的梯度微阵列电子鼻（后附彩图）[20]

(a) KAMINA 芯片 SnO_2 纳米线传感器；(b) 沿着电极阵列施加温度梯度，520（绿色区域）～600K（红色区域）芯片的 IR 图像；(c) 基于 SnO_2 纳米线的微阵列在暴露于目标气体（2～10ppm）时的电导率模式的 LDA，其中分类区域对应于 0.999 9 置信水平的正态数据分布，微阵列分别在 580K（椭圆内的恒定 T 区域）均匀加热和 520～600K（梯度 T 区域）的温度梯度下操作除空间温度梯度外,工作温度的时间变化也可以引入正交信号。Benkstein 等[21]使用微尺度电导平台整合了三种不同形貌 WO_3 材料的传感器(薄膜、纳米线 50、纳米线 100),该器件使用多晶硅加热器,通过动态脉冲温度程序进行温度控制,最短动态时间间隔可达毫秒级别。在使用脉冲程序升温,将工作温度以 30℃ 为间隔从 60℃ 升至 480℃(温度曲线如图 9.7 所示)的过程中,三种传感器在不同温度下对分析物表现出了不同的响应分布[见图 9.8(a)]。如图 9.8(b)所示,PCA 结果表明,温度梯度可以沿不同方向引入响应差异(箭头方向指示传感器温度从低到高)。图 9.8(c)中显示的正交性和相似性研究表明,随温度升高,三种传感器对不同气体有着不同的相关区间(对于 NO_2 最明显)。除使用多个传感器外,快速脉冲升温引入了另一个响应相关变量,这一新正交信号的引入更有利于观测传感器之间的响应差异。

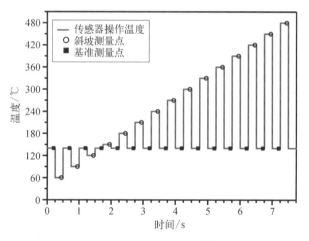

图 9.7　动态温度程序[21]

在每个空心圆(顶部)和正方形(基部)上进行电导测量

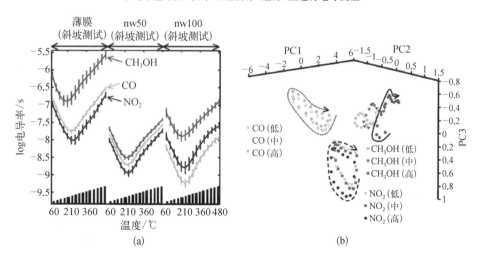

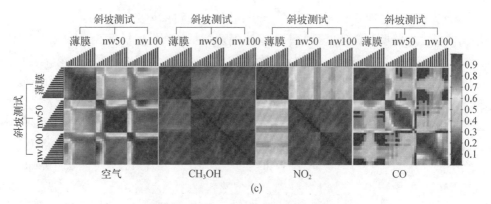

图 9.8 WO₃ 传感器传感数据分析[21]

(a) 三种不同形貌 WO₃ 传感器对三种气态分析物平均响应的周期测量,气体分别为 CH₃OH(40 μmol/mol)、CO(200 μmol/mol)和 NO₂(1 μmol/mol);(b) 六维传感器响应的相应投影,包括不同的分析物浓度,沿前三个主成分(捕获方差 = 97.12%),箭头方向指示温度增加的趋势;(c) 使用程序升温(60~480℃)进行电导测量的统计分析结果

9.4 电子鼻的集成制造

电子鼻通常由多种金属氧化物半导体气体传感器复合而成,因此要求其制造技术应尽可能具有普适性以满足多种金属氧化物的制造要求。除此之外,传感器还非常需要高灵敏度、优异的长期稳定性和高选择性,以及便捷的大规模生产技术,这些技术应适用于微电子产业,能够制造微小的、廉价的、可重复的,以及功能齐全的传感设备。化学气相沉积(CVD)、原子层沉积(ALD)和斜角沉积(OAD)技术在这方面具有很高的应用价值。其中,Hwang 等[22]通过 OAD 技术制备了近单晶的 TiO₂ 螺旋阵列,与 CVD 技术不同的是,OAD 技术制备的规则图案具有较高的比表面积,同时其近乎单晶的稳定结构有效保证了长期稳定性。在斜角沉积过程中,高度倾斜的蒸汽与受控的衬底旋转相结合,使其可以创建各种三维纳米结构阵列,如螺旋、斜杆、垂直柱、"之"字形和方形螺旋。研究进一步将 OAD 技术拓展到其他金属氧化物,包括氧化锡和氧化钨,构建了六种不同结构材料的气敏层,实现了对 H₂、CO 和 NO₂ 的出色灵敏度和选择性。

9.5 电子鼻的应用

近年来,市场上已经出现了许多商品化的电子鼻系统,并且这些电子鼻已经被应用于包括食品加工、环境监测、农业生产、工业制造、微生物病原体检测和医学诊断等各种领域。尽管有着广泛的应用范围,但不同应用场所的电子鼻都不尽相同,每个电子鼻系统都需要针对特定需求来专门进行设计制造。电子鼻的结构设计、阵列中的传感器以及数据分析模型都取决于目标分析物气体的组成和操作环境的要求,使用前须根据具体情况对系统进行评估。评估所涉及的关键考虑因素包括:样品中可能存在的目标气体种类和浓度范围、针对这些气体的各个传感器阵列的选择性(确保高准确度的同时减少冗余传感器)、传感器精度、重现性、响应-恢复速度、长期稳定性和整体性能。

9.5.1 电子鼻在食品工业中的应用

电子鼻系统应用市场中所占比例最大的是食品工业,其中电子鼻主要通过监测相关气体以进行食品质量控制。由于许多食品会释放特殊的挥发性混合气体,通过对这些气体的成分、浓度进行精确分析,可以快速地得到关于其品质和味道的许多特征信息。作为一种有效的食品质量评估工具,电子鼻系统涵盖了食品生产链中的许多环节:从原料到产品的质量检测、水果的成熟度监测、包装中的污染检测、保质期测定、水产品加工检验、发酵过程检验、生物降解检测的早期阶段、酒精饮料分类等。近年来,越来越多的科学研究集中在将电子鼻应用于食品工业,表9.2列举了一些代表性的研究工作。大量研究表明,许多挥发性有机物由水果和蔬菜组织中的酶天然形成,水果散发的气味与其成熟度直接相关。对于奶酪来说,发酵过程取决于细菌的生长、脂质降解和氧化以及蛋白质水解,其品质、味道与成熟过程密切相关。除此之外,肉类也是一类最容易由电子鼻监测的产品,其中大多数都与品质控制相关联,以监控产量、保质期、是否腐败等。

表9.2 电子鼻在食品工业领域的应用

样品	分析目标	使用传感器	数据分析方法	参考文献
花生	检测曲霉菌生长程度	Fox 3000 (Alpha MOS, Toulouse, France)	PCA, PC-LDA	[25]
苹果汁	检查品质	PEN3 (Airsense Analytics GmbH, Schwerin, Germany)	PCA	[26]
葵花籽油	监测油炸使用时间	Fox 4000 (Alpha MOS, Toulouse, France)	Fuzzy logic analysis	[27]
酱油	检查是否符合清真标准	Smart Nose 300 (Smart Nose Inc., Marin-Epagnier, Switzerland)	DFA	[28]
水稻	质量评估	TGS 880, TGS 822, TGS 826, TGS 2602, TGS 2620, TGS 2600 (Figaro Engineering, Japan)	Multiple linear regression (MLR)	[29]

续表

样品	分析目标	使用传感器	数据分析方法	参考文献
肉类	评估细菌含量	TGS 821, TGS 822, TGS 825, TGS 826, TGS 2600, TGS 2602, TGS 2610, TGS 2620 (Figaro Engineering, Japan)	PCA, back propagation neural network (BPNN)	[30]
肉类	黄曲霉毒素A菌株的快速检测	ISE Nose 2000 (SoaTec S.r.l., Parma, Italy)	DFA	[31]
人参	确定化学成分	16 TGS sensors (Figaro Engineering, Japan)	Support vector machine (SVM)	[32]
蜂蜜	分析来源植物以及是否掺假	Heracles (Alpha Mos, Toulouse, France) equipped with a GC	PCA, Partial least squares discriminant analysis (PLS-DA)	[33]
蜂蜜	分析来源植物、质量评价	Fox 4000 (ALPHA MOS, Toulouse, France)	PCA, DFA	[34]
葡萄酒	评估氧气暴露水平和多酚含量	14 MOS sensors	PLS-DA	[35]
咖啡	质量监测	SP-12A, SP-31, SP-AQ3, ST-31 (FIS Inc.); TGS-813, TGS-842, TGS-823, TGS-800 (Figaro Engineering, Japan)	PCA, ANN	[36]

近期，华中科技大学刘欢教授课题组通过合作研发智能电子鼻新技术，从气敏材料与识别算法两个层面联合攻关，提出了一种基于半导体传感器气敏响应全过程特征的嗅觉算法(all-feature olfactory algorithm，AFOA)，构建出高灵敏度、高可靠性、便携式智能电子鼻，成功提高了对复杂气味的识别准确率[23]。如图9.9所示，该团队利用自主研发的多种MOS气敏材料作为人工气味受体，通过MOS气体传感器单元模拟不同类型的嗅觉感受器细胞，将气-固界面反应引起的电荷转移转变为电阻值变化输出。其研发的电子鼻采用六个非特异性的MOS气体传感器形成阵列，为后续的识别算法提供了更多可学习的特征。这些MOS气体传感器采用的气敏材料分别为SnO_2量子点、SnO_2纳米线、SnO_2纳米颗粒、In_2O_3量子点、NiO纳米颗粒和WO_3量子点。由于MOS气体传感器对多种气味具有交叉敏感性，因此先进的气味识别算法对于电子鼻的性能提升和应用拓展至关重要。受人类嗅觉的启发，该团队利用人工神经网络模拟嗅球、大脑嗅觉皮层以及它们之间的复杂连接，自主研发的AFOA实现了MOS气体传感器阵列与多种气味分子响应恢复过程中完整信息的提取与分析。基于这一技术，集成MOS气体传感器阵列

的智能电子鼻可以高准确度(94.1%)、快速(2min 以内)识别出五种气味相近的中国白酒(CGJ,BYBJX,BYBNX,STJ,MTWZ)。该项研究工作展示了一种基于六个 MOS 气体传感器的智能电子鼻,内建了一种基于深度学习的气味识别算法：AFOA,能够在复杂环境中识别目标气味;同时,还展现了低成本、便携式、可进化的电子鼻协同设计和制造方案,应用前景广阔。

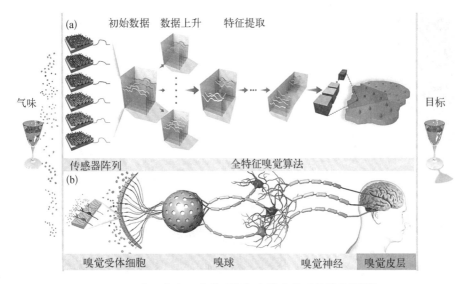

图 9.9　电子鼻人工嗅觉系统和人类嗅觉系统的比较[23]

(a) 电子鼻人工嗅觉系统；(b) 人类嗅觉系统

电子鼻等人工感官解决了单一气体传感器选择性差的问题,已引起人们对监测有害气体的浓厚研究兴趣。Ren 等[24]研究了镓对 In_2O_3 纳米管(NT)的掺杂效应,并报道了用于检测三甲胺(TMA)的四组分传感器阵列(见图 9.10)。所有镓掺杂合金化 In_2O_3($Ga-In_2O_3$)传感器在 240℃的工作温度下都对 TMA 显示出增强的灵敏度和选择性,其中 5%① Ga 掺杂合金化 In_2O_3 在 0.5～100 ppm 范围内显示出最高的响应,最低检测限为 13.83 ppb。基于气体传感特性,研究人员制作了一个四组分传感器阵列,该阵列在可变气体背景下具有独特的响应模式。利用气敏数据训练误差逆传播神经网络(BPNN)、径向基函数神经网络(RBFNN)和基于主成分分析的线性回归(PCA - LR),可高精度区分不同气体,并预测不同气体和气体混合物中目标气体的浓度。此外,研究使用该传感器阵列对六种气体(三种单一气体和三种二元气体混合物)的分类以及在不同浓度 TMA 和丙酮存在下的 TMA 浓度进行预测,准确度分别达到 92.85% 和 99.14%。

---

① 5%指摩尔分数。

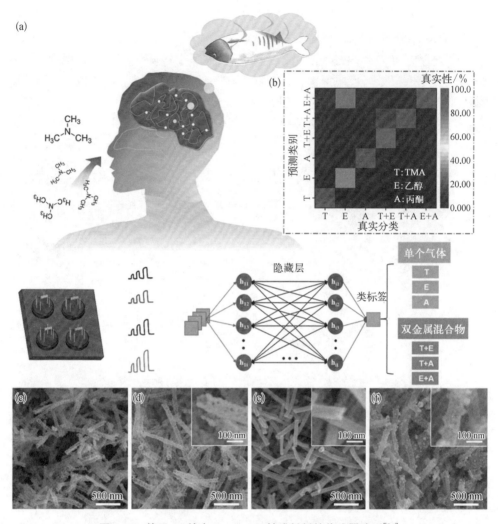

图 9.10 基于 Ga 掺杂 In_2O_3 NT 敏感材料的传感器阵列[24]

(a) 人工嗅觉系统的感知路径和用于实验的传感器阵列示意图；(b) 基于 BPNN 模型的六种气体分类结果；(c)~(f) 纯 In_2O_3 NT、1%Ga-In_2O_3 NTs、5%Ga-In_2O_3 NTs 和 10%Ga-In_2O_3 NTs 的 SEM 图，插图显示放大的图像

9.5.2 电子鼻在环境监测中的应用

空气质量问题逐渐成为人们日常生活中最受关注的焦点之一。电子鼻系统可应用在环境监测的各个方面，如室内空气质量监测、工业生产过程监控，以及汽车尾气排放检测等。由于环境氛围较为复杂，电子鼻是目前实时分类和定量检测的唯一方法。

微生物挥发性有机物（MVOC）是最常见的室内空气污染物之一。研究表明，霉菌可产生各种挥发性有机物，包括醇类、酮类、酯类和硫化合物[37]。挥发性产物的类别高度取决于微生物本身，甚至可以根据 MVOC 将真菌在物种水平上进行分类。Kuske[37] 等报道了一个由 15 个 MOS 传感器组成的电子鼻系统，以用于检测和分类细菌和真菌，该电子鼻能够区分 5 种真菌，准确率高达 96%。

Helli 等[38]使用若干 MOS 传感器组成阵列，进行了 H_2S 和 NO_2 的检测（见图 9.11）。在干燥的气氛中，传感器阵列可以根据 FDA 算法正确识别分析物组成成分[见图 9.11(a)]；然而高湿度和二氧化碳会对传感器检测精度造成很大影响[见图 9.11(b)]。这种现象反映了大多数 MOS 传感器的最大缺陷，即很容易受到环境条件变化的影响，尤其是湿度和温度。因此，降低 MOS 传感器对环境条件的依赖性已成为一个重要的发展方向。

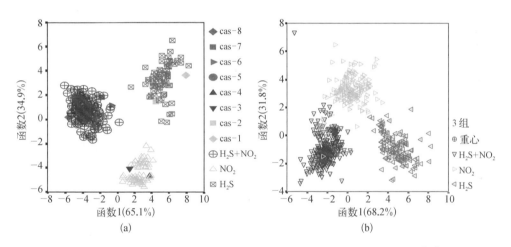

图 9.11　传感器阵列在(a)干燥组和(b)潮湿组中的气体组分分析结果[38]

除空气监测外，电子鼻还可以通过检测样品中的挥发物来分析水和土壤样品[39-40]。对液、固态样品的分析通常通过取液、固上部气体样品进行分析来实现[41]。特别地，针对极端环境无视觉输入情况下对受困人员的识别与救援需求，中国科学院上海微系统与信息技术研究所陶虎团队受自然界星鼻鼹鼠"触嗅融合"感知启发，将 MEMS 嗅觉、触觉柔性传感器阵列与多模态机器学习算法融合，构建了仿星鼻鼹鼠触嗅一体智能机械手[42]，如图 9.12 所示。得益于硅基 MEMS 气体传感器（灵敏度超越人类 1 个数量级）、压力传感器（探测限超越人类 1 个数量级）的优异性能，该机械手手指触摸物体后可准确获取其局部微形貌、材质硬度和整体轮廓等关键特征，掌心可同步嗅出物体"指纹"气味，进一步通过仿生触嗅联觉（BOT）机器学习神经网络实时处理，最终完成识别人体、确认部位、判断掩埋状态、移开障碍物、闭环救援。此外,该项研究还与应急管理部上海消防研究所合作,通过到一线消防救援单位实地调研,真实还原构建了人体被瓦砾石堆覆盖的掩埋场景,并在此环境下对包括人体在内的11种典型物体进行识别,触嗅联觉识别准确率达96.9%,较单一感觉提升了15%。相较之前报道的单一触觉(548个传感器)感知研究,该机械手通过触(70个)、嗅(6个)联觉,仅使用1/7数量的传感器,便达到了更理想的识别目的。并且缩小传感器规模和减少样本量后的传感器更适合复杂环境、资源有限条件下的快速反应和应用。除此之外,面对实际救援中经常遇见的存在干扰气体或器件部分损坏等情况,通过多模态感知的互补和神经网络的快速调节,该系统仍保持良好的准确率(>80%)。

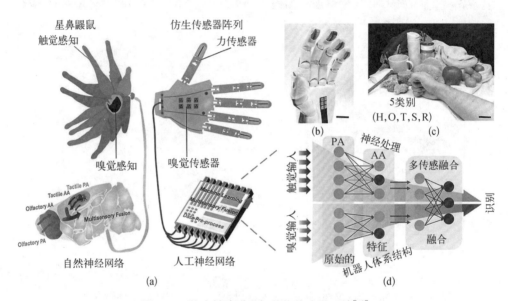

图9.12 仿生触嗅联觉相关智能感觉系统[42]

9.5.3 电子鼻在呼吸系统疾病诊断中的应用

自古代起便有记录表明,可以通过呼出气体的气味进行疾病诊断。到目前为止,已经在人类呼出气体中鉴定出超过3000种挥发性有机物,其中许多与某些疾病直接相关。这些挥发性有机物产生于人体的代谢过程,以及呼吸道或身体其他部位的疾病过程[43]。某些气体的浓度变化通常代表相应的病理变化,例如,如表9.3所示,由于碳水化合物代谢异常,糖尿病患者呼出的气体中通常含有更高浓度的丙酮[44-46];一氧化氮含量是表征哮喘和COPD的一项重要指标[47-49];癸烷、4-甲基辛烷、十一烷、醛、苯及其衍生物、1-丁醇可用于肺癌检测[50-51,17];羰基硫化物、二硫化碳和异戊二烯与肝脏疾病有关[52-53]。

表 9.3　电子鼻在部分疾病初期检测中的应用

疾　病	相　关　气　体	所使用传感器
哮喘	NH_3、NO —	SnO_2 和 WO_3 传感器[60] Cyranose 320[61]
慢性阻塞性肺疾病(COPD)	— — —	4 个 MOS 传感器(Figaro，Japan)[62] 7 个 QMB 传感器[63] Cyranose 320[64]
肺癌	癸烷、乙苯、丙基苯等 — 甲醛 癸烷、4-甲基辛烷、十一烷、乙醛、苯系物、1-丁醇等	Cyranose 320[65] Cyranose 320[66] 4 个进行不同掺杂的 SnO_2 传感器[59] 8 个 TGS-type MOS 传感器、4 个电化学传感器、催化燃烧传感器[67]
慢性肝病	—	7 个 QMB 传感器[68]
呼吸道感染	—	DiagNose, C-it, Zutphen, Netherlands[69]
糖尿病	丙酮	Cyranose 320[70] 6 个 TGS-type MOS 传感器、3 个温度调控传感器、CO_2 传感器、温度-湿度传感器[71] 12 个 TGS-type MOS 传感器[72]

气相色谱-质谱(GC-MS)被普遍认为是呼出气体分析的标准测试手段，因为它可以进行大范围、高准确度的 VOC 识别，有助于未知的病理生理过程研究[43,54-55]。然而在目前的临床检测中，GC-MS 的使用程度仍然有限，因为它价格昂贵、操作复杂且耗时，尤其不适用于早期疾病的筛查。因此，实际应用中对快速、准确、非侵入性和低成本的早期诊断手段仍有着迫切需求。作为分析复杂气体的强大工具，电子鼻系统在检测过程中无须对混合气体的成分进行逐一分析，而是通过识别混合物的指纹特征进行快速分类，从而确保了快速、低成本检测，使其成为临时诊断的有效手法。许多医学研究人员在过去十余年中发表了大量实验成果，以证明使用电子鼻诊断人类疾病的可行性，并通过检测 VOC 鉴定了许多不同的病原微生物[10]，如 Gibson 等[56]首先使用传感器阵列对 12 种不同的细菌和人类致病酵母进行分类，准确率高达 93.4%。

波兰的研究人员通过利用商用 VOC 气体传感器开发的电子鼻装置，展示了当地医院对新型冠状病毒肺炎(COVID-19)的检测结果[57]。如图 9.13 所示，研究人员考虑了研究中发现的影响检测结果的技术问题，并相信此次研究有助于推进新技术的开发,以限制 COVID-19 和类似病毒感染的传播。实验在当地医院专门治疗 COVID-19 感染者的病房进行,呼吸样本是在一个住院患者组和一个对照组中采集的。他们共调查了 56 份呼吸样本(33 份 COVID-19 重症患者呼吸样本、17 份健康对照组呼吸样本和 6 份环境空气样本),并覆盖了不同年龄、性别和有基础疾病的患者。研究结果显示老年人群的检出率更高。此外,他们注意到所分析的呼吸样本的湿度存在差异,COVID-19 感染者呼吸样本的平均湿度高于健康志愿者。在研究人员的探索性研究中,他们专注于可能的修正和针对波动环境因素的抗干扰力,以提高检测精度,最终确定了最可靠的气体传感器参数,这些参数可以应用于测试过程,大大缩短了检测时间,并进一步加强了实际检测对选择性好、灵敏度高的传感器的需求。此外,COVID-19 感染的单独 VOC 标志物的浓度为几十 ppb 水平,尽管如此,呼出气体中大量 VOC 的混合物对应用商业气体传感器检测 COVID-19 感染者仍有很大影响。

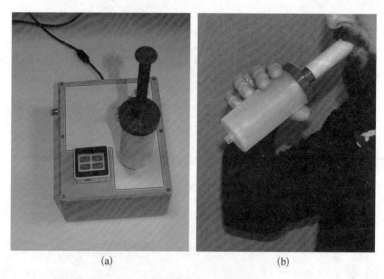

图 9.13 利用商用 VOC 气体传感器开发的电子鼻装置用于 COVID-19 检测[57]
(a) 新开发的电子鼻利用呼出气体中最后一波潮气末部分分析气体样本;
(b) 使用 BioVOC™ 收集呼气样本

Yan 和 Zhang[58] 设计了一个具有 10 个气体传感器的呼吸分析测试装置,通过测定丙酮气体的浓度实现了糖尿病患者的无创检测。研究选取了来自 36 个患者的 203 份呼吸样本,通过顺序前进法(SFS)选取了包含样品信息量最大的特征,由于样品量较少,单纯的回归模型可能出现过拟合问题,对此研究采取了全局训练与局部训练相结合的办法。全局训练可以利用尽可能多的样品,但忽略了不同患者之间的方差,因此可能会出现准确率不高的情况;而局部训练只采用单个患者的样本,训练完成后再使用其他患者的样本进行修正。最终,该研究实现了 79.3% 的检测准确率。

甲醛是一种潜在的肺癌呼吸标志物,在呼出气体中检测甲醛时,通常会受到多种气体的干扰,如丙酮、氨气和乙醇。Güntner 等[59]构建了四种元素(Pt、Si、Pd 和 Ti)掺杂的多孔二氧化锡传感器,实现了在混合气体中对甲醛的特异性检测。研究通过火焰喷雾热解的方式在硅基衬底上沉积了不同掺杂成分的二氧化锡材料,不同的掺杂成分可诱导其选择性出现差异。在干扰气体存在的情况下,回归曲线出现移位,于是该研究通过多元线性回归的方法,使得即使在较高丙酮、氨气和乙醇存在的条件下,依然能够检测到甲醛。

电子鼻作为一种非侵入性的早期医疗诊断技术,有着巨大优势,但也存在一些局限性,其明显的缺点包括缺乏标准化的样本采集和数据分析系统,以及采集的人口样本较少,难以建立标准的数据库[67,73]。此外,呼出气体的成分和浓度存在不可忽视的个体差异,可能对诊断准确性产生很大影响[59,74]。

9.6 结论与展望

近几十年来,电子鼻的发展取得了显著成功。作为检测、分类单一或复杂气体的强大工具,电子鼻系统具有传感过程快速、无创等优势,适应了气体传感应用领域的新需求,应用范围涵盖食品工业、环境监测、医学诊断等。电子鼻的主旨即集成多个传感器组成阵列,以输出多维信号,继而进行模式识别,输出结果,有效避免了传统 MOS 传感器交叉选择性的缺点。然而电子鼻技术的实际应用仍然面临挑战,主要来自一些传感器的时间漂移和低湿度耐受性问题。为了解决这些问题,可采用周期性校正,以及开发具有更好传感性能的新传感材料。随着大数据和物联网(IoT)的发展,交叉了材料、传感、微器件工艺、计算机、应用数学等领域的电子鼻技术必将对科技发展做出卓越贡献。

参考文献

[1] Rahman M, Charoenlarpnopparut C, Suksompong P, et al. A false alarm reduction method for a gas sensor based electronic nose[J]. Sensors, 2017, 17: 2089.

[2] Hsieh Y, Yao D. Intelligent gas-sensing systems and their applications[J]. J. Micromech. Microeng., 2018, 28: 093001.

[3] Zhou X, Cheng X, Zhu Y, et al. Ordered porous metal oxide semiconductors for gas sensing[J]. Chin. Chem. Lett., 2018, 29: 405-416.

[4] Wang C, Yin L, Zhang L, et al. Metal oxide gas sensors: sensitivity and influencing factors[J]. Sensors, 2010, 10: 2088-2106.

[5] Zhou X, Zhu Y, Luo W, et al. Chelation-assisted soft-template synthesis of ordered

mesoporous zinc oxides for low concentration gas sensing[J]. J. Mater. Chem. A, 2016, 4: 15064 - 15071.

[6] Lee J. Gas sensors using hierarchical and hollow oxide nanostructures: overview[J]. Sens. Actuators B, 2009, 140: 319 - 336.

[7] Wilkens W F, Hartman J D. Microcirculatory impairment as a factor in inflammatory tissue damage[J]. Ann. N. Y. Acad. Sci., 1964, 116: A2.

[8] Persaud K, Dodd G. Analysis of discrimination mechanisms in the mammalian olfactory system using a model nose[J]. Nature, 1982, 299: 5881.

[9] Gardner J W, Bartlett P N. A brief history of electronic noses[J]. Sens. Actuators B, 1994, 18: 211 - 220.

[10] Wilson A D, Baietto M. Applications and advances in electronic-nose technologies[J]. Sensors, 2009, 9: 5099 - 5148.

[11] Berna A. Metal oxide sensors for electronic noses and their application to food analysis[J]. Sensors, 2010, 10: 3882 - 3910.

[12] Dymerski T M, Chmiel T M, Wardencki W. Invited Review Article: An odor-sensing system—powerful technique for foodstuff studies[J]. Rev. Sci. Instrum. 2011, 82: 111101.

[13] Scott S M, James D, Ali Z. Data analysis for electronic nose systems[J]. Microchim. Acta., 2006, 156: 183 - 207.

[14] Buratti S, Benedetti S, Scampicchio M, et al. Characterization and classification of Italian Barbera wines by using an electronic nose and an amperometric electronic tongue[J]. Anal. Chim. Acta, 2004, 525: 133 - 139.

[15] Olsson J, Borjesson T, Lundstedt T, et al. Detection and quantification of ochratoxin A and deoxynivalenol in barley grains by GC-MS and electronic nose[J]. Int. J. Food Microbiol., 2002, 72: 203 - 214.

[16] Dutta R, Hines E L, Gardner J W, et al. Tea quality prediction using a tin oxide-based electronic nose: an artificial intelligence approach[J]. Sens. Actuators B, 2003, 94: 228 - 237.

[17] Dragonieri S, Annema J T, Schot R, et al. An electronic nose in the discrimination of patients with non-small cell lung cancer and COPD[J]. Lung Cancer, 2009, 64: 166 - 170.

[18] Gutierrez-Osuna R. Pattern analysis for machine olfaction: A review[J]. IEEE Sens. J., 2002, 2: 189 - 202.

[19] Schaller E, Bosset J O, Escher F. 'Electronic noses' and their application to food[J]. Food Sci. Technol. Leb., 1998, 31: 305 - 316.

[20] Sysoev V, Goschnick J, Schneide T, et al. A gradient microarray electronic nose based on percolating SnO_2 nanowire sensing elements[J]. Nano Lett., 2007, 7: 3182 - 3188.

[21] Benkstein K D, Raman B, Lahr D L, et al. Inducing analytical orthogonality in tungsten oxide-based microsensors using materials structure and dynamic temperature control[J]. Sens. Actuators B, 2009, 137: 48 - 55.

[22] Hwang S, Kwon H, Chhajed S, et al. A near single crystalline TiO_2 nanohelix array: enhanced gas sensing performance and its application as a monolithically integrated electronic nose[J]. Analyst., 2013, 138: 443 - 450.

[23] Fang C, Li H, Li L, et al. Smart electronic nose enabled by an all-feature olfactory algorithm[J]. Adv. Intell. Syst., 2022, 4: 2200074.

[24] Ren W, Zhao C, Niu G, et al. Gas sensor array with pattern recognition algorithms for highly sensitive and selective discrimination of trimethylamine[J]. Adv. Intell. Syst., 2022, 2200169.

[25] Shen F, Wu Q, Liu P, et al. Detection of *Aspergillus* spp. contamination levels in peanuts by near infrared spectroscopy and electronic nose[J]. Food Control, 2018, 93: 1-8.

[26] Wu H, Yue T, Xu Z, et al. Sensor array optimization and discrimination of apple juices according to variety by an electronic nose[J]. Anal. Methods, 2017, 9: 921-928.

[27] Upadhyay R, Sehwag S, Mishra H N. Electronic nose guided determination of frying disposal time of sunflower oil using fuzzy logic analysis[J]. Food Chem., 2017, 221: 379-385.

[28] Park S W, Lee S J, Sim Y S, et al. Analysis of ethanol in soy sauce using electronic nose for halal food certification[J]. Food Sci. Biotechnol., 2017, 26: 311-317.

[29] Baskar C, Nesakumar N, Balaguru R J B, et al. A framework for analysing E-nose data based on fuzzy set multiple linear regression: Paddy quality assessment[J]. Sens. Actuators A, 2017, 267: 200-209.

[30] Timsorn K, Thoopboochagorn T, Lertwattanasakul N, et al. Evaluation of bacterial population on chicken meats using a briefcase electronic nose[J]. Biosyst. Eng., 2016, 151: 116-125.

[31] Lippolis V, Ferrara M, Cervellieri S, et al. Rapid prediction of ochratoxin A-producing strains of Penicillium on dry-cured meat by MOS-based electronic nose[J]. Int. J. Food Microbiol., 2016, 218: 71-77.

[32] Miao J, Luo Z, Wang Y, et al. Comparison and data fusion of an electronic nose and near-infrared reflectance spectroscopy for the discrimination of ginsengs[J]. Anal. Methods, 2016, 8: 1265-1273.

[33] Gan Z, Yang Y, Li J, et al. Using sensor and spectral analysis to classify botanical origin and determine adulteration of raw honey[J]. J. Food Eng., 2016, 178: 151-158.

[34] Huang L, Liu H, Zhang B, et al. Application of electronic nose with multivariate analysis and sensor selection for botanical origin identification and quality determination of honey [J]. Food Bioproc. Tech., 2014, 8: 359-370.

[35] Rodriguez-Mendez M L, Apetrei C, Gay M, et al. Evaluation of oxygen exposure levels and polyphenolic content of red wines using an electronic panel formed by an electronic nose and an electronic tongue[J]. Food Chem., 2014, 155: 91-97.

[36] Rodriguez J, Duran C, Reyes A. Electronic nose for quality control of Colombian coffee through the detection of defects in "Cup Tests"[J]. Sensors, 2010, 10: 36-46.

[37] Kuske M, Romain A C, Nicolas J. Microbial volatile organic compounds as indicators of fungi. Can an electronic nose detect fungi in indoor environments? [J]. Build Environ., 2005, 40: 824-831.

[38] Helli O, Siadat M, Lumbreras M. Qualitative and quantitative identification of H_2S/NO_2 gaseous components in different reference atmospheres using a metal oxide sensor array[J].

Sens. Actuators B, 2004, 103: 403-408.

[39] Bieganowski A, Jozefaciuk G, Bandura L, et al. Evaluation of hydrocarbon soil pollution using E-nose[J]. Sensors, 2018, 18: 2463.

[40] Blanco-Rodríguez A, Camara V F, Campo F, et al. Development of an electronic nose to characterize odours emitted from different stages in a wastewater treatment plant[J]. Water Res., 2018, 134: 92-100.

[41] Capelli L, Sironi S, Del Rosso R. Electronic noses for environmental monitoring applications[J]. Sensors, 2014, 14: 19979-20007.

[42] Liu M, Zhang Y, Wang J, et al. A star-nose-like tactile-olfactory bionic sensing array for robust object recognition in non-visual environments[J]. Nat. Commun., 2022, 13: 79.

[43] Dragonieri S, Pennazza G, Carratu P, et al. Electronic nose technology in respiratory diseases[J]. Lung, 2017, 195: 157-165.

[44] Wang Z, Wang C, Lathan P. Breath acetone analysis of diabetic dogs using a cavity ringdown breath analyzer[J]. IEEE Sens. J., 2014, 14: 1117-1123.

[45] Turner C, Walton C, Hoashi S, et al. The most common methods for breath acetone concentration detection: A review[J]. J. Breath Res., 2009, 3: 11-20.

[46] Li W, Liu Y, Lu X, et al. A cross-sectional study of breath acetone based on diabetic metabolic disorders[J]. J. Breath Res. 2015, 9: 016005.

[47] Spahn J D, Malka J, Szefler S J. Current application of exhaled nitric oxide in clinical practice[J]. J. Allergy Clin. Immunol., 2016, 138: 1296-1298.

[48] Malmberg L P. Exhaled nitric oxide in childhood asthma—time to use inflammometry rather than spirometry?[J]. J. Asthma, 2004, 41: 511-520.

[49] Wou K, Feinberg J L, Wapner R J, et al. Cell-free DNA versus intact fetal cells for prenatal genetic diagnostics: what does the future hold? Expert Rev[J]. Mol. Diagn., 2015, 15: 989-998.

[50] Silva L, Freitas A C, Rocha-Santos T, et al. Breath analysis by optical fiber sensor for the determination of exhaled organic compounds with a view to diagnostics[J]. Talanta, 2011, 83: 1586-1594.

[51] Liu F L, Xiao P, Fang H L, et al. Single-walled carbon nanotube-based biosensors for the detection of volatile organic compounds of lung cancer[J]. Physica E Low Dimens., 2011, 44: 367-372.

[52] Sehnert S S, Jiang L, Burdick J F, et al. Breath biomarkers for detection of human liver diseases: preliminary study[J]. Biomarkers, 2002, 7: 174-187.

[53] Mochalski P, Wzorek B, Sliwka I, et al. Suitability of different polymer bags for storage of volatile sulphur compounds relevant to breath analysis[J]. J. Chromatogr., 2009, 877: 189-196.

[54] Adiguzel Y, Kulah H. Breath sensors for lung cancer diagnosis[J]. Biosens. Bioelectron., 2015, 65: 121-138.

[55] Wilson A D. Advances in electronic-nose technologies for the detection of volatile biomarker metabolites in the human breath[J]. Metabolites, 2015, 5: 140-163.

[56] Gibson T D, Prosser O, Lowery P, et al. Detection and simultaneous identification of

microorganisms from headspace samples using an electronic nose[J]. Sens. Actuators B, 1997, 44: 413 - 422.

[57] Kwiatkowski A, Borys S, Sikorska K, et al. Clinical studies of detecting COVID-19 from exhaled breath with electronic nose[J]. Sci. Rep., 2022, 12: 15990.

[58] Yan K, Zhang D. Blood glucose prediction by breath analysis system with feature selection and model fusion[J]. Annu Int Conf IEEE Eng Med Biol Soc., 2014, 9: 6406.

[59] Güentner A T, Koren V, Chikkadi K, et al. E-nose sensing of low-ppb formaldehyde in gas mixtures at high relative humidity for breath screening of lung cancer? [J]. ACS Sens., 2016, 1: 528 - 535.

[60] Moon H G, Jung Y, Han S D, et al. All villi-like metal oxide nanostructures-based chemiresistive electronic nose for an exhaled breath analyzer[J]. Sens. Actuators B-Chem, 2018, 257: 295 - 302.

[61] Dragonieri S. An electronic nose in respiratory disease[J]. J. Allergy Clin. Immunol. 2007, 120: 856 - 862.

[62] Vries R, Brinkman P, Schee M P, et al. Integration of electronic nose technology with spirometry: validation of a new approach for exhaled breath analysis[J]. J. Breath Res. 2015, 9: 046001.

[63] Incalzi R A, Pennazza G, Scarlata S, et al. Reproducibility and respiratory function correlates of exhaled breath fingerprint in chronic obstructive pulmonary disease[J]. PLoS One, 2012, 7: e45396.

[64] Shafiek H, Fiorentino F, Merino J L, et al. Using the electronic nose to identify airway infection during COPD exacerbations[J]. PLoS One, 2015, 10: e0135199.

[65] Thriumani R, Zakaria A, Hashim Y, et al. A study on volatile organic compounds emitted by in-vitro lung cancer cultured cells using gas sensor array and SPME - GCMS[J]. BMC Cancer, 2018, 18: 1 - 17.

[66] Tirzite M, Bukovskis M, Strazda G, et al. Detection of lung cancer in exhaled breath with an electronic nose using support vector machine analysis [J]. J. Breath Res. 2017, 11: 036009.

[67] Li W, Liu H, Xie D, et al. Lung cancer screening based on type-different sensor arrays[J]. Sci. Rep., 2017, 7: 1 - 12.

[68] De Vincentis A, Pennazza G, Santonico M, et al. Breath-print analysis by e-nose for classifying and monitoring chronic liver disease: a proof-of-concept study[J]. Sci. Rep., 2016, 6: 25337.

[69] Schnabel R M, Boumans M L, Smolinska A, et al. Electronic nose analysis of exhaled breath to diagnose ventilator-associated pneumonia[J]. Respir. Med. 2015, 109: 1454 - 1459.

[70] Arasaradnam R P, Quraishi N, Kyrou I, et al. Insights into 'fermentonomics': evaluation of volatile organic compounds (VOCs) in human disease using an electronic 'e-nose'[J]. J. Med. Eng. Technol. 2011, 35: 87 - 91.

[71] Yan K, Zhang D, Wu D, et al. Design of a breath analysis system for diabetes screening and blood glucose level prediction[J]. IEEE. Trans. Biomed. Eng. 2014, 61: 2787 - 2795.

[72] Guo D, Zhang D, Li N, et al. A novel breath analysis system based on electronic olfaction

[J]. IEEE. Trans. Biomed. Eng. 2010, 57: 2753-2763.

[73] Rocco G. Liquid biopsy—a novel diagnostic tool for management of early-stage peripheral lung cancer[J]. J. Thorac. Cardiovasc Surg. 2018, 155: 2622-2625.

[74] Dai Y, Molazemhosseini A, Liu C. In vitro quantified determination of β-amyloid 42 peptides, a biomarker of neuro-degenerative disorders, in PBS and human serum using a simple, cost-effective thin gold film biosensor[J]. Biosensors, 2017, 7: 29.

第10章
半导体金属氧化物传感器的应用

近年来,基于半导体金属氧化物(SMO)材料的气体传感器已广泛应用于居家、工厂、医院和实验室等人类生活生产的场所,为快速灵敏的半定量检测提供了一种极具前景的方法。它不仅成本低、制造简便、尺寸可设计,且具有出色的气体传感特性[包括高灵敏度、出色的选择性、快速响应/恢复和低检测限(低于 ppm 水平)],同时可检测多种气体,因而备受关注。基于半导体金属氧化物的气体传感器可检测的气体包括挥发性有机物、有毒气体、一些特定的危险气体,以及可燃气体等,这使得基于 SMO 的气体传感器成为气体报警、环境监测、食品卫生检疫和医疗诊断的有效手段。本章将根据气体种类的分类总结近期基于 SMO 的气体传感器的相关研究,对具有代表性的 SMO 传感器进行归类,并将其传感性能列于表格中,讨论灵敏度、选择性和工作温度等气体传感器性能的重要进展,读者可以从参考文献中了解更多细节。

10.1 挥发性有机物传感器

挥发性有机物是室温下具有高蒸气压的有机化学品,它们的沸点通常很低,蒸气压很高,这使得它们很容易从液体或固体中蒸发进入环境。VOC 可以自然地或人工地产生,其中一个主要来源是工厂或实验室排放。部分挥发性有机物不仅污染环境,而且可能直接危害人体健康。例如,醇和芳香烃可以刺激黏膜和上呼吸道,从而对人体健康有潜在危害。又如苯和甲醛,已被证明在人体内长期积累可能导致癌症。因此,有必要开发用于低浓度和快速检测 VOC 的气体传感器。

10.1.1 乙醇气体传感器

乙醇(C_2H_5OH)在通常情况下是一种无色、易燃、易挥发的液体,沸点为 78.3℃,广泛存在于各类酒精饮料和食品中,是最常用和最广泛使用的溶剂之一,在食品工业、制药工业和化学工业中都有广泛的应用。通常,接触乙醇蒸气并不危险,但是当吸入一定量时,它可能导致头痛、嗜睡、眼睛刺激和呼吸困难。此外,由于过量摄入酒精(乙醇)是导致交通事故的主要原因之一,因此以 ppm 水平定量检测乙醇蒸气不仅具有医学意义,而且具有社会重要性。表 10.1 总结了许多用于研究乙醇传感的金属氧化物,其中,SnO_2 和 ZnO 是最具代表性的基于 SMO 的乙醇传感器材料。例如,Chen 等[1]报道了通过水热法大批量生长的直径为 4~15 nm 的单晶 SnO_2 纳米棒。这种基于单晶 SnO_2 纳米棒的传感器在 300℃的工作温度下表现出灵敏度为 4.2~31.4(对 10~300 ppm 乙醇气体)的响应。通过燃烧化学气相沉积工艺可生成具有高度多孔纳米结构的 SnO_2 薄膜,以此制成的气体传感器在 300℃的工作温度下具有高达 1 075(对 500 ppm 乙醇蒸气)的灵敏度[2]。Li 等[3]在 Si 衬底上通过热挥发-沉积后退火处理的方法,直接合成了多层 SnO_2 纳米片。气体传感测试表明,350℃下该多层 SnO_2 纳米片对 50 ppm 乙醇气体的灵敏度响应是单层 SnO_2 纳米片的两倍以上,这是由于多层纳米片的比表面积较大。Liu 等[4]报道了核壳结构 SnO_2 微球在 260℃时对 10~50 ppm 乙醇表现出极佳的灵敏度,高于空心 SnO_2 纳米颗粒相同条件下的响应。ZnO 是另一种常用于乙醇传感的金属氧化物。通过简单的低温水热法合成的 ZnO 纳米棒,即使在 1 ppm 的低浓度下,对乙醇也显示出相当大的响应[5]。为了进一步提高传感性能,Tian 等[6]采用新颖的溶液相处理技术合成了沿陶瓷管原位取向生长的直径为 8 nm 和 (0001)晶面暴露的 ZnO 纳米棒阵列气体传感器。与基于(1010)晶面暴露的 ZnO 纳米棒的传感器相比,这种材料具有更高的灵敏度和更快的响应速度(小于 10 s)。研究提出,这主要是由于(0001)晶面的高暴露可以提高氧物种吸附。将由双溶剂制备的花状分级 ZnO 用于乙醇气体检测,检测浓度降至 10 ppb,这可能归因于特定的分级结构[7]。多孔结构有利于提高表面积和促进气体扩散,从而提高气体传感性能。Zhou 等[8]以两亲性嵌段共聚物 PEO-b-PS 为结构导向剂合成了有序介孔 ZnO 材料。制备的介孔 ZnO 具有良好的乙醇传感性能,灵敏度高(灵敏度为 66,对 50 ppm 乙醇气体,350℃)、响应速度和恢复速度快(6 s/7 s)、选择性高,如图 10.1 所示。除了经典 n 型 SMO 的 SnO_2 和 ZnO 之外,其他用于乙醇传感的金属氧化物,包括其他 n 型 SMO(如 In_2O_3[9]、V_2O_5[10]、TiO_2[11])和 p 型 SMO(如 CuO[12-13]、NiO[14])也被广泛研究。

表 10.1 半导体金属氧化物乙醇气体传感器

材料	结构形貌	气体浓度 /ppm	操作温度 /℃	灵敏度	响应/恢复 时间/s	参考 文献
SnO_2	薄膜	500	300	1 075^a	31/8	[2]
SnO_2	纳米片	50	350	3.78^a	—/—	[3]
SnO_2	纳米棒	300	300	31.4^a	1/1	[1]

续表

材料	结构形貌	气体浓度/ppm	操作温度/℃	灵敏度	响应/恢复时间/s	参考文献
SnO_2	微球	50	260	60.5[a]	9/20	[4]
SnO_2/Pt	纳米线	500	200	8 400[c]	—/—	[16]
SnO_2/Ag	纳米线	100	450	228.1[a]	5/100	[15]
$Sb-SnO_2$	纳米线	10	300	25/15[a]	1/5	[26]
$La_2O_3-SnO_2$	纳米线	100	400	57.3[a]	1/110	[29]
SiO_2-SnO_2	纳米颗粒	50	300	318[a]	—/—	[28]
ZnO	纳米棒	200	320	35[a]	54/61	[5]
ZnO	纳米棒	100	370	42[a]	22/10	[31]
ZnO	纳米棒阵列	100	370	70[a]	10/10	[6]
ZnO	介孔薄膜	50	350	66[a]	6/7	[8]
ZnO	纳米柱	100	350	18.29[a]	10/20	[32]
ZnO	介孔微球	5	450	2.2[a]	4/6	[33]
ZnO	花状微球	10	300	4[c]	17/12	[7]
ZnO	花状纳米棒	100	300	176.8[a]	—/—	[34]
$ZnO-Au$	纳米线	50	325	7[a]	5/20	[18]
$ZnO-PdO$	花状结构	100	320	35.4[a]	1/7	[19]
Fe_2O_3/ZnO	纳米棒	500	220	22.1[a]	—/—	[30]
In_2O_3	纳米棒	50	330	11.3[a]	6/11	[35]
In_2O_3	纳米线	100	370	2[a]	10/20	[9]
In_2O_3-Rh	空心球	100	371	4 748[a]	1/80	[24]
$Au@In_2O_3$	核壳结构	100	160	36.14[a]	4/2	[20]
TiO_2	薄膜	50	30	535%[b]	5/52	[11]
$Ag-TiO_2$	纳米带	500	200	46.153[a]	2/2	[23]
CuO	薄膜	12.5	180	2.2[a]	31/52	[13]
CuO	纳米棒	1 000	210	9.8[a]	42/51	[12]
NiO	半球	200	300	5[a]	—/—	[14]
V_2O_5	纳米带	100	200	1.7[a]	32/30	[10]
V_2O_5/Ti	纳米带	100	200	2.0[a]	49/85	
V_2O_5/Fe	纳米带	100	200	2.3[a]	36/64	
V_2O_5/Sn	纳米带	100	200	3.1[b]	37/126	
$Ag@\alpha-Fe_2O_3$	核壳结构	100	250	6[a]	5.5/16	[22]

注：a $(R_a/R_g$ 或 $R_g/R_a)$；b $[(R_a-R_g)/R_a,\%]$；c (I_g/I_a)。

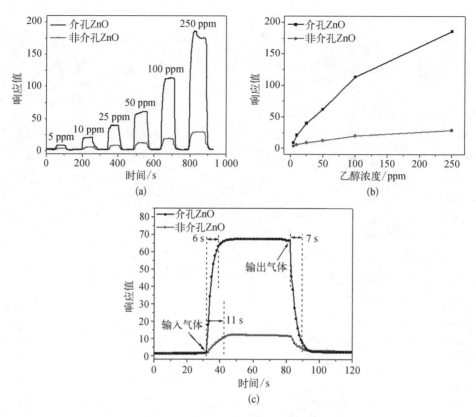

图 10.1 介孔 ZnO 对乙醇的传感性能研究[8]

(a) 介孔 ZnO 传感器和非介孔 ZnO 传感器的响应-恢复曲线;(b) 介孔 ZnO 传感器和非介孔 ZnO 传感器对不同浓度乙醇蒸气的灵敏度与浓度之间的关系;(c) 介孔 ZnO 传感器和非介孔 ZnO 传感器对 50 ppm 乙醇蒸气的动态响应-恢复曲线

负载贵金属是提高传感性能的好方法,使用对乙醇敏感的金属氧化物作为主要载体并修饰贵金属是有效提升乙醇传感性能的途径。据报道,与纯 SnO_2 纳米线相比,Ag 修饰的 SnO_2 纳米线在 450℃ 的工作温度下对 100 ppm 乙醇气体的响应增强了 3.7 倍[15]。其他例子如 Pt/SnO_2[16]、Pd/SnO_2[17] 也明显提高了 SnO_2 对乙醇气体的响应。Au 修饰的 ZnO 传感器可以在 325℃ 下对 50 ppm 的乙醇气体进行响应,响应和恢复时间分别为 5 s 和 20 s[18]。据报道,Pd 也可提高 ZnO 对乙醇气体的传感性能[19]。具有核壳纳米结构的 $Au@In_2O_3$ 复合材料在 160℃ 的低工作温度下表现出优异的灵敏度(36.14,对 100 ppm 乙醇气体)以及对乙醇的高选择性[20]。基于 $Au@In_2O_3$ 的传感器的响应比基于纯 In_2O_3 的传感器的响应高约 1.5 倍。其他贵金属/金属氧化物复合材料,如 Au/WO_3[21]、Ag/Fe_2O_3[22]、Ag/TiO_2[23]、Rh/In_2O_3[24],均已被研究用于乙醇传感。在上述复合材料中,似乎 Au 和 Ag 材料对提高乙醇传感性能更有效。Ren 等[25] 提出了一种基于两亲性嵌段共聚物

(PEO-b-PS)、多金属氧酸盐($H_4SiW_{12}O_{40}$)和有机铂配合物[$Pt(cod)(Me)_2$]的一锅法共组装方法,合成了一种 Pt 纳米颗粒(约 2.8 nm)修饰的 Si 掺杂 WO_3 纳米线阵列交织而成的三维介孔超结构材料(Pt/Si-WO_3 NWIMSs),并构筑了半导体乙醇气体传感器。该传感器在低温下(100℃)对乙醇表现出高灵敏度($S=93@$ 50 ppm)、低检出限(0.5 ppm)、快速响应/恢复速度(17 s/7 s)、高选择性以及良好的长期稳定性(见图 10.2)。研究同时利用原位红外、GC-MS、DFT 计算等深入研究了气敏机理,并基于该敏感材料,开发了高性能气体传感器模组,结合蓝牙传输实现了与智能手机通信,以及对环境中乙醇浓度的智能实时监测。

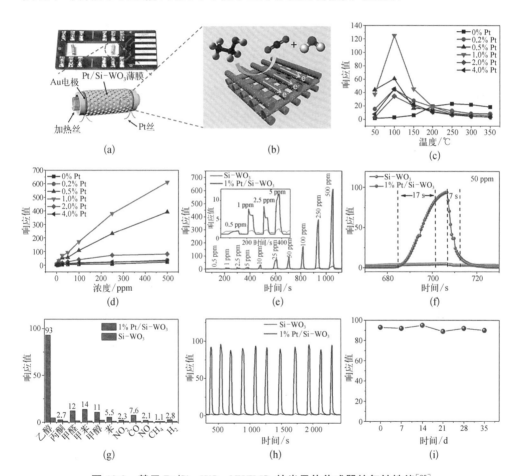

图 10.2 基于 Pt/Si-WO_3 NWIMSs 的半导体传感器的气敏性能[25]

(a) 气体传感器装置的光学图像和示意图;(b) 乙醇分子和铂纳米颗粒修饰的氧化钨纳米线之间的相互作用图;(c) 不同工作温度下 Pt/Si-WO_3 NWIMSs 对 50 ppm 乙醇的响应;(d) 在 100℃温度下,不同铂含量(原子百分比)的 Pt/Si-WO_3 NWIMSs 对不同浓度(0.5~500 ppm)乙醇的响应;(e) 在 100℃下对不同浓度(0.5~500 ppm)乙醇的动态响应-恢复曲线;(f) 50 ppm 乙醇的响应-恢复时间;(g) 对 50 ppm 不同气体的响应;(h) 在 50 ppm 乙醇中的循环;(i) 气体传感器的长期稳定性

杂原子掺杂是提高传感性能的另一种方法，In、Sb 的掺杂剂都可以通过调节金属氧化物的电子特性进而提高材料的气敏性能。Wan 和 Wang[26]报道了 Sb 掺杂 SnO_2 纳米线的合成及其在乙醇传感中的应用。基于 Sb 掺杂的 SnO_2 传感器显示出快速的响应/恢复时间（1 s/5 s，对 10 ppm 乙醇），比纯 SnO_2（长达 10 min）好得多，研究认为 Sb 掺杂有利于吸收氧分子。由 Li 等[27]制备的掺杂 In 的 ZnO 纳米线表现出优异的灵敏度（约 27，对 100 ppm 乙醇浓度）和快速的响应/恢复时间（均短于 2 s）。

此外，与其他氧化物，如 SiO_2、La_2O_3 和 Fe_2O_3 进行复合，也可提高 SnO_2 和 ZnO 对乙醇传感的性能。Tricoli 等[28]通过火焰喷雾热解（FSP）方法制备了掺杂 SiO_2 的 SnO_2 颗粒。添加 0～4% 的二氧化硅可以很好地限制烧结晶粒的颈尺寸，从而提高灵敏度并降低对乙醇的检测限（100 ppb）。Van Hieu 等[29]在 SnO_2 纳米线上负载了 La_2O_3 颗粒，以获得对 C_2H_5OH 的高灵敏度和高选择性。与纯 ZnO 相比，Fe_2O_3 修饰的 ZnO 纳米线表现出对乙醇的高响应性和高选择性。这些金属氧化物复合材料都可以形成异质结以改善传感性能[30]。

10.1.2　丙酮气体传感器

与乙醇传感器类似，丙酮传感器近年来引起了越来越多的关注。尽管丙酮在正常使用时仅表现出轻微的毒性，但它仍有可能成为呼吸分析中的标志物，比如糖尿病患者的呼出气体中丙酮含量往往较高。表 10.2 总结了不同金属氧化物的丙酮传感特性，其中，ZnO、In_2O_3、$\alpha\text{-}Fe_2O_3$ 和 WO_3 是被研究得最多的金属氧化物，因为它们具有很高的丙酮检测潜力。不过用于丙酮传感的纯金属氧化物的相关研究很少。Vomiero 等[36]合成了 In_2O_3 的单晶纳米线，其在 400℃ 时对丙酮有响应。Nguyen 和 EI-Safty[37]报道了介孔/大孔结构的结晶 Co_3O_4 纳米棒，并将其用于制备丙酮传感器，与乙醇和苯相比，基于结晶 Co_3O_4 纳米棒的传感器对丙酮的响应最高，响应速度快，恢复时间为 1 min。后来人们又发现 WO_3 是对丙酮敏感的金属氧化物。Chen 等[38]制备了 WO_3 纳米片传感器，其在 300℃ 的工作温度下对丙酮蒸气具有高且稳定的响应（对 1 000 ppm 丙酮的灵敏度为 42）、低检测限（2 ppm）和短的响应和恢复时间（3～10 s 和 12～13 s，对应 2～1 000 ppm）。

表 10.2　半导体金属氧化物丙酮气体传感器

材料	结构形貌	气体浓度/ppm	操作温度/℃	灵敏度	响应/恢复时间/s	参考文献
WO_3	纳米晶	1 000	300	42[b]	10/13	[38]
$SiO_2\text{-}WO_3$	薄膜	0.6	400	4.63[a]	—/—	[42]

续 表

材 料	结构形貌	气体浓度/ppm	操作温度/℃	灵敏度	响应/恢复时间/s	参考文献
Cu-WO$_3$	纤维	20	300	5.24[b]	15/40	[44]
SnO$_2$-WO$_3$	薄膜	5	350	3.6[b]	—/—	[45]
Au/WO$_3$	纳米棒	200	300	131.26[c]	98/91	[41]
Pd/WO$_3$	纳米棒	200	300	138.62[c]	120/76	[41]
AuPd/WO$_3$	纳米棒	200	300	152.4[c]	101/96	[41]
α-Fe$_2$O$_3$	纳米棒	100	300	10[b]	34/44	[46]
α-Fe$_2$O$_3$-Au	纳米棒	100	260	20[b]	26/24	[46]
α-Fe$_2$O$_3$-ZnO	纳米棒	100	250	43[b]	19/20	[46]
α-Fe$_2$O$_3$-ZnO-Au	纳米棒	100	225	112[b]	17/13	[46]
1%Pt/γ-Fe$_2$O$_3$	颗粒	1	250	75[d]	10	[47]
La 掺杂 α-Fe$_2$O$_3$	纳米管	50	240	26[b]	3	[48]
ZnO	纳米棒阵列	100	300	30[b]	5/15	[49]
Sn 掺杂 ZnO	颗粒	400	300	130[c]	—/—	[50]
Co$_3$O$_4$	纳米棒	74 750	300	22 360[c]	24/180	[37]
In$_2$O$_3$	纳米线	25	400	7[b]	—/—	[36]
In$_2$O$_3$/Pt	纳米颗粒	1.56	200	12[b]	25/120	[39]
SnO$_2$/Pt	纤维	3	300	2.47[a]	15/—	[40]

注: a (R_a/R_g-1); b (R_a/R_g 或 R_g/R_a); c (R_a/R_g,%); d [(R_a-R_g)/R_a,%]。

考虑到丙酮可作为呼出气体检测的标志物,近年来开发具有低检出限和耐高湿度的丙酮传感器已成为热点。用贵金属修饰金属氧化物是提高金属氧化物丙酮传感性能的一种极为有效的方法。Karmaoui 等[39] 合成的 In$_2$O$_3$ 纳米颗粒表面负载有 2%(质量百分比)的 Pt 金属纳米颗粒(2~3 nm),其对丙酮气体的检出限可低至 10 ppb 甚至更低。此外,这种基于 In$_2$O$_3$/Pt 的传感器在 200℃ 的工作温度下就可以在潮湿空气(相对湿度为 75%)中对 0.29 ppm 的丙酮气体响应,显示出在呼出气体检测中的潜在应用。Shin 等[40] 制备的 Pt 纳米颗粒修饰的分级结构 SnO$_2$ 纤维具有优异的灵敏度($R_a/R_g-1=0.72$,丙酮浓度为 120 ppb),在极低的浓度范围(0.2~1 ppm)具有快速的响应时间(<11 s)和恢复时间(<6 s)。除了对单一金属修饰,对双贵金属的修饰也有报道。Kim 等[41] 发现,与单一 Au 或 Pd 修饰的 WO$_3$ 相比,Au 和 Pd 颗粒共同修饰的 WO$_3$ 表现出更佳的丙酮感应,结果表明用双金属纳米颗粒进行修饰可能更有效。除了修饰贵金属外,Si 掺杂也是提高 WO$_3$ 对丙酮气体的传感性能的有效方法[42]。Pratsinis 团队制备的 Si 掺杂 WO$_3$ 薄膜检

测器具有超低丙酮浓度检测限（低至 20 ppb），在理想（干燥空气）和实际（高达 90%RH）环境条件下具有高信噪比。结果表明，Si 掺杂不仅可以增加对丙酮具有选择性的 ε-WO_3 相的含量，还可以提高其热稳定性，从而获得优异的传感性能（见图 10.3）。Ren 等[43]利用 Si 掺杂的亚稳态 ε-WO_3 三维正交堆叠纳米线阵列构筑了旁热式微型气敏器件，并将其用于高性能丙酮传感。由于亚稳态 ε-WO_3 纳米线阵列结构同时具有 3D 堆垛多孔结构、丰富的界面活性氧（O^-、O_2^- 等）和良好的电子传递行为，该材料展示出优异的丙酮气体传感性能（见图 10.4），具有高的灵敏度（S=216 vs 50 ppm）、低检出限（LoD<10 ppb）、良好的选择性和短的响应/恢复时间（5 s/12 s），有望应用于基于人体呼出气体检测的早期糖尿病的非侵入式检测。此外，该研究还利用 GC-MS 检测气敏催化产物研究了气敏机理，证明了丙酮气体传感的催化产物是 CO_2 和 H_2O。此外，还可以通过构建异质结来增强材料的丙酮传感性能。Bai 等[44]和 Hernández 等[45]分别使用 Cu 掺杂的 WO_3 和 WO_3/SnO_2 复合材料制备了高性能的丙酮传感器。

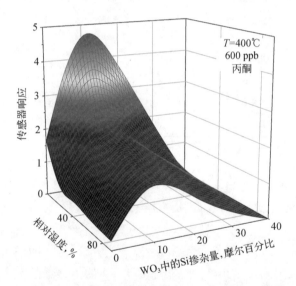

图 10.3 基于不同含量 Si 掺杂的 WO_3 的传感器在不同相对湿度下对 600 ppb 丙酮的响应（工作温度为 400℃）

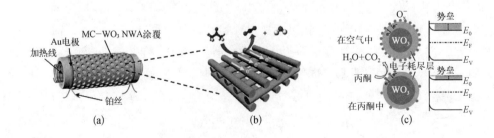

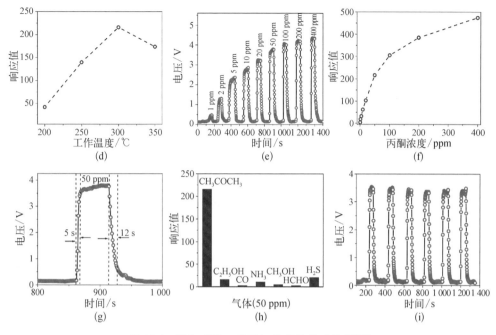

图 10.4　MC‑WO₃‑NWAs 的气体传感性能[43]

(a) 旁热式气体传感器结构示意图；(b) 丙酮分子与 Si 掺杂 ε‑WO₃ 纳米线相互作用示意图；(c) 丙酮气体传感机理示意图；(d) 传感器在不同温度下对 50 ppm 丙酮的响应；(e) 300℃时,传感器对不同浓度(1.0～400 ppm)丙酮的响应‑恢复曲线；(f) 传感器对丙酮浓度的响应($S=R_a/R_g$)；(g) 传感器在 300℃时对 50 ppm 丙酮的响应‑恢复时间；(h) 传感器对 300℃下 50 ppm 不同气体的响应；(i) 传感器对 50 ppm 丙酮的重复响应‑恢复曲线

10.1.3　甲醛气体传感器

甲醛(HCHO)是室温下无色易燃的气体,可以从人工制作的家具、装饰材料和建筑材料等中释放出来,是住宅环境和工业环境中最主要的羰基化合物之一,也是一类主要的 VOC。甲醛已被归类为人类致癌物,并可能导致肺损伤、鼻咽癌和白血病等多种健康问题。世界卫生组织(WHO)已确定甲醛暴露限值,平均 30 min 内最高值为 0.08 ppm。美国职业安全与健康管理局(OSHA)将甲醛立即威胁生命或健康的浓度(IDLH)设定为 20 ppm,工作场所的暴露限制为 0.75 ppm。虽然甲醛具有毒性和致敏性,但由于其高反应性和相对低的成本,在脲醛树脂、酚醛树脂和三聚氰胺树脂等重要树脂的生产中必不可少,它仍然广泛应用于工业过程。

目前,已经有不同的金属氧化物用于甲醛检测,见表 10.3,其中,SnO₂、ZnO、NiO 基传感器表现出较好的甲醛传感性能。Li 等[51]报道了基于多孔 SnO₂ 纳米球的传感器可以在 260℃检测到 0.5 ppm 甲醛,这低于 OSHA 规定的允许暴露限值。Lai等[52]直接以介孔二氧化硅为硬模板,通过纳米浇铸法合成了有序介孔氧化镍(NiO)材料,并构筑了一类具有高响应性能的甲醛(HCHO)气体传感器(见图10.5)。通过选择具有不同孔径的介孔二氧化硅作为模板,获得了一系列具有不同结构参数(如比表面积、孔径、孔壁厚度)的介孔NiO,并研究了NiO样品对甲醛(HCHO)的气敏特性。结果表明,这种介孔NiO即使在低浓度水平下也比块体NiO对HCHO具有更高的响应,更大的比表面积和孔径以及更薄的孔壁将有助于提高NiO的传感性能。

表10.3 半导体金属氧化物甲醛气体传感器

材料	结构形貌	气体浓度/ppm	工作温度/℃	灵敏度	响应/恢复/s	参考文献
SnO_2	纳米线	10	270	2.45[a]	90/150	[59]
SnO_2	多孔纳米球	1	260	3.14[a]	13/14	[51]
Au/SnO_2	核壳结构	50	RT	2.9[a]	80/69	[60]
$Pd-SnO_2$	薄膜	10	250	1.55[a]	50/50	[61]
$Pd-SnO_2$	纤维	0.1	190	1.7[a]	53/103	[53]
SnO_2-NiO	薄膜	0.3	300	18[c]	—/—	[62]
SnO_2-NiO	纳米纤维	10	200	6[c]	50/80	[56]
$SnO_2-In_2O_3$	纳米线	10	375	7.5[a]	—/—	[63]
SnO_2/Fe_2O_3	空心球	10	250	2[a]	—/—	[64]
ZnO	纳米棒	50	RT(紫外光活化)	11.7[c]	—/—	[58]
ZnO	厚膜	0.001	210	7.4[a]	—/—	[65]
CdO-Sn掺杂ZnO	纳米颗粒	200	200	2 000[a]	—/—	[57]
MnO_2-ZnO	阵列结构	200	320	27[a]	27/12	[66]
Ga-ZnO	薄膜	205	400	13[b]	30/210	[67]
Au/ZnO	核壳结构	5	RT	10.57[a]	138/104	[68]
NiO	薄膜	10	300	1.5[a]	—/—	[69]
NiO	薄膜	20	340	4.3[a]	—/—	[70]
NiO	薄膜	5	340	12.65[c]	—/—	[71]
Co_3O_4	介晶	100	200	2.8[a]	—/—	[72]
In_2O_3	八面体线串	100	400	1.7[a]	48/58	[73]

注:a(R_a/R_g或R_g/R_a);b(R_a/R_g,%);c[$(R_a-R_g)/R_a$,%]。

为了进一步提高金属氧化物的甲醛传感性能,研究人员开始采用贵金属修饰和氧化物复合的策略。由Tian等[53]合成的Pd修饰的SnO_2纤维具有超低的甲醛检测限(低至50 ppb),优于其他还原性气体的选择性,以及在190℃相对低的工作

图 10.5　基于有序介孔 NiO 材料的甲醛传感器[52]

温度下对甲醛较短的响应/恢复时间(分别为 53 s 和 103 s,对 100 ppb 甲醛),这远优于纯的 SnO$_2$ 材料(见图 10.6)。在 Pt/SnO$_2$ 体系中也可以看到类似的性能提升,其检测限低至 ppb 级别[54]。Cai 等[55]基于负载双金属 PdPt 敏化剂的 SnO$_2$ 多壳空心微球(PdPt/SnO$_2$ - M),开发了一种低成本集成微机电系统(MEMS)传感器。MEMS 传感器对 HCHO 具有高灵敏度($S=83.7$@1 ppm)、50 ppb 的超低检测限和超短的响应/恢复时间(5.0/7.0 s@1 ppm)(见图 10.7)。这些优异的 HCHO 传感性能归因于其独特的多壳中空结构,具有大而易接近的表面积、丰富的界面、合适的介孔结构以及双金属 PdPt 的协同催化效应。因此,PdPt 双金属纳米颗粒可用于在这种多壳中空结构上构建具有高含量和良好分散性的协同敏化剂,进一步降低工作温度和促进 HCHO 的超灵敏检测。这种基于 PdPt/SnO$_2$ - M 的 MEMS 传感器提供了一种独特且高度灵敏的检测 HCHO 的方法,为其在环境监测中的潜在应用奠定了良好的基础。Zheng 等[56]制备的 NiO 掺杂的 SnO$_2$ 纳米纤维在 200℃ 的工作温度下对甲醛具有良好的传感性能,其对 10 ppm 甲醛的响应时间和恢复时间分别为 50 s 和 80 s,最小检测限降至 0.08 ppm。Han 等[57]采用共沉淀法制备了 Sn、Ni、Fe、Al 掺杂的 ZnO 和纯 ZnO,并测试了其对甲醛的传感性能。结果表明,2.2%①的 Sn 掺杂剂可以使 ZnO 对甲醛的灵敏度增加 2 倍以上,而其他掺杂剂的反应灵敏度几乎没有增加甚至降低。此外,将 10%② CdO 用于活化 2.2% Sn

---

① 2.2%指摩尔百分比。
② 10%指摩尔百分比。

掺杂的 ZnO，其最优工作温度为 200℃，低于纯 ZnO 的最优工作温度（400℃），并且比其他还原气体具有更高的选择性。除进行复合或者掺杂外，Peng 等[58]还报道了紫外光照射可以显著提高直径约为 40 nm 的 ZnO 纳米棒对甲醛的传感性能。

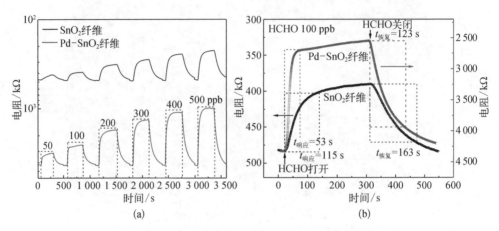

图 10.6　Pd 修饰的 SnO_2 纳米纤维对 HCHO 的传感性能[53]

(a) 纯 SnO_2 和 Pd 修饰的 SnO_2 纳米纤维传感器在 190℃下对不同浓度 HCHO 气体的动态响应；
(b) 基于纯 SnO_2 纤维的传感器和 Pd-SnO_2 纤维传感器对 100 ppb HCHO 气体的响应曲线放大图

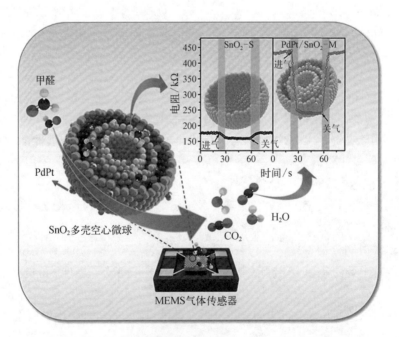

图 10.7　基于双金属 PdPt 敏化的 SnO_2 多壳空心微球的 MEMS 甲醛传感器[58]

10.1.4 苯系物(苯、甲苯、二甲苯)传感器

苯、甲苯和二甲苯(苯系物,BTX)是典型的室内芳香族污染物,会对人类健康构成威胁,因此,开发具有高选择性和灵敏度的BTX气体传感器是非常重要的。一些用于检测BTX的相关金属氧化物传感器列于表10.4中。

表10.4 半导体金属氧化物苯系物(BTX)传感器

气体	材料	气体浓度/ppm	工作温度/℃	灵敏度	响应/恢复时间/s	参考文献
苯	TiO_2-ZnO	100	370	27[a]	10/5	[74]
	WO_3膜	20	200	12.5[a]	70/120	[87]
	Cu_2O纳米颗粒	50	230	9.7[a]	3/4	[76]
	Cr掺杂ITO	1 000	210	72[a]	38/53	[77]
	Au/SnO_2	0.005	415	300[b]	50/—	[75]
甲苯	Co_3O_4空心球	10	100	3[a]	—/—	[88]
	介孔Co_3O_4	100	190	23.55[a]	233/165	[78]
	PdO修饰ZnO	100	240	10.9[a]	1/9	[79]
	四足形ZnO	100	380	7[a]	50/20	[89]
	ZnO微球	20	300	24.5[a]	0.3/3	[90]
	α-Fe_2O_3/NiO	100	300	18.68[a]	—/—	[80]
	Fe_3O_4-NiO	100	280	13[a]	—/—	[91]
	SnO_2/α-Fe_2O_3空心球	10	350	2.3[a]	—/—	[64]
	SnO_2/Fe_2O_3纳米管	50	260	25.3[a]	—/—	[92]
二甲苯	Cr掺杂NiO	5	400	11.6[a]	—/—	[81]
	Cr掺杂NiO	5	220	20.9[a]	—/—	[93]
	Pd负载SnO_2	5	350	17.5[a]	—/—	[83]
	Ni掺杂Al_2O_3	10	340	5.5[a]	20/300	[94]
	Co_3O_4纳米方块	100	200	6.45[a]	—/—	[95]

注:a (R_a/R_g);b $[(I_g-I_a)/I_a]$,%。

苯来自石油化工生产、工业排放和机动车尾气,具有毒性,致癌。200 ppm 的苯即可对人体造成麻醉作用,在2%的苯气体中暴露5~10 min 可致人死亡,苯也可以诱导白血病和淋巴瘤的发生。Zhu 等[74]通过混合 ZnO-TiO_2 纳米颗粒和烧结处理制备了 VOC 传感器。基于 ZnO-TiO_2 的传感器在370℃的工作温度下对 100 ppm 苯气体的灵敏度(R_a/R_g)为27,响应时间为10 s,恢复时间为5 s。深入研

究表明，随着 TiO_2 含量的增加（5%～10%），ZnO 的晶粒尺寸减小，从而提高了传感性能。Elmi 等[75]制备了 5%～10%（质量百分比）Au 修饰的 SnO_2 薄膜，并将其用于制备 VOC 传感器。获得的传感器在 415℃ 的工作温度下具有 300% 对 5 ppb 苯的响应，响应时间为 50 s。Wang 等[76]制备了 Cu_2O 八面体纳米结构和纳米棒。基于 Cu_2O 八面体材料的传感器具有灵敏度（R_a/R_g）接近 10 的高响应，以及非常短的响应时间和恢复时间（分别为 3 s 和 4 s）。结果表明，八面体结构有助于提高材料对苯的传感特性。Vaishnav 等[77]制备了 Cr/ITO 薄层，并将其应用于苯气体的检测，其灵敏度和浓度（30～100 ppm 苯）之间显示出良好的线性关系，与甲苯相比对苯具有良好的选择性。

与苯和二甲苯相比，甲苯传感器已被广泛研究。Co_3O_4 已被证明具有甲苯传感特性，ZnO 基材料和 Fe_2O_3 基材料在甲苯传感方面也显示出巨大的潜力。Liu 等[78]采用介孔二氧化硅（SBA-15）作为硬模板，用纳米浇筑法合成了介孔 Co_3O_4。结果表明，与甲醛、乙醇、丙酮、甲醇和氨相比，基于介孔 Co_3O_4 的传感器对 100 ppm 甲苯具有较高的响应性（在 190℃ 下 $R_a/R_g=23.55$）和良好的选择性。Lou 等[79]制备了具有花状结构的 PdO 修饰的 ZnO 材料，花状 ZnO 纳米结构由许多直径约为 200 nm 的棒聚集组成。基于 PdO-ZnO 的传感器在 240℃ 下对 100 ppm 甲苯具有灵敏度（R_a/R_g）为 10.9 的响应和超短的响应/恢复时间（1 s/9 s）。Wang 等[80]报道了通过简便的水热法合成了具有中空纳米结构的分级 α-Fe_2O_3/NiO 复合材料。基于 α-Fe_2O_3/NiO 的传感器在 300℃ 的工作温度下对 100 ppm 甲苯具有高响应性（$R_a/R_g=18.68$），比纯 NiO 高出 13.18 倍。

与甲苯一样，二甲苯也会引起眼睛、鼻黏膜和皮肤的刺激，造成头痛、疲劳等不适。然而二甲苯传感器却很少，通过使用基于金属氧化物的传感器来区分二甲苯和甲苯是一项挑战。有研究报道，具有分级纳米结构的 1.15%① Cr 掺杂的 NiO 材料可以通过简单的溶剂热反应合成，并用于检测各种气体[81]。气体传感测试表明，1.15% Cr 掺杂的 NiO 基传感器对 5 ppm 邻二甲苯和甲苯具有高响应（R_g/R_a 分别为 11.61 和 7.81），对 5 ppm 苯、甲醛、乙醇、氢气和一氧化碳的交叉响应可忽略不计，以及具有较好的选择性。Guo 等[82]采用超声湿法化学腐蚀和热解法制备了由不规则二维纳米片组装而成的 CuO/WO_3 分级空心微球，并评价了基于 CuO/WO_3 分级结构的微机电系统（MEMS）二甲苯气体传感器的传感性能（见图 10.8）。研究发现，CuO/WO_3 MEMS 传感器较其他 MEMS 传感器表现出增强的气敏性能。CuO/WO_3-3（CuO 与 WO_3 的质量比为 3%）传感器具有更快的响应/恢复速度、对二甲苯的最高响应值，以及更高的选择性和长期稳定性。由于 CuO-WO_3 独特的三维（3D）分层结构和 p-n 异质结，具有良好传

---

① 1.15% 指原子百分比。

感性能的传感器在二甲苯的快速检测和监测方面具有很大的潜力。贵金属修饰金属氧化物用于二甲苯和甲苯传感的研究也有报道。Hong 等[83]用具有蛋黄-蛋壳结构的 Pd 负载的 SnO_2 材料检测甲苯。结果表明,基于 Pd 负载的 SnO_2 的传感器对甲苯具有优异的选择性,对所有其他室内污染物如苯、H_2、C_2H_5OH、HCHO、甲苯的交叉响应可忽略不计,如图 10.9 所示。此外,该传感器的检出限可低至 0.1 ppm。

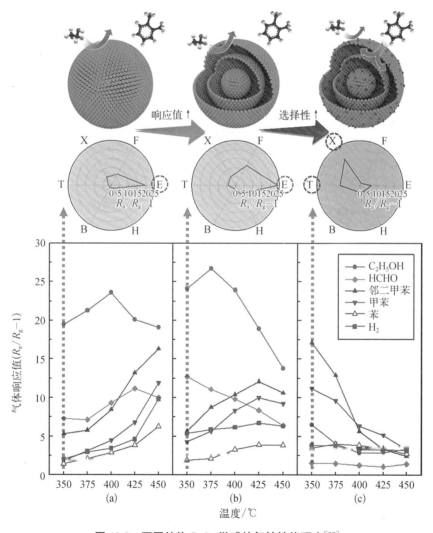

图 10.8 不同结构 SnO_2 微球的气敏性能研究[82]

(a) 致密 SnO_2 球;(b)、(c) 蛋黄-蛋壳结构 SnO_2 球和负载 Pd 的蛋黄-蛋壳结构 SnO_2 球在 350~450℃下对各种分析气体的气体响应,其中 B 表示苯,H 表示氢气,E 表示乙醇,F 表示甲醛,X 表示邻二甲苯,T 表示甲苯

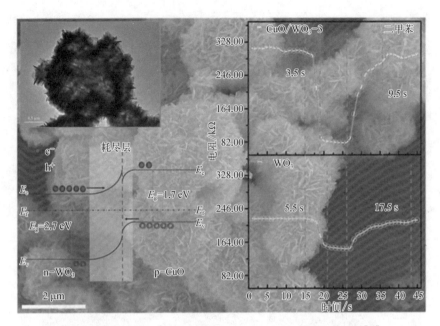

图 10.9 基于 CuO/WO₃ 分级空心微球材料的二甲苯传感器[83]

使用传统的金属氧化物半导体(MOS)气体传感器检测惰性气体化学品,如苯系物(即苯、甲苯、乙苯和二甲苯)蒸气通常比较困难。CeO_2 是一种广泛用于苯及其同系物裂解的催化剂。然而,由于其绝缘特性,CeO_2 催化剂不能直接添加到用于传感应用的 MOS 基材料中。Wang 等[84]提出了一种用于痕量 BTEX 检测的新型双层传感结构,其中绝缘 CeO_2 催化剂涂覆在 ZnO 传感层的上层。他们的研究证实,CeO_2 催化剂的顶层对底部 ZnO 传感层电阻的影响可以忽略不计。为了实现对痕量 BTEX 的高传感性能,使用两种纳米材料,即三角形 CeO_2 纳米片和纳米多孔 ZnO 构建了双层气体传感器。制作的 ZnO-CeO_2 传感器对 BTEX 蒸气具有优异的传感性能,其甲苯检测限(LoD)达到 10 ppb。原位质谱进一步揭示了 CeO_2 纳米片的催化裂化机理(见图 10.10)。

Cao 等[85]以 Mn 掺杂 Co_3O_4 作为传感材料,采用原位漫反射傅里叶变换红外光谱技术研究了甲苯气体传感过程中传感材料界面处甲苯分子的动力学演变过程,发现甲苯气体分子在传感材料表面逐级氧化生成了一系列中间态物种,推断甲苯分子氧化过程由数个基元反应构成,而非一步完全氧化过程。进一步地,该团队[86]更新了传感材料的制备方法,获得了对甲苯气体响应具有显著提高的 Ti 掺杂 Co_3O_4 传感材料。通过测试传感器对甲苯、邻二甲苯、间二甲苯、对二甲苯、乙基苯的响应-恢复曲线,并从相应的气体响应-恢复曲线中提取特征因子,构成气体传感响应分析数据集,以及结合机器学习算法,实现了对甲苯、二甲苯异构体、乙基苯气体的准确定性分析(见图 10.11)。

图 10.10　基于 CeO_2 - ZnO 双层结构的 BTEX 传感器[84]

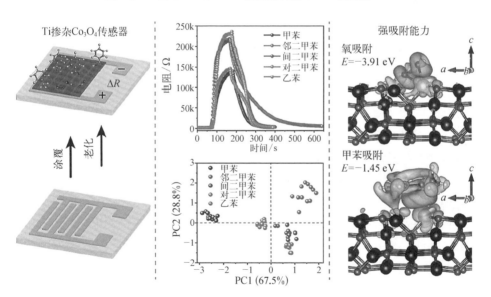

图 10.11　基于 Ti 掺杂 Co_3O_4 传感材料的 BTEX 传感器(后附彩图)[86]

10.2　环境气体传感器

10.2.1　二氧化碳传感器

二氧化碳是大气的组成成分之一,测定室内环境中的二氧化碳含量具有重要意义。实验表明,在空气中氧含量充足的情况下,二氧化碳浓度为5%时,它对人体无害;但是如果氧浓度低于17%,则空气中含有4%的二氧化碳,就会导致人体中毒。表10.5总结了一些半导体金属氧化物二氧化碳传感器。

表10.5 半导体金属氧化物二氧化碳传感器

材　料	结构形貌	气体浓度/ppm	工作温度/℃	参考文献
Gd_2O_3	微球	1%～4%	470	[96]
La_2O_3	薄膜	200 ppm	RT	[98]
Nd_2O_3	微球	100 ppm	400	[97]
$LaOCl\text{-}SnO_2$	纳米线	250 ppm	400	[99]
$BaTiO_3\text{-}CuO$	薄膜	500 ppm	300	[101]
$BaTiO_3\text{-}CuO\text{-}Ag$	薄膜	500 ppm	200	[100]

考虑到空气中的二氧化碳含量为0.03%,二氧化碳传感器的检测限通常不需要非常低,但使用典型的SMO(如SnO_2、ZnO、WO_3等)来检测CO_2是很困难的。Michel等[96]通过共沉淀法合成了直径为0.7～2.5 μm的Gd_2O_3空心微球。基于Gd_2O_3空心微球的传感器可以在470℃的工作温度下检测空气中1%的二氧化碳,且具有良好的重复性。Michel等[97]又用类似的方法制备了Nd_2O_3微球材料。基于Nd_2O_3微球的传感器可以在400℃的最佳工作温度下在短的响应时间(3.6 s)内检测到100 ppm的CO_2。Jinesh等[98]则报道了使用包含原子层沉积的La_2O_3薄膜的金属-绝缘体-硅(MIS)电容器在室温进行CO_2检测,传感器可以对200 ppm的CO_2响应。Trung等[99]使用p型LaOCl修饰n型SnO_2纳米线制备了二氧化碳传感器。与基于纯SnO_2纳米线的传感器相比,基于$LaOCl\text{-}SnO_2$纳米线的传感器在400℃下检测CO_2气体时具有更高的灵敏度($R_a/R_g=6.8$,对4 000 ppm CO_2)和更快的响应、恢复速度,其优异的性能可归结于LaOCl和SnO_2纳米线中p-n异质结的形成及LaOCl的催化作用。Herrán等[100]报道了负载2.26% Ag的$BaTiO_3\text{-}CuO$可用作CO_2的敏感材料,其在300℃时的响应为18%[$(R_g-R_a)/R_a$]。

10.2.2 氧气传感器

氧气(O_2)是空气的主要成分之一,是人类生存最基本的需要。从化学角度来看,氧气参与许多氧化反应,因此,氧气的含量会影响燃料燃烧、金属腐蚀和食品保鲜等。监测氧含量在废气处理、工业生产和医疗保健中具有重要意义。用于氧传感的半导体金属氧化物尚未被广泛研究,已有的报道主要基于氧化锆、氧化钛、氧化锌,关于SMO的氧气传感器列于表10.6。

表 10.6　半导体金属氧化物氧气传感器

材　料	结构形貌	气体浓度	操作温度/℃	参考文献
ZrO_2	纳米膜	0.21～1 ppm	450	[107]
$Ce_{1-x}Zr_xO_2$	纳米膜	10^{-7}～10^5	600～800	[102]
Nb/TiO_2	纳米膜	10 ppm	300	[104]
Pd/TiO_2	纳米膜	100 ppm	300	[103]
$SrTiO_3$	纳米膜	20%	40	[108]
ZnO	纳米线	8%～11%	50	[109]
Cr‑ZnO	纳米膜	25%	250	[105]
Mn‑ZnO	纳米棒	5～15 ppm	RT	[110]

Izu 等[102]报道了使用新沉淀法制备 Zr 掺杂二氧化铈纳米粉末(约 100 nm)。所有样品(0～20%① ZrO_2)都具有多孔结构,由粉末制备的基于 CeO_2‑ZrO_2 厚膜的电阻氧气传感器都表现出较大的氧分压依赖性,即氧气分压范围为 10^{-17}～10^5 Pa 时都具有灵敏度。实验无法观察到 ZrO_2 浓度对响应时间的变化,但在 1 073 K 时氧分压从高到低变化的响应时间为 9～20 ms,而从低到高变化的响应时间则为 1～2 ms。

实际上,高的工作温度是氧气传感器应用的障碍。Castañeda[103]将 Pd 负载在 TiO_2 薄膜上制成了氧气传感器。气体传感实验结果表明,在 300℃的零级空气中,O_2 浓度为 100 ppm,灵敏度达到 1.18[$S=(R_g-R_a)/R_a$]的稳定值,这与采用更昂贵、更复杂的技术制备的传感器检测氧气的报告结果相同。Sotter 等[104]通过溶胶‑凝胶法制备了 Nb 掺杂的 TiO_2 纳米薄膜,并将其在 600～900℃的温度下煅烧。与纯 TiO_2 相比,Nb 掺杂的 TiO_2 在 10 ppm O_2 存在下显示出较低的工作温度。这种优良的性能提升可归因于 Nb 掺杂提供的缺陷,其提高了吸附氧的量。后来,Al‑Hardan 等[105]通过射频反应共溅射合成了 1%(原子百分比)Cr 掺杂的 ZnO,并用于氧气传感应用。Cr 掺杂的 ZnO 传感器显示出较低的工作温度(250℃),而未掺杂的 ZnO 传感器的工作温度约为 350℃(见图 10.12)。

Liang 等[106]报道了一种可伸缩、自加热、自黏和室温氧气传感器,具有良好的重复性、全浓度检测范围(0～100%)、低理论检测限(5.7 ppm)、高灵敏度(0.2%/ppm)、良好的线性度。该研究采用聚丙烯酰胺‑壳聚糖(PAM‑CS)双网络(DN)有机水凝胶作为新型换能器材料制备了耐湿性材料。PAM‑CS DN 有机水凝胶采用简单的浸泡和溶剂替换策略,由 PAM‑CS 复合水凝胶转化而来。与原始水

---

① 0～20%指摩尔百分比。

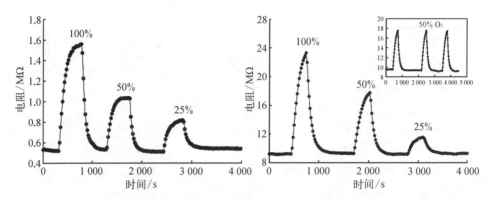

图 10.12 对于不同的 O_2 浓度，未掺杂的 ZnO 传感器(左)在 350℃ 的工作温度下的动态响应和 Cr 掺杂的 ZnO 传感器(右)在 250℃ 的工作温度下的动态响应[105]

插图显示了 Cr 掺杂的 ZnO 传感器在测试 50% O_2 时的可重复性

凝胶相比，DN 有机水凝胶的机械强度、保水性、抗冻性和对氧气的敏感性都大大增强。值得注意的是，施加拉伸应变可以提高基于有机水凝胶的氧气传感器的灵敏度和响应速度。此外，自愈前后对相同浓度氧气的反应基本相同。重要的是，他们提出了一种电化学反应机制来解释氧气传感器的正电流漂移，并通过合理设计的实验证实了这种传感机制。该有机水凝胶氧气传感器可用于实时监测人体呼吸，验证了其实用性（见图 10.13）。

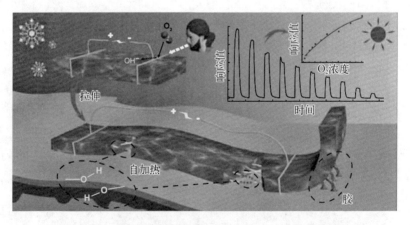

图 10.13 室温下高性能自愈、自黏和稳定的有机水凝胶基可拉伸氧气传感器[106]

10.2.3 二氧化硫传感器

二氧化硫（SO_2）是一种无色但有害的气体，闻起来像烧焦的火柴。二氧化硫可被氧化成三氧化硫，在水蒸气存在下很容易转化成硫酸雾，也可被氧化形成酸性气溶胶。二氧化硫是硫酸盐的前体,硫酸盐则是大气中可吸入颗粒的主要成分之一。暴露于高浓度的 SO_2 可导致若干健康问题,包括呼吸系统疾病和心血管疾病。此外,二氧化硫和氮氧化物是酸雨的主要前体,会导致严重的环境破坏。

用于 SO_2 检测的传感器中,WO_3 和 SnO_2 是最有希望的金属氧化物。半导体金属氧化物 SO_2 传感器汇总于表 10.7 中。纯的金属氧化物对 SO_2 的传感性能并不能满足现代社会的要求,因此需采用贵金属负载或掺杂来提高 WO_3 和 SnO_2 的传感性能。Shimizu 等[111]测试了不同金属氧化物的 SO_2 传感特性,发现 WO_3 在 400℃时的灵敏度最高。该项工作还研究了不同贵金属修饰的效果,发现添加 1.0% Ag 可有效提高 WO_3 在 450℃下的灵敏度。一些研究表明,其他原子掺杂可以改善 SnO_2 的 SO_2 传感特性。Hidalgo 等[112]使用 Ni 掺杂的 SnO_2 制备了 SO_2 传感器,1%① Ni 掺杂的 SnO_2 基传感器呈现线性曲线,最大极限响应为 32 ppm SO_2。Das 和同事则首次报道钒掺杂的 SnO_2 基半导体传感器可用于 SO_2 泄漏监测[112]。这种传感器对 SO_2 具有出色的灵敏度,检测限低(低至 5 ppm),并且在存在其他气体如一氧化碳、甲烷和丁烷时具有良好的选择性。除了 WO_3 和 SnO_2 外,用于 SO_2 检测的金属氧化物并不多。Liang 等[113]报道了基于 NASICON(钠超离子导体)和 V_2O_5 掺杂 TiO_2 传感电极的紧凑型管状传感器可用于 SO_2 传感。在 600℃下烧结的 5% V_2O_5 掺杂的 TiO_2 传感器在 200～400℃的工作温度下对空气中 1～50 ppm SO_2 具有出色的传感性能。Das 等报道基于 V_2O_5 掺杂 SnO_2 的传感器对 SO_2 与其他干扰气体,如 NO、NO_2、CH_4、CO、NH_3 和 CO_2 具有极好的选择性和快速的响应/恢复速度(10 s/35 s,对 50 ppm SO_2),如图 10.14 所示[114]。

表 10.7 半导体金属氧化物二氧化硫传感器

材料	结构形貌	气体浓度/ppm	工作温度/℃	灵敏度	响应/恢复时间/s	参考文献
V - SnO_2	粉末	100	350	70[b]	—/—	[114]
Ni - SnO_2	颗粒	32	RT	0.95[c]	—/—	[112]
Pt - WO_3	膜	1	200	6[c]	—/—	[115]
Ag - WO_3	粉末	500	450	11[a]	—/—	[111]
V_2O_5 - TiO_2	膜	50	300		10/35	[113]

注:a (R_a/R_g);b $[(R_a-R_g)/R_a \times 100]$;c $[(R_a-R_g)/R_a]$。

---

① 1%指摩尔百分比。

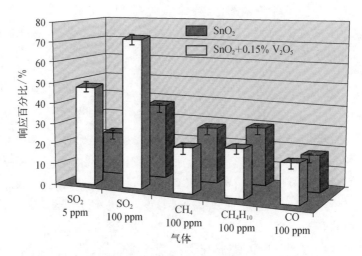

图 10.14 工作温度为 350℃ 时 SnO_2 和 $SnO_2+0.15\%$ V_2O_5 传感器对不同气体的灵敏度[114]

10.2.4 臭氧传感器

臭氧(O_3)是一种淡蓝色气体,具有明显的刺激性气味,因其为地球屏蔽了来自太阳的大量有害紫外线而闻名。臭氧是一种强氧化剂,被广泛应用于工业和日常生活的杀菌消毒。但是,如果浓度高于正常水平,它仍然对人体健康有害。因此,越来越多的半导体气体传感器被开发来检测 O_3 的浓度。用于检测 O_3 的半导体金属氧化物传感器总结在表 10.8 中。

表 10.8 半导体金属氧化物臭氧传感器

材料	结构形貌	气体浓度/ppm	工作温度/℃	灵敏度	响应/恢复时间/s	参考文献
Ag_2WO_4	纳米棒	0.95	300	2[a]	6/13	[118]
WO_3	膜	0.03	250	16[a]	—/—	[120]
WO_3	膜	0.2	350	2[a]	—/—	[116]
SnO_2	膜	1	200~350	$10^2 \sim 10^4$[a]	5~10	[117]
$CuCrO_2$	膜	100	RT	—	240/120	[119]

注:a (R_g/R_a)。

在 O_3 传感材料中,WO_3 受到了较多的关注。Bendahan 等[116]通过 RF 溅射将 WO_3 沉积在硅基板上。基于 WO_3 膜的传感器在 350℃ 的工作温度下对低浓度臭氧(0.2～0.4 ppm)表现出良好的稳定性和快速响应。Korotcenkov 等[117]报道了通过喷雾热解沉积的方法制备基于 SnO_2 薄膜的臭氧传感器。在 35%～45% RH 的空气中臭氧浓度约为 1 ppm 时,传感器在暴露于 10^2～10^4 范围内的臭氧时显示出大的响应(R_{ozone}/R_{air})。即使在约为 200℃ 的低工作温度下,响应时间仍然保持在 5～10 s。Silva[118]合成了具有 1-D 纳米棒状结构的 Ag_2WO_4,并将其用于 O_3 检测。所获得的传感器在 300℃ 的工作条件下,即使臭氧浓度低至 80 ppb,仍可显示出良好的灵敏度和响应/恢复性能。然而,大多数臭氧传感器的工作温度通常接近甚至高于 200℃,因此近年来,用新材料降低工作温度受到关注。Deng 等[119]通过溶胶-凝胶和脉冲激光沉积方法制备了 $CuCrO_2$ 薄膜。如图 10.15 所示,半导体 $CuCrO_2$ 薄膜材料在室温下对臭氧具有选择性和可逆响应,这意味着 $CuCrO_2$ 有希望成为高性能臭氧传感器的新选择。

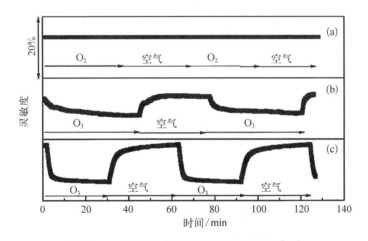

图 10.15　$CuCrO_2$ 薄膜室温下的臭氧传感[119]

(a) $CuCrO_2$ 薄膜在 300 K 时对 O_2-空气-O_2 的响应;(b)、(c) $CuCrO_2$ 薄膜和烧结块体材料在 300 K 时对 100 ppm 臭氧-空气-100 ppm 臭氧的响应

10.2.5　氨气传感器

氨(NH_3)是一种无色气体,具有强烈而刺激性的气味。作为食物和肥料的前体,氨对陆地生物的营养需求做出了重大贡献。氨也是许多药物合成的基础。尽管氨是非常有用的,但氨对人体具有腐蚀性和危害性,特别是当其浓度超过 25 ppm 时。由于氨气的广泛使用,基于半导体金属氧化物的气体传感器对氨气(NH_3)的传感引起了大量的研究关注。目前已有一些研究报道了各种传感材料,如 ZnO、WO_3、SnO_2 和 In_2O_3 作为 NH_3 传感器。与其他气体传

感器不同，许多 NH₃ 传感器可以在室温下工作。表 10.9 列出了基于 SMO 的氨气传感器。

表 10.9 半导体金属氧化物氨气传感器

材　料	结构形貌	气体浓度/ppm	工作温度/℃	灵敏度	响应/恢复时间/s	参考文献
ZnO	空心球	0.95	220	7.9^a	—/—	[126]
ZnO	纳米棒	50～800	RT	QCM	—/—	[121]
Fe₂O₃-ZnO	颗粒	0.4	RT	10^{4b}	20/20	[122]
SnO₂/MWCNTs	薄膜	60～800	RT	2～32^a	<300	[124]
WO₃	纳米线	300～1 500	250	2～10^a	7/8	[123]
WO₃/Pt	介孔	200	125	13.61^a	43/272	[127]
In₂O₃	纳米管	25	RT	2 500^a	<20/<20	[125]
SnO₂/CuO	纳米线	1‰	RT	300%^{c}	125/500	[128]
PANI/TiO₂	薄膜	23	RT	1.2	—/—	[129]

注：a (R_g/R_a)；b (I_g/I_a)；c $(R_a-R_g)/R_a$。

Minh 等[121]通过简单的低温水热法在石英晶体微天平(QCM)的金电极上直接合成了垂直排列的 ZnO 纳米棒，并将其用于氨的检测。研究人员测试了具有不同长度 ZnO 纳米棒传感器的 NH₃ 传感特性，在室温下可检测的 NH₃ 浓度范围为50～800 ppm。同时，增加 ZnO 纳米棒的长度可以增加气体传感响应。Tang 等[122]通过溶胶-凝胶法制备了 Fe₂O₃-ZnO 纳米复合材料，并测试了不同 Fe∶Zn 组分的传感特性。测试结果表明，Fe∶Zn＝2‰的传感器在室温下对 NH₃ 具有高灵敏度和选择性，并具有超短的响应时间和恢复时间(均小于 20 s)。据报道，碳纳米管(CNT)可增强金属氧化物对 NH₃ 的传感性能。Hieu 等[123]在多孔单壁碳纳米管(SWCNTs)薄膜的基底上沉积了钨金属，得到了 WO₃ 的大规模纳米线状结构。这种材料具有多孔结构，平均直径约为 70 nm，而长度则达到 μm 级别。传感器在 250℃时可对 300～1 500 ppm 的 NH₃ 响应，其快速响应时间和恢复时间可降至 7 s 和 8 s。此外，Hieu 等[124]还使用 SnO₂/MWCNTs 复合材料制造了 NH₃ 传感器。传感器在室温下显示出高的响应速度和恢复性能(小于 5 min)，可检测的 NH₃ 浓度范围为 60～800 ppm。CNT 也可以用作模板。Du 等[125]通过在 CNT 模板上使用逐层组装以及随后的煅烧合成了平均直径为 20～60 nm 的多孔多晶 In₂O₃ 纳米管。这种 In₂O₃ 纳米管在室温下具有优异的 NH₃ 敏感性，并且由于超高的表面积、多晶性和多孔结构的特点，其具有良好的重现性和快速响应/恢复时间(小于 20 s)(见图 10.16)。

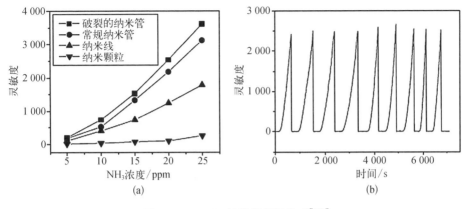

图 10.16　In_2O_3 纳米管氨气传感[125]

(a) 基于 In_2O_3 纳米结构的四种类型气体传感器在室温下的灵敏度响应与 NH_3 浓度（5~25 ppm）曲线，包括破裂的 In_2O_3 纳米管、常规 In_2O_3 纳米管、In_2O_3 纳米线和 In_2O_3 纳米颗粒；(b) 基于破裂的 In_2O_3 纳米管的气体传感器在室温下对 20 ppm NH_3 的灵敏度循环测试曲线

10.3　有毒气体传感器

10.3.1　一氧化碳传感器

一氧化碳（CO）是一种无色、无味且无刺激性的气体，仅依靠人的感官难以检测。它是由燃料的不完全燃烧产生的，并且通常出现在汽车排放的尾气中。研究表明，CO 可以不可逆地结合血液中作为氧运输分子的血红蛋白的铁中心。由于不可逆的结合，氧气不再被吸收，这会导致细胞缺氧从而造成人体受损。高浓度的 CO 暴露甚至会导致人死亡。英国政府于 2000 年 1 月通过的国家空气质量标准认为，暴露于浓度为 10 ppm 的 CO 气体中 8 h 即可对生命和健康有直接危害。即使浓度低于该值，CO 也可能对人体健康有害。因此，考虑到安全因素，迫切需要对空气中的 CO 浓度进行监控。

基于半导体金属氧化物的 CO 传感器总结在表 10.10 中，其中，基于 SnO_2 的传感器和基于 ZnO 的传感器是报道较多的两类传感器。贵金属修饰的 SnO_2 传感器也被广泛应用于 CO 传感。Wang 等[130]合成了用于 CO 气体传感器的 Pt 负载 SnO_2 多孔纳米材料（见图 10.17）。气体传感测试表明，在室温下 Pt 负载的 SnO_2 对 100 ppm CO 的最高响应达到 64.5（R_a/R_g），远高于之前的报道结果。而且，这种传感器对 H_2 的干扰具有良好的 CO 选择性。Wang 和 Chen[131]通过共沉淀法制备了氧化钒锡纳米颗粒，对一系列样品进行了测试。结果表明，较低含量的钒物种（V∶Sn=0.05 和 0.1）可以提高 SnO_2 传感器的性能。这可归因于还原的钒阳离子(V^{3+}和V^{4+})作为电子供体在传感器表面上增加的氧吸附。Wu等[132]报道,在室温下,0.1%CNT掺杂的Co_3O_4-SnO_2复合材料对CO具有较好响应,检测浓度范围为20~1 000 ppm。

表10.10 半导体金属氧化物一氧化碳传感器

材料	结构形貌	气体浓度/ppm	工作温度/℃	灵敏度	响应/恢复时间/s	参考文献
ZnO	介孔粉体	10	350	0.5[b]	—/—	[137]
ZnO	纳米棒	200	400	176[d]	46/27	[133]
CeO_2-ZnO	薄膜	10 000	380	450[c]	44/40	[134]
ZnO-CuO/Pt	颗粒	1 000	RT	2.64[a]	80/80	[135]
Pd-SnO_2	粉末	300	150	3[a]	5/22	[138]
Pt-SnO_2	多孔粉体	100	RT	64.5[a]	144/—	[130]
V掺杂SnO_2	粉末	250	175	4[a]	—/—	[131]
Co_3O_4	纳米棒	50	250	6.55[a]	4/6	[139]
Co_3O_4-SnO_2 CNT	薄膜	20~1 000	RT			[132]
WO_3/Pt	介孔粉体	100	125	10[a]	1/16	[136]

注:a (R_g/R_a);b $(G/G_0)-1$;c $(R_a-R_g)/R_g\times 100$;d $(R_a/R_g)\times 100$。

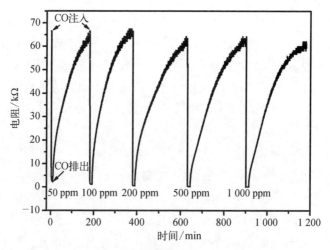

图10.17 室温下SnO_2/Pt PNS传感器(厚度为0.5 mm)电阻随时间变化的曲线[130]

将不同浓度的CO气体重复注入腔室并从腔室中排出

ZnO 基 CO 传感器是另一种被广泛报道的类型。Khoang 等[133]制备了在晶片级上生长的 ZnO 纳米棒平面型传感器用于 CO 检测,其中最好的 ZnO NR 传感器可以在 400℃下响应 CO,灵敏度为 0.37%/ppm。Al-Kuhaili 等[134]报道了一种用于 CO 传感器的氧化锌(ZnO)和氧化铈(CeO_2)混合膜。CeO_2-ZnO 膜在 380℃的工作温度下可对 500~10 000 ppm 的 CO 表现出良好的敏感性。Yu 等[135]对 ZnO-CuO 和 Pt/ZnO-CuO 进行了一氧化碳(CO)的传感测试。基于优化的 ZnO 和 CuO 比例(1∶1,质量比)的传感器在室温下对 1 000 ppm CO 的响应(R_{co}/R_{air})最高为 1.28,响应时间和恢复时间分别为 41 s 和 86 s。然而,在 1∶1 的 ZnO-CuO 负载了 0.4% Pt 后,传感器在相同的 CO 浓度下响应可达到 2.64(R_{co}/R_{air})。传感器(0.4% Pt/ZnO-CuO)对 CO(100~1 000 ppm)表现出良好的线性关系,其响应时间和恢复时间为 81 s。

最近,Ma 等[136]利用两亲性嵌段共聚物 PEO-b-PS、疏水性铂前体和亲水性钨前体的共组装合成了 Pt 敏化的高度有序介孔 WO_3 复合材料。基于 WO_3/Pt 的传感器在低温(125℃,57%RH)下对 100 ppm 的 CO 气体表现出很高的灵敏度($R_{air}/R_{gas}=10\pm1$),超快的响应/恢复速度(16 s/1 s)以及良好的选择性,其优异的性能归因于材料具有的高度有序介孔结构和 Pt 纳米粒子的敏化作用。

10.3.2 H_2S 传感器

硫化氢(H_2S)是一种无色、有毒、易燃的气体,具有臭鸡蛋气味,被归类为化学窒息气体,因为它可以与人体血液中的血红蛋白结合,阻止氧气被带到身体的重要组织和器官,从而对人体造成伤害。硫化氢通过有机物(如粪肥)的厌氧分解产生,在低浓度下由于其特有的臭鸡蛋味而易于被察觉,在较高浓度下,H_2S 会导致嗅觉麻痹,给人一种虚假的安全感;当其浓度超过阈值极限值(TLV)时,它可能导致短暂的麻痹和死亡。因此,从安全角度来看,H_2S 的监测在石油化工和煤炭化工中非常重要。

表 10.11 总结了用于 H_2S 的半导体金属氧化物传感器。在这些传感器中,CuO 扮演着重要的角色,许多不同形貌的 CuO 材料已经被开发为 H_2S 传感器。Ramgir 等[140]使用 CuO 薄膜在室温下制备了亚 ppm 级别的 H_2S 传感器。在低浓度范围(100~400 ppb),其响应曲线是高度可逆的,具有短的响应时间(60 s)和恢复时间(90 s)。然而,对于高浓度(>50 ppm)的 H_2S,材料表面上的 CuO 颗粒可以转化为 CuS,导致不可逆的响应曲线。Zhang 等[141]通过一种简便、低成本、无表面活性剂的方法合成了 CuO 纳米片材料。传感特性测试表明,传感器具有 30 ppb~1.2 ppm 的宽线性测量范围、2 ppb 的低检测限和短响应/恢复时间(4 s/9 s)。此外,传感器显示出优异的 H_2S 选择性,即使暴露于 100 倍浓度的其他气体

[氮气(N_2)、氧气(O_2)、一氧化氮(NO)、一氧化碳(CO)、二氧化氮(NO_2)、氢气(H_2)]也无大的响应。Li 等[142]通过模板辅助电沉积法合成了取向排列的 CuO 纳米线。基于 CuO 纳米线的传感器对 H_2S 具有良好的响应和可重复性、2.5 ppb 的低检测限和 10~100 ppb 的良好线性范围。除了纯的 CuO 之外,CuO 基纳米复合材料也被研究以增强 H_2S 传感。Kim 等[143]报道了 Pd 修饰的 CuO 不仅可以提高灵敏度,还可以缩短响应时间和恢复时间。Xue 等[144]合成了均匀涂覆 CuO 的 SnO_2 纳米棒。CuO-SnO_2 核/壳 p-n 结纳米棒在 60℃ 的工作温度下表现出超高响应(高达 $9.4×10^6$)。

表 10.11 半导体金属氧化物硫化氢传感器

材　料	结构形貌	气体浓度/ppm	工作温度/℃	灵敏度	响应/恢复时间/s	参考文献
CuO	纳米线	0.1	180	0.237^b	—/—	[142]
CuO	空心球	1	190	350^c	3/9	[148]
CuO	薄膜	0.4	RT	20^c	60/90	[140]
CuO	纳米线陈列	0.5	160	1^d	—/—	[149]
CuO	纳米片	1	240	320^c	4/9	[141]
CuO-SnO_2	纳米棒	10	60	$9.4×10^{4a}$	—/—	[144]
Pd-CuO	纳米棒	100	300	$31,243^c$	690/80	[143]
$ZnSnO_3$	纳米笼	50	310	17.6^a	20/—	[150]
In_2O_3	颗粒	50	268.5	124.9^a	—/—	[151]
SnO_2	介孔	50	350	170^a	11/165	[146]
WO_3	介孔	50	250	250^a	2/38	[145]

注：a (R_g/R_a)；b $(R_a-R_g)/R_a$；c $(R_a-R_g)/R_a×100\%$；d $\Delta I/I_0$。

Li 等[145]以两亲性嵌段共聚物 PEO-b-PS 为模板,通过溶剂蒸发诱导组装方法合成了具有大孔径和结晶骨架的高度有序介孔 WO_3。得益于其高度有序的介孔结构和连续的晶体框架,介孔 WO_3 基 H_2S 气体传感器在低浓度(0.25 ppm)下表现出优异的响应,具有快速响应时间(2 s)和恢复时间(38 s)(见图 10.18)。使用类似的方法,Xiao 等[146]合成了有序介孔 SnO_2,其对 H_2S 也具有优异的传感性能。Wu 等[147]报道了微机电系统(MEMS)器件上原位生长的 H_2S 气体传感器,该传感器以 Pt 纳米粒子(NP)修饰的 SnO_2-ZnO 作为敏感材料,对 H_2S 传感显示出灵敏度高、选择性好、重复性好、功耗低的优异性能(见图 10.19)。

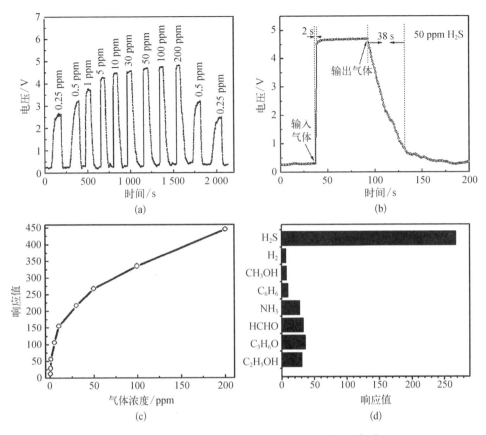

图 10.18 介孔 WO₃ 传感器的 H₂S 传感性能研究[145]

(a) 在 250℃的工作温度下,结晶介孔 WO₃ 传感器对不同浓度(0.25~200 ppm) H₂S 的响应和恢复曲线;
(b) 基于介孔 WO₃ 的传感器对 50 ppm H₂S 的响应和恢复曲线;(c) 灵敏度与 250℃下 H₂S 浓度之间的关系;(d) WO₃ 传感器对 50 ppm 不同气体的响应

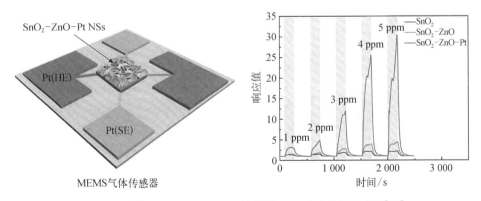

图 10.19 基于 SnO₂-ZnO-Pt 材料的 MEMS 硫化氢传感器[147]

10.3.3 二氧化氮传感器

氮氧化物(NO_2 和 NO)是汽车尾气和化学工厂排放产生的有毒气体。二氧化氮(NO_2)是最危险的气体之一,阈值极限值(TLV)低至 3 ppm。NO_2 也是光化学烟雾的主要成分,它在产生近地面臭氧的大气反应中起主要作用,并且刺激眼睛,可导致肺部损伤。尽管 NO 与 NO_2 相比毒性相对较低,但它也会引起酸雨、光化学烟雾和近地面臭氧的产生。对氮氧化物的检测对于减少它们对环境和人类的有害影响非常重要。

对 NO_2 传感的研究远远多于 NO,因为与 NO 相比,NO_2 被认为是更危险的气体。表 10.12 为半导体金属氧化物 NO_2 传感器的总结。从研究来看,WO_3 是对 NO_2 气体最敏感的半导体金属氧化物,具有不同形貌结构的 WO_3 已被用于 NO_2 检测。Zhang 等[152]采用大气等离子喷涂技术合成了 WO_3 敏感层,并用于 NO_2 传感。气体传感测试表明,其对 NO_2(0～450 ppb)具有良好的敏感性,工作温度范围为 95～240℃。Heidari 等[153]使用二氧化硅作为模板,通过硬模板法制备了介孔氧化钨。介孔 WO_3 在工作温度为 200℃时对 75～500 ppb 的气体具有良好的灵敏度和线性工作区间。Liu 等[154]合成了氧化钨纳米棒组装的微球,并将其用于 NO_2 检测。经 350℃退火,该材料在 200℃的工作温度下对 NO_2 气体具有高灵敏度和快速的响应速度,这得益于其三维网络具有高的有效比表面积。You 等[155]通过对不规则 WO_3 纳米片的水热处理合成了方形 WO_3 纳米片。基于方形 WO_3 纳米片的传感器在 125℃的最佳工作温度下表现出高灵敏度(R_g/R_a = 92,对 100 ppb NO_2),远高于不规则 WO_3 纳米片(在最佳工作温度 150℃下 R_g/R_a = 7,对 100 ppb NO_2),如图 10.20 所示。该结果表明结构在气体传感性能中起重要作用。此外,这种基于方形 WO_3 的传感器可以检测低至 40 ppb 的 NO_2。Kida 等人[156]通过酸化处理的方式合成了 WO_3 纳米片。基于 WO_3 纳米片的传感器在 200℃下表现出较高的灵敏度(R_g/R_a=150,对 50 ppb NO_2),同时对 NO 具有较好的抗干扰性。

表 10.12 半导体金属氧化物二氧化氮传感器

材料	结构形貌	气体浓度/ppm	工作温度/℃	灵敏度	响应/恢复时间/s	参考文献
WO_3	粉末	0.45	130	77[a]	—/—	[152]
WO_3	纳米片	0.1	125	92[a]	—/—	[155]
WO_3	介孔	0.05～0.5	150～300	<50[a]	7～22.5/4～22.5	[153]
WO_3	薄膜		150	1.6[b]	—/—	[160]

续表

材　料	结构形貌	气体浓度/ppm	工作温度/℃	灵敏度	响应/恢复时间/s	参考文献
WO_3	微球	20	200	500[b]	—/—	[154]
WO_3/Ag	介孔	0.1	75	44[a]	5.5 min/2.46 min	[162]
ZnO	纳米棒	1	350	1.8[a]	3 min	[158]
ZnO	纳米棒	1	200	0.075[b]	20 s	[163]
Co_3O_4-ZnO	纳米线	5	200	45.4[a]	—/—	[159]
Pd 掺杂 TiO_2	纳米纤维	2.1	180	38[a]	—/—	[164]

注：a (R_g/R_a)；b $(R_a-R_g)/R_a$；c $(R_a-R_g)/R_a\times 100\%$。

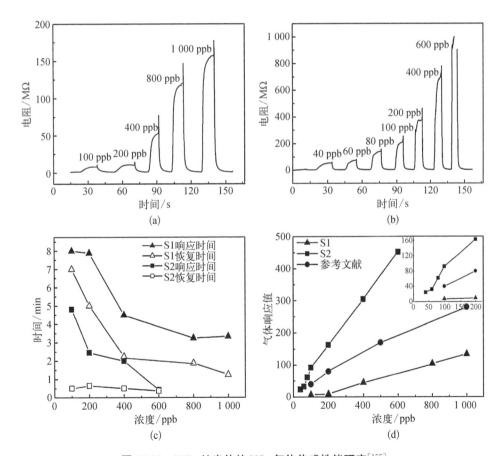

图 10.20　WO_3 纳米片的 NO_2 气体传感性能研究[155]

(a)、(b) S1 和 S2 对不同浓度 NO_2 的动态响应性；(c) S1 和 S2 对不同浓度 NO_2 的响应和恢复特性；(d) 在每个最佳工作温度下 S1、S2 和参考文献样品对于不同 NO_2 浓度气体的响应(插图显示在低浓度 NO_2 下的气体响应)，其中 S1 表示不规则的 WO_3 纳米片，S2 表示方形 WO_3 纳米片

除 WO_3 外,ZnO 是半导体金属氧化物中另一类报道较多的用于 NO_2 传感的气敏材料。Min 等[157]在硅晶片上制作了 ZnO 薄膜微阵列,发现传感器对 NO_2 具有极好的敏感性。Cho 等[158]通过水热法制备了 ZnO 纳米棒,良好分散的 ZnO 纳米棒在 1 ppm NO_2 下的响应值为 1.8,而在 50 ppm CO 下电阻没有显著响应。结果表明,这种 ZnO 纳米棒可应用于汽车通风系统中,以检测 NO_2 存在而避免 CO 干扰。Na 等[159]报道,使用 Co_3O_4 修饰 ZnO 纳米线可以实现 C_2H_5OH 和 NO_2 的选择性检测,这可归因于纳米晶 Co_3O_4 的催化作用和 p-n 异质结的形成。

以氧化铟(In_2O_3)作为敏感材料构筑 NO_2 传感器也有相关研究。Ren 等[161]以大分子量($M_n = 24\ 600$ g/mol)的两亲性嵌段共聚物聚环氧乙烷-b-聚苯乙烯(PEO-b-PS)为模板剂,以 $InCl_3$ 为无机前驱体,采取 CaO_2 释放氧气加速模板剂降解的策略,用溶剂挥发诱导自组装的方法,合成了具有晶化骨架、大孔径(约 14.5 nm)、高比表面积(48 m^2/g)的有序介孔片层 In_2O_3 材料,其在 NO_2 气体传感的应用中表现出优异的性能,具有低的检测限(50 ppb)、对 250 ppb NO_2 显示出高的灵敏度(10.5)和短的响应/恢复时间(2 min/2.5 min)(见图 10.21)。

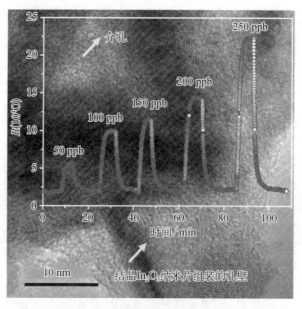

图 10.21　基于有序介孔片层氧化铟(In_2O_3)材料的 NO_2 传感器[161]

10.4 可燃气体传感器

10.4.1 甲烷传感器

甲烷(CH_4)是通过自然和人类活动产生的一种无色、无味、无毒的可燃温室气体。它通过天然材料的腐烂形成,在垃圾填埋场、沼泽、化粪池系统和下水道中很常见,可用作燃料,也可用于制造有机化学品。CH_4在空气中可形成爆炸性混合物,含量低至5%。如今,以甲烷为主要成分的天然气已作为燃料进入千家万户。因此,甲烷传感器的开发具有重要的安全意义。

已有很多金属氧化物材料用于CH_4传感研究,但是只有少数半导体金属氧化物对CH_4气体敏感,见表10.13。贵金属修饰的ZnO是一种对甲烷气体敏感的材料。Basu等[165]制备了基于Pd修饰ZnO薄膜的MIM(金属-绝缘体-金属)传感器。即使温度低于100℃,传感器也能响应CH_4,并在氮气(1%甲烷)中和空气(1%甲烷)中保持稳定。Bhattacharyya等[166]研究了不同贵金属催化剂(Pd、Ag、Rh、Pt)修饰的ZnO的甲烷传感性能。结果表明,Pd-Ag和Rh修饰可以提供比Pt修饰(300℃)更低的最佳操作温度(250℃)。Rh修饰比Pd-Ag(74.3%)和Pt(69.3%)修饰具有更高的灵敏度响应(83.6%),但Pd-Ag修饰显示出最短的响应时间和恢复时间。所有贵金属修饰的ZnO在适当的温度下都能响应0.1%的CH_4。Kim等[167]报道,与氧化铝负载的Pd催化剂混合可以促进氧化锡基传感器的甲烷传感性能。传感器可以在500~10 000 ppm范围内响应甲烷,且在385℃的工作温度下具有足够高的灵敏度。Prasad等[168]发现单相二氧化钒在50℃时可以响应50 ppm CH_4,从而降低CH_4传感器的工作温度,如图10.22所示。Dayan等[169]发现ZnO∶Sb薄膜在360℃对1 000 ppm甲烷响应值可以达到20。

表10.13 半导体金属氧化物甲烷传感器

材料	结构形貌	气体浓度	工作温度/℃	灵敏度	响应/恢复时间/s	参考文献
VO_2	膜	50 ppm	50	1.4[a]	—/—	[168]
ZnO∶Sb	膜	1 000 ppm	360	20[b]	10 s/—	[169]
SnO_2/Os	膜	1 000 ppm	270	1.25[c]	—/—	[171]
Pd-Ag/ZnO	膜	1%	70	48[d]	4.6/22.7	[165]
Rh/ZnO	膜	1%	250	74.3[a]	23/73	[166]
SnO_2(Ca,Pt) Pd/Al_2O_3	膜	500 ppm	385	90[a]	—/—	[167]

注:a $(R_0-R_g)/R_0\times100$,%;b $(G_g-G_a)/G_a$;c I_g/I_0;d $(I_g-I_0)/I_0$。

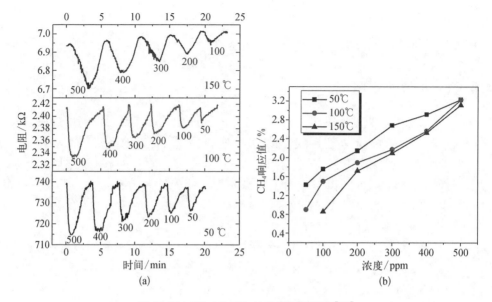

图 10.22　VO₂ 对 CH₄ 的传感性能研究[168]

(a) VO₂ 对 CH₄ 在 50℃、100℃和 150℃的动态响应曲线（曲线下方的数字表示浓度）；
(b) 传感器响应(S)-浓度曲线在各种工作温度下的比较

检测低浓度的 CH₄ 在工业制造中具有重要意义，例如石化工业和天然气催化，但由于其高度对称和稳定的结构，检测低浓度的 CH₄ 并不容易。Chen 等[170]通过水热合成和 H₂ 处理合成了具有丰富缺陷的 ZnO_{1-x} 纳米花（NFs），缺陷增强了气敏性能。为了实现对 CH₄ 的低浓度检测，又合成了超薄钯团簇（1～2 nm）敏化剂，并将其修饰到 ZnO_{1-x} 表面。研究发现 Pd-Cs-2/ZnO_{1-x} 气体传感器对 CH₄ 显示出增强的气体传感特性，即使在 ppm 浓度水平下也是如此。其在最佳工作温度 260℃下对 50 ppm CH₄ 的气体响应可以达到 5.0，具有良好的气体选择性，响应时间和恢复时间分别仅为 16.2 s 和 13.8 s（见图 10.23）。

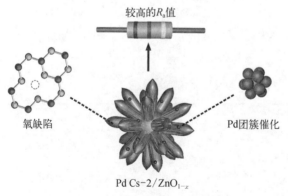

图 10.23　基于 Pd-ZnO_{1-x} 的传感器的甲烷气体传感机理示意图[170]

10.4.2　氢气传感器

氢气（H₂）是一种无色、无味的可燃气体，具有低的最小点火能量（0.017 mJ）、

第 10 章 半导体金属氧化物传感器的应用

高燃烧热(142 kJ/g H_2)和宽易燃范围(4%～75%),以及高燃烧速度和爆轰感度,点火温度为 560℃。氢气还具有强大的还原性,因此可以作为强还原剂。由于氢气的高可燃性,监测氢气浓度对于安全而言至关重要。

如今,越来越多的半导体金属氧化物被用作 H_2 检测的传感材料,并且得到了深入的研究。其中,SnO_2 基和 ZnO 基材料是主要的两种传感材料。用于 H_2 检测的 SMO 传感器见表 10.14。Wang 等[172]制备了用于 H_2 检测的 p 型 NiO/n 型 SnO_2 异质结纳米复合纤维,该传感器在 320℃的工作温度下获得了极快的响应恢复行为(约 3 s),检测限约 5 ppm。Liu 等[173]通过静电纺丝法合成了纯 SnO_2 纳米纤维和掺杂的 SnO_2 纳米纤维。气体传感实验表明,掺杂可以提高 SnO_2 的 H_2 传感性能。1% Co 掺杂的 SnO_2 纳米纤维表现出最高的响应(在 330℃的工作温度下对 100 ppm 的 H_2 达到 24),具有非常短的响应/恢复时间(2 s/3 s)。降低氢气工作温度是氢气传感器研发的一项挑战。Wang 等[174]使用 Pd 催化剂修饰 SnO_2 纳米纤维,所制备的传感器具有超低检测限(20 ppb)、高响应、快速响应/恢复性能,甚至在室温下也具有选择性(见图 10.24)。Das 等[175]将 Pt/ZnO 纳米线肖特基二极管作为氢气传感器,该传感器也可以在室温下工作。

表 10.14 半导体金属氧化物氢气传感器

材 料	结构形貌	气体浓度	工作温度/℃	灵敏度	响应/恢复时间/s
TiO_2	纳米管	1 000 ppm	290	—	150/—[177]
ZnO/Pt	纳米线	2 500 ppm	RT	80[b]	—/—[175]
p-NiO/n-SnO_2	纳米纤维	5 ppm	320	3[a]	3[172]
Co 掺杂 SnO_2	纳米纤维	100 ppm	330	24[a]	2/3[173]
30% Pd^0 负载的 SnO_2	纳米纤维	0.02 ppm	RT	1.09[a]	3[174]
Pd-In_2O_3	颗粒	1%①	RT	4.6×10^{7a}	28/32[179]
MoO_3	纳米片	1%	200	2.3[a]	7/24[176]

注:a R_g/R_0;b $(I_g-I_0)/I_0$,%;① 指体积分数。

其他金属氧化物用于 H_2 检测也有报道。Alsaif 等[176]通过一种基于液体的有机溶剂辅助研磨和超声处理方法合成了二维水合的 α-MoO_3 纳米薄片,基于 α-MoO_3 纳米薄片的传感器在 200℃时对 1%的 H_2 表现出良好的响应,响应时间和恢复时间分别为 7 s 和 24 s。Varghese 等[177]合成了 TiO_2 纳米管阵列。基于

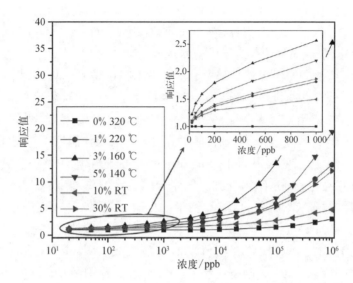

图 10.24 基于负载 Pd^0 的 SnO_2 纳米纤维的传感器在不同工作温度下对 20 ppb~1 000 ppm H_2 的响应关系图[174]

插图表示它们在低 H_2 浓度下的可接受响应;0、1%、3%、5%、10%、30%均指原子百分比

TiO_2 的传感器对 1 000 ppm 的 H_2 表现出高达 1 000 的响应值。

Luo 等[178]开发了一种简便、有效和可扩展的 H_2 还原方法,用于合成具有丰富氧空位的 n 型金属氧化物半导体材料 MOSs-D(SnO_2-D、ZnO-D 和 In_2O_3-D)。得到的 SnO_2-D 传感器显示出比基于 SnO_2 的传感器更高的响应,气敏测试结果表明,与原始 SnO_2 传感器(1.13)相比,SnO_2-D4 传感器仍然对 6 ppm H_2 表现出增强响应(2.3)(见图 10.25)。此外,相比于原始的 SnO_2 传感器,SnO_2-D4 还具有更低的工作温度和更快的响应/恢复速度。该研究不仅提供了一种简单、通用的 H_2 还原方法来合成具有氧空位的各种 n 型 MOS-D,而且还证明了氧空位结构是提高 n 型 MOSs 气敏性的可行途径。

10.4.3 LPG 传感器

液化石油气(LPG)主要由丙烷或丁烷组成,是一种易燃的碳氢化合物气体混合物,用于加热器具、烹饪和车辆燃料。LPG 具有潜在的危险性,因为其在意外或错误泄漏时可能会引发爆炸事故。因此,检测液化石油气是一项巨大的需求和挑战。

用于 LPG 检测的 SMO 传感器总结在表 10.15 中,其中基于 ZnO 的传感器得到了深入的研究。Patil 等[180]制造了用于高灵敏度 LPG 气感的纳米晶 ZnO 薄膜。ZnO 薄膜在 300℃时对 LPG 表现出高响应(见图 10.26),并且对 CO_2、H_2、NH_3、

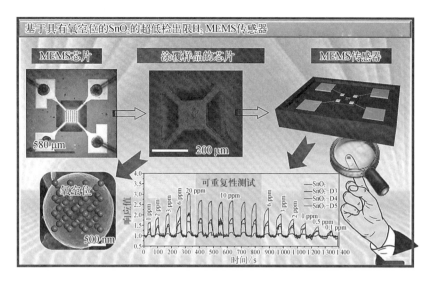

图 10.25 具有丰富氧空位的 SnO_2 微球的超低检出限的 MEMS 氢气传感器[178]

C_2H_5OH 和 Cl_2 具有优异的选择性。Shinde 等[181]也报道了 ZnO 在 LPG 传感中的应用,发现 ZnO 气体传感器的气敏性取决于其粒度。Waghulade 等[182]制备了纳米级氧化镉,基于 CdO 的传感器对 75 ppm LPG 的灵敏度在 450℃ 的工作温度下达到了 341%,响应时间和恢复时间分别为 3~5 s 和 8~10 s。传感器在 LPG 浓度范围内(25~75 ppm)表现出可靠的性能。Sen 等[183]发现 PANI/Fe_2O_3(3%)纳米复合材料对浓度为 50~200 ppm 的 LPG 敏感,室温下响应时间为 60 s,这证明了 LPG 监测的潜在应用价值。

表 10.15 半导体金属氧化物 LPG 传感器

材料	结构形貌	气体浓度	工作温度/℃	灵敏度	响应/恢复时间/s
PANI/Fe_2O_3	复合结构	50 ppm	28	0.5[b]	60/—[183]
$NdFeO_3$	膜	5%	RT	8 225%[a]	90/—[184]
ZnO	膜	1 000 ppm	300	1 727[a]	2/8[180]
ZnO	膜	0.8%	350	82%[b]	90/105[185]
ZnO	膜	0.8%	400	43%[b]	—/—[181]
CdO	粉末	785 ppm	450	341[b]	3~5/8~10[182]

注:a R_g/R_0;b $(R_g-R_0)/R_0$。

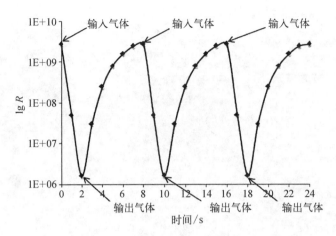

图 10.26 基于 ZnO 的传感器在 300℃下对 1 000 ppm LPG 的响应和恢复[181]

10.5 其他气体传感器

除了上述常见的气体传感器,还有一些气体传感器是针对人们生活生产中的特殊气体而被开发研究的,比如李斯特菌的呼出气体 3-羟基-2-丁酮、有机物腐败气体三甲胺等,检测这些气体对于食品安全、卫生医疗非常重要;一些工业车间生产中会产生的气体,如氯气、正丁醇、乙炔气等,监测这些气体对于安全生产十分关键;一些化合物如硝基爆炸物的痕量快速响应,对于现代安保、反恐同样意义深远。本节将进行简要的介绍。

李斯特菌是最具毒性的食源性病原体之一,其生长过程中产生的 3-羟基-2-丁酮是微生物特异性挥发性有机物。因此,3-羟基-2-丁酮可以作为检测食品、水等中李斯特菌的生物标志物。Zhu 等[186]应用高度有序的介孔氧化钨检测了 3-羟基-2-丁酮,这种基于有序介孔 WO_3 的气体传感器由于其介孔结构和晶体框架而显示出优异的灵敏度、快速响应和对 3-羟基-2-丁酮的高选择性响应,如图 10.27 所示。通过使用这种基于介孔 WO_3 的气体传感器可以实现有效且快速的细菌检测,具有良好的灵敏度、宽范围内的线性关系以及对李斯特氏菌的高选择性。

三甲胺(TMA)是一种无色、易吸湿和易燃的叔胺,在低浓度下具有强烈的鱼腥味,在较高浓度下具有类似氨的气味。TMA 是植物和动物分解的产物,有毒,对人体健康有害。因此,对 TMA 的检测非常重要。例如,Zhang 等[187]制备了分层异质纳米结构 Fe_2O_3/ZnO 作为 TMA 传感器。该传感器在 260℃的工作温度下对 TMA 具有高灵敏度,响应时间和恢复时间分别为 0.7 s 和 7.1 s(见图 10.28)。

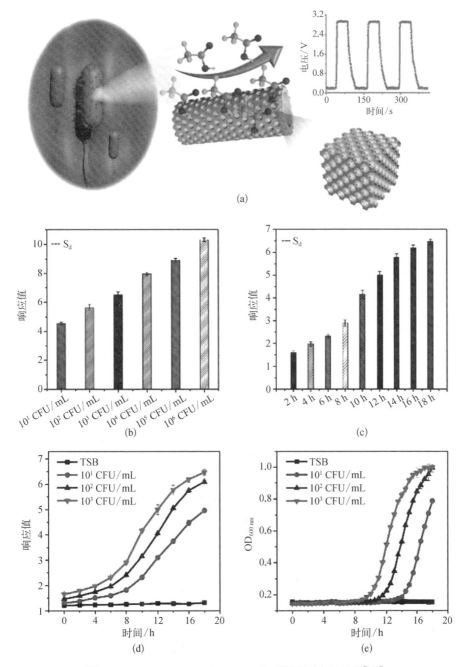

图 10.27　WO_3 - PEO_{117} - b - PS_{186} 传感器检测李斯特菌[186]

(a) 使用基于 WO_3 - PEO_{117} - b - PS_{186} 的传感器 S_d 检测李斯特菌的原理示意图;(b) 在 30℃孵育 18 h 后,S_d 对各种浓度李斯特菌的典型响应;(c) S_d 对 10^3 CFU·mL^{-1} 李斯特菌的典型响应(30℃孵育 2~18 h);(d) S_d 对各种浓度的李斯特菌(分别为 10^1、10^2、10^3 CFU·mL^{-1})和胰蛋白胨大豆肉汤的对照样品(TSB)的灵敏度(30℃孵育 18 h,间隔 2 h);(e) 使用浊度测量的方法,在接种 200 μL 初始浓度分别为 0、10^1、10^2、10^3 CFU·mL^{-1} 的细菌悬浮液的孔中,每 30 min 获得的李斯特菌生长的变异性(持续 18 h)

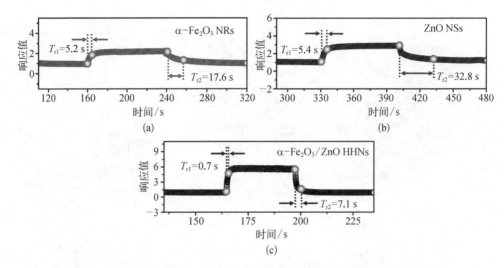

图 10.28 三种传感器在工作温度为 260℃ 时对 50 ppm 三甲胺的响应[187]

Shen 等[188]提出了一种新的金属氧化物半导体气体传感器，利用双金属 Au@Pt 纳米晶修饰的 α-Fe_2O_3 空心纳米立方体（NCs）作为传感材料检测了三甲胺（TMA）（见图 10.29）。该研究通过多种分析方法对 Au@Pt/α-Fe_2O_3 结构进行了表征，并对其气敏性能进行了研究。与原始 α-Fe_2O_3 NC 传感器相比，Au@Pt/

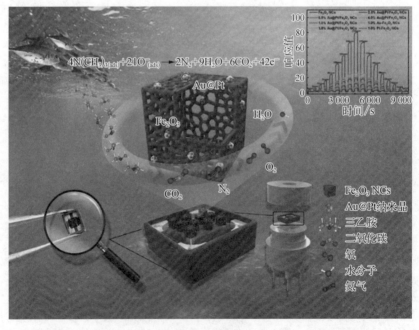

图 10.29 基于 Au@Pt-α-Fe_2O_3 空心纳米立方体的 TMA 传感器[188]

α-Fe_2O_3 NCs 在较低温度(150℃)下对 100 ppm TMA 气体表现出更快的响应时间(5 s)和更高的响应(R_a/R_g=32)。此外,研究还评估了 Au@Pt/α-Fe_2O_3-NC 传感器用顶空固相微萃取和气相色谱-质谱法检测大黄鱼新鲜度的能力,其在现场检测海鲜新鲜度方面表现出良好的应用前景。该传感器的高性能归因于 Au@Pt/α-Fe_2O_3 纳米晶具有的高比表面积(212.9 m^2/g)、特殊中空形态和 Au@Pt 双金属纳米晶的协同效应。

Feng 等[189]提出了一种自模板合成策略,以无毒、可再生、低成本的植物多酚为主要螯合剂,锡离子和氯金酸为金属源,制备了介孔 Au-SnO_2 纳米球。由于多酚的强螯合能力,研究人员获得了稳定的金/锡多酚甲醛球并将其用作前驱体。经过直接煅烧,成功制备了具有介孔晶体骨架、均匀直径(约 120 nm)和高比表面积(105.2 m^2/g)的 Au-SnO_2 材料。介孔 Au-SnO_2 纳米球被进一步用于 TEA 传感,结果表明,它们在低温(50℃)下对三乙胺(TEA)表现出高响应(5.16),检测限为 0.11 ppm,金的修饰可以有效地降低活化能和工作温度。与蓝牙信号传输系统集成后,该气体传感器可用于快速响应(约 30 s)的无线监测低浓度(10~50 ppm) TEA。该策略可用于合成介孔贵金属-氧化物复合材料,进一步提高气体传感器的传感性能(见图 10.30)。

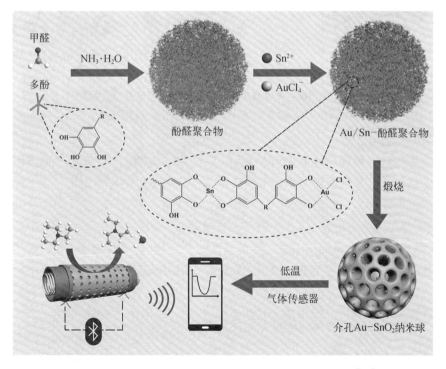

图 10.30　基于介孔 Au-SnO_2 纳米球材料的 TEA 传感器[189]

氯气(Cl_2)是一种常温常压下呈黄绿色、有剧毒且有强烈刺激性气味的气体。Cl_2可以通过呼吸道进入人体并溶解在黏膜所含的水分里,从而对人体的呼吸道黏膜造成严重伤害。它是氯碱工业的主要产品之一,具有强氧化性,因此常作为强氧化剂,用于农药杀菌、消毒剂、漂白剂溶剂等。此外,它在工业上也可用于生产塑料(如PVP)、合成纤维、染料以及各种氯化物。考虑到其应用的广泛性,开发高灵敏度的氯气传感器具有重要的安全价值。Li 等[190]报道了一种利用室温固相反应合成的In_2O_3/SnO_2异质结纳米材料。基于该材料的传感器在260℃的工作温度下对50 ppm Cl_2的最大灵敏度(R_a/R_g)可达(1 034.3 ± 67.7),响应/恢复时间为2 s/9 s,可连续工作30天。其机理主要是由于n-n异质结的形成,在In_2O_3接触SnO_2的表面生成了累积层,氯气作为一种强氧化性气体可以更多地从材料中接受电子,导致电阻急剧上升,获得了较大的灵敏度,如图10.31所示。

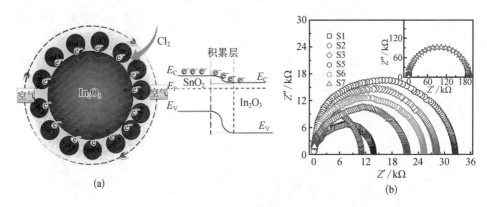

图 10.31 In_2O_3/SnO_2异质结传感机理及交流阻抗研究(后附彩图)[190]

(a) In_2O_3/SnO_2异质微观结构机制的示意图;(b) 基于S1、S2、S3、S4(插图)、S5、S6和S7的传感器的交流阻抗谱

正丁醇是一种刺激性和麻醉性液体,广泛用作溶剂、有机合成中间体和提取剂。暴露于正丁醇气体环境中可引起头晕、头痛、嗜睡和皮炎等症状。正丁醇检测的常用传统方法是气相色谱(GC),然而这种方式成本较高、耗时且操作复杂。目前,随着气体传感器的发展,一些研究人员发现,金属氧化物半导体也是制备高灵敏度、高选择性和可靠性正丁醇气体传感器的良好材料。Wang等[191]制备了蠕虫状的介孔SnO_2材料,并发现其在150℃下对正丁醇具有高度选择性和灵敏度。最佳的气体传感器对10 ppm 正丁醇的灵敏度(R_a/R_g)高达600,响应时间仅约11 s,并且在常见的VOC气体(甲醛、甲醇、乙醇、丙酮、苯、甲苯等)中具有很高的选择性(见图10.32)。

乙炔(C_2H_2),俗称风煤和电石气,是一种室温下无色、极易燃的气体。C_2H_2是最有效和多功能的燃气,可以应用于焊接、切割、矫直和其他局部加热。从理论

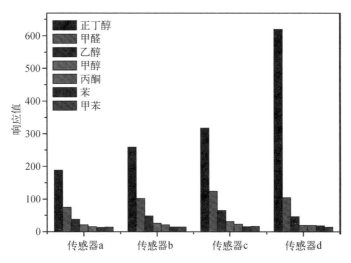

图 10.32 SnO$_2$ 气体传感器在 150℃ 下对不同气体的响应(后附彩图)[191]

上讲,乙炔具有高度可燃性和爆炸性,在空气中具有宽泛的可燃性范围(标准温度和压力下为 2.4%~83%),因此有必要对乙炔气体进行检测。目前基于半导体金属氧化物的乙炔传感器的相关报道较少。Zhang 等[192]报道了用柠檬酸辅助水热的方法合成外表面修饰颗粒的 ZnO 纳米盘材料。该材料在 420℃ 的工作温度下对 200 ppm 乙炔气体的灵敏度(R_a/R_g)高达 52,响应/恢复时间为 15 s/19 s,具有较宽的测试范围(1~4 000 ppm)、较好的线性工作区间(见图 10.33)、较低的检测限,以及在 CH$_4$、H$_2$、LPG、CO 和 NO 等易燃气体中具有非常好的选择性。

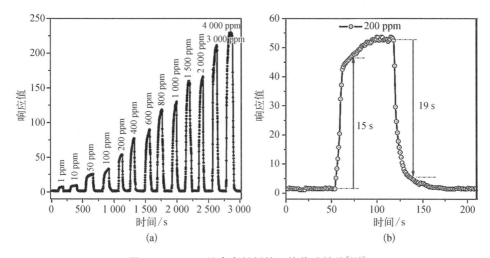

图 10.33 ZnO 纳米盘材料的乙炔传感性能[192]

(a) 传感器对乙炔的典型响应恢复曲线(420℃,1~4 000 ppm 气体中);(b) 放大的响应和恢复特征曲线(200 ppm 乙炔)

硝基爆炸物，如著名的三硝基甲苯（TNT）、环三次甲基三硝胺（黑索金，RDX）、二硝基甲苯（DNT）、对硝基甲苯（PNT）和三硝基苯酚（PA）等，是具有一定爆炸力的可用于制作炸药的化合物，在军工、采矿等领域有重要应用，故对它们的痕量监测是保障其用得其所的必要手段，具有安保、反恐的重要现实意义。然而，由于硝基爆炸物的超低饱和蒸气压，实现对其超灵敏和实时的检测仍然是一个挑战，例如，TNT 和 RDX 的饱和蒸气压在室温下分别仅为几 ppb（十亿分之一）和 ppt（万亿分之一）。Guo 等[193]提出，基于肖特基结的传感器可以对微量的气体变化产生可检测的信号变化，因此有望在痕量监测方面发挥重要作用。Dou 等将 n 型的半导体金属氧化物 ZnO 与具有 p 型半导体行为的还原氧化石墨烯（rGO）进行复合，构筑了 rGO 包覆的 ZnO 核壳结构材料。该材料会形成众多的"微肖特基结"，以此构建的传感器对 TNT、DNT、PNT、RDX 和 PA 等硝基爆炸物的灵敏度分别可达到 56.8%、58.4%、20.2%、80% 和 16.3%，其响应时间都小于 5 s，具有高灵敏度和快速响应检测的性能。同时，该气体传感器可以忍受 ppm 级别的干扰气体，如 EtOH（100 ppm）、NO_2（1 ppm）、NH_3（1 ppm），具有非常好的选择性（见图 10.34）。研究同时对纯的 ZnO 和纯的 rGO 进行了硝基爆炸物的检测，结果显示其性能都不如复合材料高，从而证明了复合所形成的肖特基结对传感器性能具有关键作用。

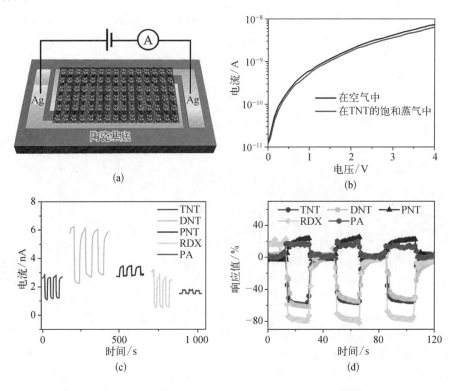

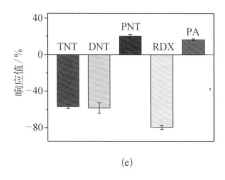
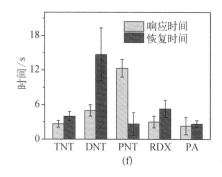

图 10.34　ZnO/rGO 对 TNT 的传感性能研究[193]

(a) 基于 ZnO/rGO 核/壳结构的传感器的示意图；(b) 传感器在空气和 TNT 饱和蒸气中的典型 I-V 曲线；(c)、(d) 电流变化和传感器对 TNT、DNT、PNT、RDX 和 PA 的室温饱和蒸气的响应；(e)、(f) 相应的响应值、响应时间和恢复时间

参考文献

[1] Chen Y, Xue X, Wang Y, et al. Synthesis and ethanol sensing characteristics of single crystalline SnO$_2$ nanorods[J]. Appl. Phys. Lett. 2005, 87: 233503-233507.

[2] Liu Y, Koep E, Liu M. A highly sensitive and fast-responding SnO$_2$ sensor fabricated by combustion chemical vapor deposition[J]. Chem. Mater., 2005, 17: 3997-4000.

[3] Li K, Li Y, Lu M, et al. Direct conversion of single-layer SnO$_2$ nanoplates to multi-layer SnO$_2$ nanoplates with enhanced ethanol sensing properties[J]. Adv. Funct. Mater. 2009, 19: 2453-2456.

[4] Liu S, Xie M, Li Y, et al. Novel sea urchin-like hollow core-shell SnO$_2$ superstructures: facile synthesis and excellent ethanol sensing performance[J]. Sens. Actuators B, 151: 229-235.

[5] Wang L, Kang Y, Liu X, et al. ZnO nanorod gas sensor for ethanol detection[J]. Sens. Actuators B, 2012, 162: 237-243.

[6] Tian S, Yang F, Zeng D, et al. Solution-processed gas sensors based on ZnO nanorods array with an exposed (0001) facet for enhanced gas-sensing properties[J]. Phys. Chem. C., 2012, 116: 10586-10591.

[7] Meng F, Ge S, Jia Y, et al. Interlaced nanoflake-assembled flower-like hierarchical ZnO microspheres prepared by bisolvents and their sensing properties to ethanol[J]. J. Alloy Compd. 2015, 632: 645-650.

[8] Zhou X, Zhu Y, Luo W, et al. Chelation-assisted soft-template synthesis of ordered mesoporous zinc oxides for low concentration gas sensing[J]. J. Mater. Chem. A., 2016, 4: 15064-15071.

[9] Huang F, Yang W, He F, et al. Controlled synthesis of flower-like In$_2$O$_3$ microrods and their highly improved selectivity toward ethanol[J]. Sens. Actuators B, 2016, 235: 86-93.

[10] Liu J, Wang X, Peng Q, et al. Preparation and gas sensing properties of vanadium oxide nanobelts coated with semiconductor oxides[J]. Sens. Actuators B, 2006, 115: 481-487.

[11] Pandeeswari R, Karn R K, Jeyaprakash B G. Ethanol sensing behaviour of sol-gel dip-coated TiO_2 thin films[J]. Sens. Actuators B, 194: 470-477.

[12] Yang C, Su X, Xiao F, et al. Gas sensing properties of CuO nanorods synthesized by a microwave-assisted hydrothermal method[J]. Sens. Actuators B, 2011, 158: 299-303.

[13] Zoolfakar A S, Ahmad M Z, Rani R A, et al. Nanostructured copper oxides as ethanol vapour sensors[J]. Sens. Actuators B, 2013, 185: 620-627.

[14] Cho N G, Hwang I, Kim H, et al. Gas sensing properties of p-type hollow NiO hemispheres prepared by polymeric colloidal templating method[J]. Sens. Actuators B, 2011, 155: 366-371.

[15] Hwang I, Choi J, Woo H, et al. Facile control of C_2H_5OH sensing characteristics by decorating discrete Ag nanoclusters on SnO_2 nanowire networks[J]. ACS Appl. Mater. Inter., 2011, 3: 3140-3145.

[16] Lin Y, Hsueh Y, Lee P, et al. Fabrication of tin dioxidenanowires with ultrahigh gas sensitivity by atomic layer deposition of platinum[J]. J. Mater. Chem., 2011, 21: 10552-10556.

[17] Van N, Duc N, Trung T, et al. Gas-sensing properties of tin oxide doped with metal oxides and carbon nanotubes: a competitive sensor for ethanol and liquid petroleum gas[J]. Sens. Actuators B, 2010, 144: 450-456.

[18] Ramgir N S, Kaur M, Sharma P K, et al. Ethanol sensing properties of pure and Au modified ZnO nanowires[J]. Sens. Actuators B, 2013, 187: 313-318.

[19] Lou Z, Deng J, Wang L, et al. Toluene and ethanol sensing performances of pristine and PdO-decorated flower-like ZnO structures[J]. Sens. Actuators B, 2013, 176: 323-329.

[20] Wang Y, Lin Y, Jiang D, et al. Special nanostructure control of ethanol sensing characteristics based on Au@In_2O_3 sensor with good selectivity and rapid response[J]. RSC Adv., 2015, 5: 9884-9989.

[21] Vallejos S, Stoycheva T, Umek P, et al. Au nanoparticle-functionalised WO_3 nanoneedles and their application in high sensitivity gas sensor devices[J]. Chem. Commun. 2011, 47: 565-567.

[22] Mirzaei A, Janghorban K, Hashemi B, et al. Synthesis, characterization and gas sensing properties of Ag@α-Fe_2O_3 core-shell nanocomposites[J]. Nanomaterials., 2015, 5: 737-749.

[23] Hu P, Du G, Zhou W, et al. Enhancement of ethanol vapor sensing of TiO_2 nanobelts by surface engineering[J]. ACS Appl. Mater. Inter., 2010, 2: 3263-3269.

[24] Kim S, Hwang I, Na C W, et al. Ultrasensitive and selective C_2H_5OH sensors using Rh-loaded In_2O_3 hollow spheres[J]. J. Mater. Chem., 2011, 21: 18560-18567.

[25] Ren Y, Xie W, Li Y, et al. Dynamic coassembly of amphiphilic block copolymer and polyoxometalates in dual solvent systems: an efficient approach to heteroatom-doped semiconductor metal oxides with controllable nanostructures[J]. ACS Cent. Sci., 2021, 7: 1885-1897.

[26] Wan Q, Wang T H. Single-crystalline Sb-doped SnO_2 nanowires: synthesis and gas sensor application[J]. Chem. Commun., 2005, 30: 3841-3486.

[27] Li L, Li C, Zhang J, et al. Bandgap narrowing and ethanol sensing properties of In-doped ZnO nanowires[J]. Nanotechnology, 2007, 18: 225504-225507.

[28] Tricoli A, Graf M, Pratsinis S E. Optimal doping for enhanced SnO_2 sensitivity and thermal stability[J]. Adv. Funct. Mater., 2008, 18: 1969-1976.

[29] Van Hieu N, Kim H, Ju B, et al. Enhanced performance of SnO_2 nanowires ethanol sensor by functionalizing with La_2O_3[J]. Sens. Actuators B, 2008, 133: 228-234.

[30] Zhu C, Chen Y, Wang R, et al. Synthesis and enhanced ethanol sensing properties of α-Fe_2O_3/ZnO heteronanostructures[J]. Sens. Actuators B, 2009, 140: 185-189.

[31] Yin M, Liu M, Liu S. Development of an alcohol sensor based on ZnO nanorods synthesized using a scalable solvothermal method[J]. Sens. Actuators B, 2013, 185: 735-742.

[32] Bie L, Yan X, Yin J, et al. Nanopillar ZnO gas sensor for hydrogen and ethanol[J]. Sens. Actuators B, 2007, 126: 604-608.

[33] Rao J, Yu A, Shao C, et al. Construction of hollow and mesoporous ZnO microsphere: a facile synthesis and sensing property[J]. ACS Appl. Mater. Inter., 2012, 4: 5346-5352.

[34] Chen Y, Zhu C, Xiao G. Reduced-temperature ethanol sensing characteristics of flower-like ZnO nanorods synthesized by a sonochemical method[J]. Nanotechnology, 2006, 17: 4537-4541.

[35] Xu J, Chen Y, Shen J. Ethanol sensor based on hexagonal indium oxide nanorods prepared by solvothermal methods[J]. Mater. Lett., 2008, 62: 1363-1365.

[36] Vomiero A, Bianchi S, Comini E, et al. In_2O_3 nanowires for gas sensors: morphology and sensing characterisation[J]. Thin Solid Films, 2007, 515: 8356-8359.

[37] Nguyen H, El-Safty S A. Meso- and macroporous Co_3O_4 nanorods for effective VOC gas sensors[J]. J. Phys. Chem. C., 2011, 115: 8466-8474.

[38] Chen D, Hou X, Li T, et al. Effects of morphologies on acetone-sensing properties of tungsten trioxide nanocrystals[J]. Sens. Actuators B, 2011, 153: 373-381.

[39] Karmaoui M, Leonardi S G, Latino M, et al. Pt-decorated In_2O_3 nanoparticles and their ability as a highly sensitive (<10 ppb) acetone sensor for biomedical applications[J]. Sens. Actuators B, 2016, 230: 697-705.

[40] Shin J, Choi S, Lee I, et al. Thin-wall assembled SnO_2 fibers functionalized by catalytic Pt nanoparticles and their superior exhaled-breath-sensing properties for the diagnosis of diabetes[J]. Adv. Funct. Mater., 2013, 23: 2357-2367.

[41] Kim S, Park S, Park S, et al. Acetone sensing of Au and Pd-decorated WO_3 nanorod sensors[J]. Sens. Actuators B, 2015, 209: 180-185.

[42] Righettoni M, Tricoli A, Pratsinis S E. Si:WO_3 sensors for highly selective detection of acetone for easy diagnosis of diabetes by breath analysis[J]. Anal. Chem., 2010, 82: 3581-3587.

[43] Ren Y, Zou Y, Liu Y, et al. Synthesis of orthogonally assembled 3D cross-stacked metal oxide semiconducting nanowires[J]. Nat. Mater., 2020, 19: 203-211.

[44] Bai X, Ji H, Gao P, et al. Morphology, phase structure and acetone sensitive properties of copper-doped tungsten oxide sensors[J]. Sens. Actuators B, 2014, 193: 100-106.

[45] Hernández P T, Naik A T, Newton E J, et al. Assessing the potential of metal oxide semiconducting gas sensors for illicit drug detection markers[J]. J. Mater. Chem. A., 2014, 2: 8952-8960.

[46] Kaneti Y V, Moriceau J, Liu M, et al. Hydrothermal synthesis of ternary α-Fe_2O_3-ZnO-Au nanocomposites with high gas-sensing performance[J]. Sens. Actuators B, 2015, 209: 889-897.

[47] Biswal R C. Pure and Pt-loaded gamma iron oxide as sensor for detection of sub ppm level of acetone[J]. Sens. Actuators B, 2011, 157: 183-188.

[48] Shan H, Liu C, Liu L, et al. Highly sensitive acetone sensors based on La-doped α-Fe_2O_3 nanotubes[J]. Sens. Actuators B, 2013, 184: 243-247.

[49] Zeng Y, Zhang T, Yuan M, et al. Growth and selective acetone detection based on ZnO nanorod arrays[J]. Sens. Actuators B, 2009, 143: 93-98.

[50] Li X, Chang Y, Long Y. Influence of Sn doping on ZnO sensing properties for ethanol and acetone[J]. Mater. Sci. Eng. C., 2012, 32: 817-821.

[51] Li Z, Zhao Q, Fan W, et al. Porous SnO_2 nanospheres as sensitive gas sensors for volatile organic compounds detection[J]. Nanoscale, 2011, 3: 1646-1652.

[52] Lai X, Shen G, Xue P, et al. Ordered mesoporous NiO with thin pore walls and its enhanced sensing performance for formaldehyde[J]. Nanoscale, 2015, 7: 4005-4012.

[53] Tian S, Ding X, Zeng D, et al. A low temperature gas sensor based on Pd-functionalized mesoporous SnO_2 fibers for detecting trace formaldehyde[J]. RSC Adv., 2013, 3: 1823-11831.

[54] Tang Y, Han Z, Qi Y, et al. Enhanced ppb-level formaldehyde sensing performance over Pt deposited SnO_2 nanospheres[J]. J. Alloys Compd., 2022, 899: 163230.

[55] Cai H, Luo N, Hu Q, et al. Multishell SnO_2 hollow microspheres loaded with bimetal PdPt nanoparticles for ultrasensitive and rapid formaldehyde MEMS sensors[J]. ACS Sens., 2022, 7: 1484-1494.

[56] Zheng Y, Wang J, Yao P. Formaldehyde sensing properties of electrospun NiO-doped SnO_2 nanofibers[J]. Sens. Actuators B, 2011, 156: 723-730.

[57] Han N, Wu X, Zhang D, et al. CdO activated Sn-doped ZnO for highly sensitive, selective and stable formaldehyde sensor[J]. Sens. Actuators B, 2011, 152: 324-329.

[58] Peng L, Zhai J, Wang D, et al. Size and photoelectric characteristics-dependent formaldehyde sensitivity of ZnO irradiated with UV light[J]. Sens. Actuators B, 2010, 148: 66-73.

[59] Castro-Hurtado I, Herrán J, Ga M G, et al. SnO_2-nanowires grown by catalytic oxidation of tin sputtered thin films for formaldehyde detection[J]. Thin Solid Films, 2012, 520: 4792-4796.

[60] Chung F, Wu R, Cheng F. Fabrication of a Au@SnO_2 core-shell structure for gaseous formaldehyde sensing at room temperature[J]. Sens. Actuators B-Chem, 2014: 190: 1-7.

[61] Wang J, Zhang P, Qi J, et al. Silicon-based micro-gas sensors for detecting formaldehyde [J]. Sens. Actuators B, 2009, 136: 399-404.

[62] Lv P, Tang Z A, Yu J, et al. Study on a micro-gas sensor with SnO_2-NiO sensitive film

for indoor formaldehyde detection[J]. Sens. Actuators B, 2008, 132: 74-80.

[63] Du H, Wang J, Su M, et al. Formaldehyde gas sensor based on SnO_2/In_2O_3 heteronanofibers by a modified double jets electrospinning process[J]. Sens. Actuators B, 2012, 166: 746-752.

[64] Sun P, Zhou X, Wang C, et al. Hollow $SnO_2/\alpha-Fe_2O_3$ spheres with a double-shell structure for gas sensors[J]. J. Mater. Chem. A., 2014, 2: 1302-1308.

[65] Chu X, Chen T, Zhang W, et al. Investigation on formaldehyde gas sensor with ZnO thick film prepared through microwave heating method[J]. Sens. Actuators B, 2009, 142: 49-54.

[66] Xie C, Xiao L, Hu M, et al. Fabrication and formaldehyde gas-sensing property of $ZnO-MnO_2$ coplanar gas sensor arrays[J]. Sens. Actuators B, 2010, 145: 457-463.

[67] Han N, Tian Y, Wu X, et al. Improving humidity selectivity in formaldehyde gas sensing by a two-sensor array made of Ga-doped ZnO[J]. Sens. Actuators B, 2009, 138: 228-235.

[68] Chung F, Zhu Z, Luo P, et al. Fabrication of a $Au@SnO_2$ core-shell structure for gaseous Au@ZnO core-shell structure for gaseous[J]. Sens. Actuators B, 2014, 199: 314-319.

[69] Lee C, Chiang C, Wang Y, et al. A self-heating gas sensor with integrated NiO thin-film for formaldehyde detection[J]. Sens. Actuators B, 2007, 122: 503-510.

[70] Castro-Hurtado I, Herrán J, Mandayo G G, et al. Studies of influence of structural properties and thickness of NiO thin films on formaldehyde detection[J]. Thin Solid Films, 2011, 520: 947-952.

[71] Castro-Hurtado I, Malagù C, Morandi S, et al. Properties of NiO sputtered thin films and modeling of their sensing mechanism under formaldehyde atmospheres[J]. Acta. Mater., 2013, 61: 1146-1153.

[72] Liu Y, Zhu G, Ge B, et al. Concave Co_3O_4 octahedral mesocrystal: polymer-mediated synthesis and sensing properties[J]. Cryst. Eng. Comm., 2012, 14: 6264-6627.

[73] Yang W, Wan P, Zhou X, et al. Self-assembled In_2O_3 truncated octahedron string and its sensing properties for formaldehyde[J]. Sens. Actuators B-Chem, 2014, 201: 228-233.

[74] Zhu B, Xie C, Wang W, et al. Improvement in gas sensitivity of ZnO thick film to volatile organic compounds (VOCs) by adding TiO_2[J]. J. Mater. Lett., 2004, 58: 624-629.

[75] Elmi I, Zampolli S, Cozzani E, et al. Development of ultra-low-power consumption MOX sensors with ppb-level VOC detection capabilities for emerging applications[J]. Sens. Actuators B, 2008, 135: 342-351.

[76] Wang L, Zhang R, Zhou T, et al. Concave Cu_2O octahedral nanoparticles as an advanced sensing material for benzene (C_6H_6) and nitrogen dioxide (NO_2) detection[J]. Sens. Actuators B-Chem, 2016, 223: 311-317.

[77] Vaishnav V S, Patel S G, Panchal J N. Development of ITO thin film sensor for detection of benzene[J]. Sens. Actuators B, 2015, 206: 381-388.

[78] Liu S, Wang Z, Zhao H, et al. Ordered mesoporous Co_3O_4 for high-performance toluene sensing[J]. Sens. Actuators B, 2014, 197: 342-349.

[79] Lou Z, Deng J, Wang L, et al. Toluene and ethanol sensing performances of pristine and PdO-decorated flower-like ZnO structures[J]. Sens. Actuators B, 2013, 176: 323-329.

[80] Wang C, Cheng X, Zhou X, et al. Hierarchical α-Fe$_2$O$_3$/NiO composites with a hollow structure for a gas sensor[J]. ACS Appl. Mater. Inter., 2014, 6: 12031-12037.

[81] Kim H, Yoon J, Choi K, et al. Ultraselective and sensitive detection of xylene and toluene for monitoring indoor air pollution using Cr-doped NiO hierarchical nanostructures[J]. Nanoscale, 2013, 5: 7066-7070.

[82] Guo M, Luo N, Chen Y, et al. Fast-response MEMS xylene gas sensor based on CuO/WO$_3$ hierarchical structure[J]. J. Hazard. Mater., 2022, 429: 127471-127476.

[83] Hong Y J, Yoon J, Lee J, et al. One-pot synthesis of Pd-loaded SnO$_2$ yolk-shell nanostructures for ultraselective methyl benzene sensors[J]. Chem-Eur. J., 2014, 20: 2737-2741.

[84] Wang D, Yin Y, Xu P, et al. The catalytic-induced sensing effect of triangular CeO$_2$ nanoflakes for enhanced BTEX vapor detection with conventional ZnO gas sensors[J]. J. Mater. Chem. A., 2020, 8: 11188-11194.

[85] Cao Z, Wang W, Ma H, et al. Porous Mn-doped Co$_3$O$_4$ nanosheets: gas sensing performance and interfacial mechanism investigation with in situ DRIFTS[J]. Sens. Actuators B, 2022, 353: 131155-131158.

[86] Cao Z, Ge Y, Wang W, et al. Chemical discrimination of benzene series and molecular recognition of the sensing process over Ti-doped Co$_3$O$_4$[J]. ACS Sens., 2022, 7: 1757-1765.

[87] Ke M, Lee M, Lee C, et al. A MEMS-based benzene gas sensor with a self-heating WO$_3$ sensing layer[J]. Sensors, 2009, 9: 2895-2906.

[88] Park J, Shen X, Wang G. Solvothermal synthesis and gas-sensing performance of Co$_3$O$_4$ hollow nanospheres[J]. Sens. Actuators B, 2009, 136: 494-498.

[89] Bai Z, Xie C, Zhang S, et al. Microstructure and gas sensing properties of the ZnO thick film treated by hydrothermal method[J]. Sens. Actuators B, 2010, 151: 107-113.

[90] Wang L, Lou Z, Fei T, et al. Zinc oxide core-shell hollow microspheres with multi-shelled architecture for gas sensor applications[J]. J. Mater. Chem., 2011, 21: 19331-19336.

[91] Qu F, Wang Y, Liu J, et al. Fe$_3$O$_4$-NiO core-shell composites: hydrothermal synthesis and toluene sensing properties[J]. Mater. Lett., 2014, 132: 167-170.

[92] Shan H, Liu C, Liu L, et al. Excellent toluene sensing properties of SnO$_2$-Fe$_2$O$_3$ interconnected nanotubes[J]. ACS Appl. Mater. Inter., 2013, 5: 6376-6380.

[93] Cao J, Wang Z, Wang R, et al. Electrostatic sprayed Cr-loaded NiO core-in-hollow-shell structured micro/nanospheres with ultra-selectivity and sensitivity for xylene[J]. Cryst. Eng. Comm., 2014, 16: 7731-7735.

[94] Akiyama T, Ishikawa Y, Hara K. Xylene sensor using double-layered thin film and Ni-deposited porous alumina[J]. Sens. Actuators B, 2013, 181: 348-352.

[95] Sun C, Su X, Xiao F, et al. Synthesis of nearly monodisperse Co$_3$O$_4$ nanocubes via a microwave-assisted solvothermal process and their gas sensing properties[J]. Sens. Actuators B, 2011, 157: 681-685.

[96] Michel C R, López-Contreras N L, Martínez-Preciado A H. Gas sensing properties of

Gd_2O_3 microspheres prepared in aqueous media containing pectin[J]. Sens. Actuators B, 2013, 177: 390-396.

[97] Michel C R, Martínez-Preciado A H, Contreras N L. Gas sensing properties of Nd_2O_3 nanostructured microspheres[J]. Sens. Actuators B, 2013, 184: 8-14.

[98] Jinesh K B, Dam V A, Swerts J, et al. Room-temperature CO_2 sensing using metal-insulator-semiconductor capacitors comprising atomic-layer-deposited La_2O_3 thin films[J]. Sens. Actuators B, 2011, 156: 276-282.

[99] Trung D D, Toan L D, Hong H S, et al. Selective detection of carbon dioxide using LaOCl-functionalized SnO_2 nanowires for air-quality monitoring[J]. Talanta, 2012, 88: 152-159.

[100] Herrán J, Mandayo G, Pérez N, et al. On the structural characterization of $BaTiO_3$-CuO as CO_2 sensing material[J]. Sens. Actuators B, 133: 315-320.

[101] Herrán J, Mandayo G, Castaño E. Semiconducting $BaTiO_3$-CuO mixed oxide thin films for CO_2 detection[J]. Thin Solid Films, 2009, 517: 6192-6197.

[102] Izu N, Oh-hori N, Itou M, et al. Resistive oxygen gas sensors based on $Ce_{1-x}Zr_xO_2$ nano powder prepared using new precipitation method[J]. Sens. Actuators B, 2005, 108: 238-243.

[103] Castañeda L. Effects of palladium coatings on oxygen sensors of titanium dioxide thin films[J]. Mater. Sci. Eng. B., 2007, 139: 149-154.

[104] Sotter E, Vilanova X, Llobet E, et al. Thick film titania sensors for detecting traces of oxygen[J]. Sens. Actuators B, 2007, 127: 567-579.

[105] Al-Hardan N, Abdullah M J, Abdul A A, et al. Low operating temperature of oxygen gas sensor based on undoped and Cr-doped ZnO films[J]. Appl. Surf. Sci., 2010, 256: 3468-3471.

[106] Liang Y, Wu Z, Wei Y, et al. Self-healing, self-adhesive and stable organohydrogel-based stretchable oxygen sensor with high performance at room temperature[J]. Nano-Micro Lett., 2022, 14: 52-55.

[107] Kaneko H, Okamura T, Taimatsu H, et al. Performance of a miniature zirconia oxygen sensor with a Pd-PdO internal reference[J]. Sens. Actuators B, 2005, 108: 331-334.

[108] Hu Y, Tan O K, Pan J S, et al. The effects of annealing temperature on the sensing properties of low temperature nano-sized $SrTiO_3$ oxygen gas sensor[J]. Sens. Actuators B, 2005, 108: 244-249.

[109] Minaee H, Mousavi S H, Haratizadeh H, et al. Oxygen sensing properties of zinc oxide nanowires, nanorods, and nanoflowers: the effect of morphology and temperature[J]. Thin Solid Films, 2013, 545: 8-12.

[110] Ahmed F, Arshi N, Anwar M S, et al. Mn-doped ZnO nanorod gas sensor for oxygen detection[J]. Curr. Appl. Phys., 2013, 13: S64-S68.

[111] Shimizu Y, Matsunaga N, Hyodo T, et al. Improvement of SO_2 sensing properties of WO_3 by noble metal loading[J]. Sens. Actuators B, 2001, 77: 35-40.

[112] Hidalgo P, Castro R R, Coelho A V, et al. Surface segregation and consequent SO_2 sensor response in SnO_2-NiO[J]. Chem. Mater., 2005, 17: 4149-4153.

[113] Liang X, Zhong T, Quan B, et al. Solid-state potentiometric SO_2 sensor combining NASICON with V_2O_5 - doped TiO_2 electrode[J]. Sens. Actuators B, 2008, 134: 25-30.

[114] Das S, Chakraborty S, Parkash O, et al. Vanadium doped tin dioxide as a novel sulfur dioxide sensor[J]. Talanta, 2008, 75: 385-389.

[115] Stankova M, Vilanova X, Calderer J, et al. Detection of SO_2 and H_2S in CO_2 stream by means of WO_3 - based micro-hotplate sensors[J]. Sens. Actuators B, 2004, 102: 219-225.

[116] Bendahan M, Boulmani R, Seguin J, et al. Characterization of ozone sensors based on WO_3 reactively sputtered films: influence of O_2 concentration in the sputtering gas, and working temperature[J]. Sens. Actuators B, 2004, 100: 320-324.

[117] Korotcenkov G, Blinov I, Ivanov M, et al. Ozone sensors on the base of SnO_2 films deposited by spray pyrolysis[J]. Sens. Actuators B, 2007, 120: 679-686.

[118] Silva L F, Catto A C, Avansi W, et al. A novel ozone gas sensor based on one-dimensional (1D) α - Ag_2WO_4 nanostructures[J]. Nanoscale, 2014, 6: 4058-4062.

[119] Deng Z, Fang X, Li D, et al. Room temperature ozone sensing properties of p-type transparent oxide $CuCrO_2$[J]. J. Alloy Compd., 2009, 484: 619-621.

[120] Vallejos S, Khatko V, Aguir K, et al. Ozone monitoring by micro-machined sensors with WO_3 sensing films[J]. Sens. Actuators B, 2007, 126: 573-578.

[121] Minh V A, Tuan L A, Huy T Q, et al. Enhanced NH_3 gas sensing properties of a QCM sensor by increasing the length of vertically orientated ZnO nanorods[J]. Appl. Surf. Sci., 2013, 265: 458-464.

[122] Tang H, Yan M, Zhang H, et al. A selective NH_3 gas sensor based on Fe_2O_3 - ZnO nanocomposites at room temperature[J]. Sens. Actuators B, 2006, 114: 910-915.

[123] Hieu N V, Quang V V, Hoa N D, et al. Preparing large-scale WO_3 nanowire-like structure for high sensitivity NH_3 gas sensor through a simple route[J]. Curr. Appl. Phys., 2011, 11: 657-661.

[124] Hieu N, Thuy L, Chien N D. Highly sensitive thin film NH_3 gas sensor operating at room temperature based on SnO_2/MWCNTs composite[J]. Sens. Actuators B, 2008, 129: 888-895.

[125] Du N, Zhang H, Chen B D, et al. Porous indium oxide nanotubes: layer-by-layer assembly on carbon-nanotube templates and application for room-temperature NH_3 gas sensors[J]. Adv. Mater., 2007, 19: 1641-1645.

[126] Zhang J, Wang S, Wang Y, et al. ZnO hollow spheres: preparation, characterization, and gas sensing properties[J]. Sens. Actuators B, 2009, 139: 411-417.

[127] Wang Y, Liu J, Cui X, et al. NH_3 gas sensing performance enhanced by Pt-loaded on mesoporous WO_3[J]. Sens. Actuators B, 2017, 238: 473-481.

[128] Mashock M, Yu K, Cui S, et al. Modulating gas sensing properties of CuO nanowires through creation of discrete nanosized p - n junctions on their surfaces[J]. ACS Appl. Mater. Inter., 2012, 4: 4192-4199.

[129] Tai H, Jiang Y, Xie G, et al. Influence of polymerization temperature on NH_3 response of PANI/TiO_2 thin film gas sensor[J]. Sens. Actuators B, 2008, 129: 319-326.

[130] Wang K, Zhao T, Lian G, et al. Room temperature CO sensor fabricated from Pt-loaded SnO_2 porous nanosolid[J]. Sens. Actuators B, 2013, 184: 33-39.

[131] Wang C, Chen M. Vanadium-promoted tin oxide semiconductor carbon monoxide gas sensors[J]. Sens. Actuators B-Chem, 2010, 150: 360-366.

[132] Wu R, Wu J, Yu M, et al. Promotive effect of CNT on $Co_3O_4-SnO_2$ in a semiconductor-type CO sensor working at room temperature[J]. Sens. Actuators B, 2008, 131: 306-312.

[133] Khoang N D, Hong H S, Trung D D, et al. On-chip growth of wafer-scale planar-type ZnO nanorod sensors for effective detection of CO gas[J]. Sens. Actuators B, 2013, 181: 529-536.

[134] Al-Kuhaili M F, Durrani S A, Bakhtiari I A. Carbon monoxide gas-sensing properties of CeO_2-ZnO thin films[J]. Appl. Surf. Sci., 2008, 255: 3033-3039.

[135] Yu M, Wu R, Chavali M. Effect of "Pt" loading in ZnO-CuO hetero-junction material sensing carbon monoxide at room temperature[J]. Sens. Actuators B, 2011, 153: 321-328.

[136] Ma J, Ren Y, Zhou X, et al. Pt nanoparticles sensitized ordered mesoporous WO_3 semiconductor: gas sensing performance and mechanism study[J]. Adv. Funct. Mater, 2018, 28: 1705268.

[137] Wagner T, Waitz T, Roggenbuck J, et al. Ordered mesoporous ZnO for gas sensing[J]. Thin Solid Films, 2007, 515: 8360-8363.

[138] Li W, Shen C, Wu G, et al. New model for a Pd-doped SnO_2-based CO gas sensor and catalyst studied by online in-situ X-ray photoelectron spectroscopy[J]. J. Phys. Chem. C, 2011, 115: 21258-21263.

[139] Patil D, Patil P, Subramanian V, et al. Highly sensitive and fast responding CO sensor based on Co_3O_4 nanorods[J]. Talanta, 2010, 81: 37-43.

[140] Ramgir N S, Ganapathi S K, Kaur M, et al. Sub-ppm H_2S sensing at room temperature using CuO thin films[J]. Sens. Actuators B, 2010, 151: 90-96.

[141] Zhang F, Zhu A, Luo Y, et al. CuO nanosheets for sensitive and selective determination of H_2S with high recovery ability[J]. J. Phys. Chem. C., 2010, 114: 19214-19219.

[142] Li X, Wang Y, Lei Y, et al. Highly sensitive H_2S sensor based on template-synthesized CuO nanowires[J]. RSC Adv., 2012, 2: 2302-2305.

[143] Kim H, Jin C, Park S, et al. H_2S gas sensing properties of bare and Pd-functionalized CuO nanorods[J]. Sens. Actuators B, 2012, 64: 594-599.

[144] Xue X, Xing L, Chen Y, et al. Synthesis and H_2S sensing properties of $CuO-SnO_2$ core/shell pn-junction nanorods[J]. J. Phys. Chem. C, 2008, 112: 12157-12160.

[145] Li Y, Luo W, Qin N, et al. Highly ordered mesoporous tungsten oxides with a large pore size and crystalline framework for H_2S sensing[J]. Angew. Chem. Int. Ed., 2014, 53: 9035-9040.

[146] Xiao X, Liu L, Ma J, et al. Ordered mesoporous tin oxide semiconductors with large pores and crystallized walls for high-performance gas sensing[J]. ACS Appl. Mater. Inter., 2018, 10: 1871-1880.

[147] Wu X, Zhu L, Sun J, et al. Pt nanoparticle-modified SnO_2 - ZnO core-shell nanosheets on microelectromechanical systems for enhanced H_2S detection[J]. ACS Appl. Nano Mater., 2022, 5: 6627-6636.

[148] Qin Y, Zhang F, Chen Y, et al. Hierarchically porous CuO hollow spheres fabricated via a one-pot template-free method for high-performance gas sensors[J]. J. Phys. Chem. C, 2012, 116: 11994-12000.

[149] Chen J, Wang K, Hartman L, et al. H_2S detection by vertically aligned CuO nanowire array sensors[J]. J. Phys. Chem. C, 2008, 112: 16017-16021.

[150] Zeng Y, Zhang K, Wang X, et al. Rapid and selective H_2S detection of hierarchical $ZnSnO_3$ nanocages[J]. Sens. Actuators B-Chem, 2011, 159: 245-250.

[151] Xu J, Wang X, Shen J. Hydrothermal synthesis of In_2O_3 for detecting H_2S in air[J]. Sens. Actuators B-Chem, 2006, 115: 642-646.

[152] Zhang C, Debliquy M, Boudiba A, et al. Sensing properties of atmospheric plasma-sprayed WO_3 coating for sub-ppm NO_2 detection[J]. Sens. Actuators B-Chem, 2010, 144: 280-288.

[153] Heidari E K, Zamani C, Marzbanrad E, et al. WO_3-based NO_2 sensors fabricated through low frequency AC electrophoretic deposition[J]. Sens. Actuators B-Chem, 2010, 146: 165-170.

[154] Liu Z, Miyauchi M, Yamazaki T, et al. Facile synthesis and NO_2 gas sensing of tungsten oxide nanorods assembled microspheres[J]. Sens. Actuators B-Chem, 2009, 140: 514-519.

[155] You L, Sun Y F, Ma J, Highly sensitive NO_2 sensor based on square-like tungsten oxide prepared with hydrothermal treatment[J]. Sens. Actuators B-Chem, 2011, 157: 401-407.

[156] Kida T, Nishiyama A, Yuasa M, et al. Highly sensitive NO_2 sensors using lamellar-structured WO_3 particles prepared by an acidification method[J]. Sens. Actuators B-Chem, 2009, 135: 568-574.

[157] Min Y, Tuller H L, Palzer S, et al. Gas response of reactively sputtered ZnO films on Si-based micro-array[J]. Sens. Actuators B-Chem, 2003, 93: 435-441.

[158] Cho P, Kim K, Lee J. NO_2 sensing characteristics of ZnO nanorods prepared by hydrothermal method[J]. J. Electroceram., 2006, 17: 975-978.

[159] Na C W, Woo H, Kim I, et al. Selective detection of NO_2 and C_2H_5OH using a Co_3O_4-decorated ZnO nanowire network sensor[J]. Chem. Commun., 2011, 47: 5148-5150.

[160] Breedon M, Spizzirri P, Taylor M, et al. Synthesis of nanostructured tungsten oxide thin films: a simple, controllable, inexpensive, aqueous sol-gel method[J]. Cryst. Growth Des., 2010, 10: 430-439.

[161] Ren Y, Zhou X, Luo W, et al. Amphiphilic block copolymer templated synthesis of mesoporous indium oxides with nanosheet-assembled pore walls[J]. Chem. Mater., 2016, 28: 7997-8005.

[162] Wang Y, Cui X, Yang Q, et al. Preparation of Ag-loaded mesoporous WO_3 and its enhanced NO_2 sensing performance[J]. Sens. Actuators B-Chem, 2016, 225: 544-552.

[163] Öztürk S, Kılınç N, Öztürk Z Z. Fabrication of ZnO nanorods for NO_2 sensor

[164] Moon J, Park J, Lee S, et al. Pd-doped TiO$_2$ nanofiber networks for gas sensor applications[J]. Sens. Actuators B-Chem, 2010, 149: 301-305.

[165] Basu P K, Jana S K, Saha H, et al. Low temperature methane sensing by electrochemically grown and surface modified ZnO thin films[J]. Sens. Actuators B-Chem, 2008, 135: 81-88.

[166] Bhattacharyya P, Basu P K, Lang C, et al. Noble metal catalytic contacts to sol-gel nanocrystalline zinc oxide thin films for sensing methane[J]. Sens. Actuators B-Chem, 2008, 129: 551-557.

[167] Kim J C, Jun H K, Huh J, et al. Tin oxide-based methane gas sensor promoted by alumina-supported Pd catalyst[J]. Sens. Actuators B-Chem, 1997, 45: 271-277.

[168] Prasad A K, Amirthapandian S, Dhara S, et al. Novel single phase vanadium dioxide nanostructured films for methane sensing near room temperature[J]. Sens. Actuators B-Chem, 2014, 191: 252-256.

[169] Dayan N J, Sainkar S R, Karekar R N, et al. Formulation and characterization of ZnO: Sb thick-film gas sensors[J]. Thin Solid Films, 1998, 325: 254-258.

[170] Chen Y, Zhang W, Luo N, et al. Defective ZnO nanoflowers decorated by ultra-fine Pd clusters for low-concentration CH$_4$ sensing: controllable preparation and sensing mechanism analysis[J]. Coatings, 2022, 12: 677.

[171] Quaranta F, Rella R, Siciliano P, et al. A novel gas sensor based on SnO$_2$/Os thin film for the detection of methane at low temperature[J]. Sens. Actuators B-Chem, 1999, 58: 350-355.

[172] Wang Z, Li Z, Sun J, et al. Improved hydrogen monitoring properties based on p-NiO/n-SnO$_2$ heterojunction composite nanofibers[J]. J. Phys. Chem. C, 2010, 114: 6100-6105.

[173] Liu L, Guo C, Li S, et al. Improved H$_2$ sensing properties of Co-doped SnO$_2$ nanofibers [J]. Sens. Actuators B-Chem, 2010, 150: 806-810.

[174] Wang Z, Li Z, Jiang T, et al. Ultrasensitive hydrogen sensor based on Pd$_0$-loaded SnO$_2$ electrospun nanofibers at room temperature[J]. ACS Appl. Mater. Inter., 2013, 5: 2013-2021.

[175] Das S N, Kar J P, Choi J, et al. Fabrication and characterization of ZnO single nanowire-based hydrogen sensor[J]. J. Phys. Chem. C, 2010, 114: 1689-1693.

[176] Alsaif M, Balendhran S, Field M, et al. Two dimensional α-MoO$_3$ nanoflakes obtained using solvent-assisted grinding and sonication method: application for H$_2$ gas sensing[J]. Sens. Actuators B-Chem, 2014, 192: 196-204.

[177] Varghese O K, Gong D, Paulose M, et al. Hydrogen sensing using titania nanotubes[J]. Sens. Actuators B-Chem, 2003, 93: 338-344.

[178] Luo N, Wang C, Zhang D, et al. Ultralow detection limit MEMS hydrogen sensor based on SnO$_2$ with oxygen vacancies[J]. Sens. Actuators B-Chem, 2022, 354: 130982.

[179] Liu B, Cai D, Liu Y, et al. High-performance room-temperature hydrogen sensors based on combined effects of Pd decoration and Schottky barriers[J]. Nanoscale, 2013, 5: 2505.

[180] Patil L A, Bari A R, Shinde M D, et al. Ultrasonically prepared nanocrystalline ZnO thin films for highly sensitive LPG sensing[J]. Sens. Actuators B-Chem, 2010, 149: 79-86.

[181] Shinde V R, Gujar T P, Lokhande C D. LPG sensing properties of ZnO films prepared by spray pyrolysis method: effect of molarity of precursor solution[J]. Sens. Actuators B-Chem, 2007, 120: 551-559.

[182] Waghulade R, Patil P, Pasricha R. Synthesis and LPG sensing properties of nano-sized cadmium oxide[J]. Talanta, 2007, 72: 594-599.

[183] Sen T, Shimpi N G, Mishra S, et al. Polyaniline/γ-Fe_2O_3 nanocomposite for room temperature LPG sensing[J]. Sens. Actuators B-Chem, 2014, 190: 120-126.

[184] Singh S, Singh A, Yadav B C, et al. Fabrication of nanobeads structured perovskite type neodymium iron oxide film: its structural, optical, electrical and LPG sensing investigations[J]. Sens. Actuators B-Chem, 2013, 177: 730-739.

[185] Mishra D, Srivastava A, Srivastava A, et al. Bead structured nanocrystalline ZnO thin films: synthesis and LPG sensing properties[J]. Appl. Surf. Sci., 2008, 255: 2947-2950.

[186] Zhu Y, Zhao Y, Ma J, et al. Mesoporous tungsten oxides with crystalline framework for highly sensitive and selective detection of foodborne pathogens[J]. J. Am. Chem. Soc., 2017, 139: 10365-10373.

[187] Zhang R, Wang L, Deng J, et al. Hierarchical structure with heterogeneous phase as high performance sensing materials for trimethylamine gas detecting[J]. Sens. Actuators B-Chem, 2015, 220: 1224-1231.

[188] Shen J, Xu S, Zhao C, et al. Bimetallic Au@Pt nanocrystal sensitization mesoporous α-Fe_2O_3 hollow nanocubes for highly sensitive and rapid detection of fish freshness at low temperature[J]. ACS Appl. Mater. Interfaces, 2021, 13: 57597-57608.

[189] Feng B, Wu Y, Ren Y, et al. Self-template synthesis of mesoporous Au-SnO_2 nanospheres for low-temperature detection of triethylamine vapor[J]. Sens. Actuators B-Chem, 2022, 356: 131358.

[190] Li P, Fan H, Cai Y. In_2O_3/SnO_2 heterojunction microstructures: facile room temperature solid-state synthesis and enhanced Cl_2 sensing performance[J]. Sens. Actuators B-Chem, 2013, 185: 110-116.

[191] Wang H, Qu Y, Chen H, et al. Highly selective n-butanol gas sensor based on mesoporous SnO_2 prepared with hydrothermal treatment[J]. Sens. Actuators B-Chem, 2014, 201: 153-159.

[192] Zhang L, Zhao J, Zheng J, et al. Hydrothermal synthesis of hierarchical nanoparticle-decorated ZnO microdisks and the structure-enhanced acetylene sensing properties at high temperatures[J]. Sens. Actuators B-Chem, 2011, 158: 144-150.

[193] Guo L, Yang Z, Li Y, et al. Sensitive, real-time and anti-interfering detection of nitro-explosive vapors realized by ZnO/rGO core/shell micro-Schottky junction[J]. Sens. Actuators B-Chem, 2017, 239: 286-294.

索 引

3H-2B气体传感器　10
3-羟基-2-丁酮(3H-2B)　10
3-羟基-2-丁酮　145,324
Arrhenius 方程　185
Avogadro 常数　188
BET 理论　169
Brown 动力学(Brownian dynamics,BD)　190
Brunauer-Emmett-Teller(BET)理论　168
CH_4 传感　319
Co_3O_4-ZnO 复合材料　92
CO 传感器　311
Cu_2O 半导体　129
CuO 传感器的　176
Debye 半径　172
Debye 长度　172
Drude 模型　253
DR 耦合　195,197,210
DR 耦合效应　202
Einstein 方程　192
Eley-Rideal 模型　170
E-R 模型　170
F127　78
Fe_2O_3 基材料　154
Fermi 能级　173
FET 气体传感器　208
Fick 第二定律　187
Fick 第一定律　187
Fick 扩散　188,190
Fick 扩散定律　187
Freundlich 方程　168
Freundlich 吸附等温式　168

Gardner 模型　167
Gauss 位移　192
Grotthuss H^+ 跳跃模型　192
Grotthuss 质子跳跃机理　179
H^+ 离子簇　192
H^+ 迁移　192
H_2O 分子　178
H_2O 分子吸附　179
H_2S 传感器　7,313
H_2 检测　321
Hatta 数　196,199
Henry 规则　168
Henry 吸附常数　168
Henry 吸附等温式　168
In_2O_3 纳米线 FET 气体传感器　256
Kirkendall 孔洞　194
Kirkendall 效应　194
KIT-6　143
KIT-6 型　133
Knudsen 扩散　143,189,190,210
Knudsen 扩散方程　199
Knudsen 扩散模型　207
Knudsen 扩散系数　189,199
Knudsen 流　189
Knudsen 数(Kn)　189
Langmuir-Hinshelwood 模型　169
Langmuir 等温方程　184
Langmuir 等温线　170
Langmuir 方程　168,169
Langmuir 吸附机理　185
L-H 模型　169

LPG 传感器　322
LPG 检测　322
Mars-van Krevelen 机制　151
Mears 假定(C_m)　195
MEMS 基底　98
MEMS 微加热器　245
MEMS 嗅觉　277
MIS(金属-绝缘体-半导体)电容器　241
MOS 传感器阵列　263
MOS 气体传感器　210
MOS 气体传感器阵列　274
MvK 机理　170,175
Mxene　180
NH_3 传感器　310
n-n 型复合传感器　34
n-n 异质结　34,91,96,148
NO_2 传感　316
NO_2 传感器　4,316
NO_2 气体传　228
n-p-n 三元异质结构　150
n-p-n 异质结　148,149
n 场效应晶体管　237
n 型半导体　2,90,91,128
O_2 传感器　180
O_3 传感　308
p+n 场效应晶体管电路　236
PEO-b-PS　79
Pluronic P123　78
p-NiO/n-SnO_2 复合物　91
p-n 结　3
p-n 异质结　13,20,39,91,148
p-n 异质结界面　148
Poiseuille 流　189
ppb 级别　13
ppb 水平　61
ppm 水平　226
ppt 级别　208,235
p-p 半导体金属氧化物　38
p-p 异质结　91,96,101,148
p 场效应晶体管　237

p 型半导体　90,91,128,129,150
p 型金属氧化物半导体　2
SBA-15　133,143
Schottky 层　181
SMOs@MOF 传感器　6
SnO_2　16
SnO_2 敏感材料　2
SnO_2 纳米线 FET 气体传感器　254
SO_2 检测　307
SPR 传感器　230
SPR 气体传感器　231
Taguchi 型传感器　242
TGS 系列气体传感器　265
Thiele 模数　195
van der Waals(范德瓦尔斯)力　168
Weisz-Prater 假定(C_{WP})　195
WO_3/NiO 基气敏元件　94
WO_3　130
Yamazoe 模型　167
ZnO　16
ZnO-Cr_2O_3 异质结　95
ZnO 纳米线　255
α-Fe_2O_3　154
β-Fe_2O_3　154
γ-Fe_2O_3　155
ε-Fe_2O_3　155
ε-WO_3　134
η-Fe_2O_3　155

A

氨(NH_3)　309
氨气传感器　8,309

B

半导体材料　1
半导体金属氧化物(SMO)　1,241,287
半导体金属氧化物　128,241
半导体金属氧化物气体传感器　30,241
半导体纳米线　251

索引

半导体纳米线场效应晶体管(NW-FET)型气体传感器 251
半导体气体传感器 241
半导体势垒 21
半导体型气体传感器 1
半导体载体 19
暴露时间 196
背栅 252
本征半导体 133
本征电子 21
本征电阻 2
苯 299,300
苯、甲苯和二甲苯(苯系物,BTX) 299
苯系物(苯、甲苯、二甲苯)传感器 299
比表面积 1,8,70,90,147
边界阈值温度 60
便携式电子鼻系统 265
便携式气体传感器 235
便携式智能电子鼻 274
标准互补金属氧化物半导体(CMOS) 244
标准状况 169
表面/界面效应 151
表面粗糙度 191
表面等离子体(SP) 229
表面等离子体波(SPW) 230
表面等离子体共振(SPR) 225,229
表面等离子体共振(surface plasmon resonance, SPR)加 208
表面等离子体共振效应 230
表面等离子体效应 230
表面电导率 70
表面电势 21
表面电子参数 68
表面电子耗尽层 244
表面反应 145
表面覆盖度 168
表面改性 57,166,210
表面改性剂 41
表面更新模型 196
表面耗竭模型 32

表面耗尽层理论 209
表面化学反应(surface chemical reaction, SCR) 205
表面化学状态 12
表面活性剂 72,76
表面活性氧 55
表面积 128
表面碱度 115,116,134
表面阶梯 69
表面控制型 30
表面扩散 143,190,191,210
表面扩散系数 192
表面敏感型传感器 241
表面敏感型电阻传感器 244
表面模型 189
表面声波(SAW)传感器 264
表面势垒高度 47
表面受体 2
表面酸碱特性 15
表面态密度 21,68,141
表面吸附/解吸理论 1
表面吸附催化反应 117
表面吸附理论 225
表面吸附态浓度 68
表面吸附氧 2,15,47,106,187
表面吸附氧物种 55
表面吸脱附 166
表面氧化还原反应 21
表面氧空位 69,233
表面氧物种 1,150
表面重组系数 182
丙酮传感 8
丙酮传感器 292
丙酮气体传感器 292
波长 225
玻璃态 66
薄膜-渗透模型 196
不饱和配位 49

C

测定限(limit of quantitation, LoQ) 184

层次聚类分析(HCA) 268
掺杂 111
掺杂剂 31
长期稳定性 6,8
场效应管(field-effect transistor,FET) 180
场效应晶体管(FET)气体传感器 225
场效应晶体管(FET)型 252
场效应晶体管 236
场效应晶体管参数 252
场效应晶体管型气体传感器 251,252
超大规模集成电路(VLSI)技术 241
超晶格 76
超顺磁性 75
程序升温脱附 174
尺寸限制的扩散 189
尺寸效应 66,67
赤铁矿 154
臭氧(O_3) 308
臭氧传感器 308
触觉柔性传感器阵列 277
传感层 4
传感层厚 182
传感层厚度 199
传感机制 129
传感器功能 242
传感器集成 265
传感器小型化 246
传感器阵列 252,263,275
传质 135
串行路径 129
磁赤铁矿 155
磁化强度 75
磁铁矿纳米晶组 75
催化层 4
催化过滤效应 6
催化活性 6,35,60,90
催化剂 102
催化敏化效应 249
催化氧化 151

D

大孔 143,190
大孔-介孔 WO_3 147
带弯曲 39
带隙 12,97
带隙能量(E_g) 17
单壁碳纳米管 118
单层化学吸附 178
单层吸附 169
单档扩散 190
单核细胞增生李斯特菌 9
单晶 16,66,152
单晶金属氧化物 152
单向导电性 20
单原子催化剂 49
氮掺杂 40
氮氧化物(NO_2 和 NO) 316
导带 2,17,96
导带电子 18
导电聚合物(CP)传感器 264
导电聚合物 13
导电性 38
德拜长度 31,41,66
等离子体 172,230
点缺陷 16,156
电导率 2,54,102,204,238
电导型 252
电导型传感器 252
电荷补偿机制 114
电荷传输 15
电荷俘获效率 133
电荷耗尽层 90,148
电荷交换 21
电荷扩散长度 18
电荷载体 15
电荷转移 90
电荷转移相互作用 52
电化学传感 225
电极电势 241

电离度 178
电离作用 37
电子/空穴 225
电子/空穴受体密度 15
电子 15,128,156
电子鼻 13,263,265,272
电子鼻技术 263
电子鼻系统 273
电子传导 96,157
电子传导率 157
电子电荷补偿 114
电子给体 175,178
电子给体缺陷 117
电子供体 35
电子构型 15
电子函数 21
电子耗尽层 7,10,17,30,102,129,138,150,172
电子和空穴复合 91,97
电子核壳结构 30
电子积累层(electron accumulation layer, EAL) 175
电子结构 14,17,20,90,134
电子-空穴对 15,102
电子密度 36
电子密度 n_e 253
电子敏化 47,90,102
电子敏化剂 45
电子敏化效应 114
电子能级跃迁 18
电子迁移率 2,54,182
电子亲和力 175,178
电子受 228
电子受体 128
电子陷阱 2
电子效应 90,91
电子跃迁 133
电子云重叠 134
电子重组 96
电子主导型 150

电子转移 43,49
电阻 91
电阻率 14,18
电阻气体传感器 166
电阻器 241
电阻式传感器 31
电阻式气敏元件 241
电阻式气体传感器 1
顶栅 252
动力学直径 188
动态脉冲温度程序 271
短路光电流 232
煅烧温度 112
对流性流 189
对硝基甲苯(PNT) 330
多壁碳纳米管 118
多层感知器 269
多层前向神经元网络 269
多层物理吸附 178
多层吸附 169
多级介孔 TiO_2-SnO_2 复合材料 97
多金属氧酸盐(POM) 228
多晶 16,66,152
多晶半导体 152
多晶金属氧化物导电性 66
多孔材料 70,189
多模态机器学习算法 277
多维数据分析(MDA) 267
多维数据分析(MDA)技术 268
多相催化 50
多元线性回归 281
多组分共组装 211

E

二次判别分析(QDA) 269
二级动力学 184
二极管 241
二甲苯 300
二甲苯传感器 300
二维(two-dimensional,2D)材料 180

二维材料 118
二维六方 133
二维六方 $P6mm$ 型 143
二维六方结构 143
二维六方介孔 143
二硝基甲苯(DNT) 330
二氧化氮(NO_2) 316
二氧化氮传感器 316
二氧化硫(SO_2) 306
二氧化硫传感器 306
二氧化碳 303
二氧化碳传感器 303,304
二氧化锡 6
二元金属氧化物异质结 90

F

发散型的介孔孔道 98
反相微乳液 80
反应动力学 54
反应速率常数 199
反应位点 128,142
范德瓦耳斯力 75
仿生触嗅联觉(BOT)机器学习神经网络 277
非化学计量比 14
非金属元素掺杂 111
非晶态 66
非线性DR模型 198
费米能级 20,34,68,90,91
费米能级介导的电荷转移 148
分级结构 6
分散性 75
分子动力学(molecular dynamics, MD)模拟 188
分子界面模型 166
分子扩散 143,190,210
封闭膜 245
负载电阻(R_L) 31
负载电阻器(R_L) 243
负载贵金属 166,290

G

改性剂 57
干扰气体 4
各向异性形态 74
工作温度 54,70,269
工作温度梯度 270
功函数 1,37,45,97,241,249
功率强度 225
供体 15
供体能级 36
供体缺陷 157
共价键 75
共结晶 151
共吸附 179,209
构型扩散 190
固-气界面的相互作用 130
固态传感器 253
固态电阻型半导体金属氧化物气体传感器 128
固态扩散 190
固态气体传感器 66
光催化 228
光电响应 14
光激活(如紫外光)气体传感 225
光激活化学电阻气体传感器 225
光激活气体传感 225
光强度 227
光生空穴 226
光生载流子 228
光响应行为 225
光学气体传感器 208
光诱导氧离子 226
光致氧离子 226
光子 225
贵金属 45,69,131
贵金属-金属氧化物纳米复合物 233
贵金属敏化 49
贵金属纳米颗粒 3,49
贵金属纳米粒子 102

贵金属修饰 90
贵金属原位负载 211
过渡金属 131,133
过渡流 189

挥发性有机物 208,287
挥发性有机物传感器 287
恢复动力学 40
混晶 152
活化 168
活化能 4,54,68,90,185
活性位 41
活性位点 48,69,128,175
活性氧物种 172
火焰离子化检测器(FID) 264
霍尔迁移率 154

H

耗尽层 97,148,149
耗尽层宽度 140
合金 130
核-壳半导体材料 18
痕量 BTEX 检测 302
呼出气分析 7
互扩散 187
滑移流 189
化学电阻传感器 10
化学电阻式气体传感器 59
化学电阻型传感器 7,241
化学键 168
化学键合 69
化学键相互作 171
化学敏化 47,90,102
化学敏化剂 47
化学气相沉积(CVD) 82,272
化学气相沉积 20
化学稳定性 59
化学吸 168
化学吸附 21,167,168,209
化学吸附能 191
化学吸附氧 35,47,69
化学吸脱附 171,174,176
化学效应 90
还原气体 2,129
还原型半导体 14
还原性气氛 15
还原氧化反应速率 96
还原氧化石墨烯 139
环绕栅 FET 252
环三次甲基三硝胺 330
换能器 241
挥发性有机化合物 5,182

J

机械合金法 20
机械稳定性 245
基线电阻 45,60
基线漂移 227
基于主成分分析的线性回归(PCA‑LR) 275
极化相互作用 168
极性 113
极坐标图 268
疾病诊断 278
集成电路(IC)技术 241
集成化 14
几何效应 90
甲苯 300
甲苯传感 94
甲苯传感器 300
甲苯检测 235
甲醛 295
甲醛传感 295
甲醛检测 295
甲醛气体传感器 295
甲烷(CH_4) 319
甲烷传感器 319
价带 17,101
价态平衡 134
间接间隙半导体 18
间隙原子 16
监督技术 267

检测极限　4
检测下限　3
检测限　60
检出限(limit of detection, LoD)　181
检出限(LoD)　225
键合作用　113
交叉敏感性　263
交叉选择性　281
胶束　78
胶体纳米晶体阵列　77
接收器　241
结构导向剂　106,207
结构-活性关系　166
结晶度　60,152,154,207
结晶性　15,128
解离氧　172
解吸速率　226
介电常数　14,241
介观结构　145
介孔　190
介孔 Co_3O_4 纳米针阵列　129
介孔 n 型 WO_3 半导体　129
介孔 SnO_2　32
介孔 WO_3/Pt 复合材料　106
介孔 WO_3 材料　145
介孔材料　143
介孔结构　78
介孔晶态氧化钨　32
介孔氧化钨　8
界面催化位点　8
界面反应行为　166
界面能　155
界面势垒　90
金红石相　152
金属-电介质界面　133
金属-绝缘体-半导体场效应晶体管(MISFET)　241
金属空位　156
金属硫化物　128
金属氧化物/石墨烯复合材料　118

金属氧化物/碳纳米管复合材料　118
金属氧化物　1,128
金属氧化物 SO_2 传感器　307
金属氧化物半导体(MOS)　166,263
金属氧化物半导体(MOS)气体传感器阵列　265
金属氧化物半导体　13,30,166
金属氧化物半导体场效应晶体管(MOSFET)气体传感器　241,252
金属氧化物传感器阵列　13
金属氧化物纳米片　151
金属氧化物纳米线　252,253
金属氧化物前驱体　72
金属氧化物微球　74
金属有机骨架(MOF)　73
金属元素掺杂　113
金属载体界面相互作用　49
金属-载体强相互作用　106
禁带　19
禁带宽度　2,90,170
晶胞　16
晶畴尺寸　134
晶格畸变　154
晶格匹配效应　138
晶格取向　128
晶格氧　151,172,175,187,209
晶格氧成　176
晶化孔壁　143
晶界控制　141
晶界势垒　21
晶粒尺寸　21,51,66,67,90,115,128,140,182
晶粒间接触　68
晶粒间接触面积　66,68
晶粒控制　141
晶粒团聚　66
晶面　15,68,152
晶面取向　66
晶体　15
晶体结构　15,66,90,152

晶体缺陷　16,238
晶体台阶　69
晶形　152
晶型　15
晶圆级　248
颈部控制　141
径向基函数神经网络(RBFNN)　275
竞争效应　200
竞争性吸附　2
静电纺丝　81,114
静电屏蔽效应　172
静电相互作用　31
局部表面等离子体效应　232
聚吡咯(PPy)　138
聚合物　241
聚集程度　71
聚集阻力　71
聚类分析(CA)　268
决速步骤　185
均方位移(mean-square-displacement, MSD)　192

K

抗干扰性　21
抗湿度特性　43
颗粒聚集　71
可穿戴智能电子设备　238
空间电荷层　1,67,142
空间电荷的宽度　67
空间电荷区　20
空间电荷区域　67
空间温度梯度　271
空间自约束方法　151
空穴　15,91,129,156
空穴耗尽层　102
空穴积累层(HAL)　31
空穴积累层　38,102,174
空穴累积层(HAL)　129
空穴主导型　150
孔壁厚度　207

孔道半径　202
孔道对称性　207
孔道工程策略　108
孔道结构　210
孔道形貌　204
孔结构　143
孔径　143,189
孔径大小　204,207
孔容　207
孔体积　128
孔隙度　71
孔隙率　15,21,66,70,128,142,143
宽禁带　228
宽频带　230
扩散　21,187,209
扩散尺度　189
扩散机理　189
扩散理论　181
扩散路径　187
扩散模型　143,167
扩散深度　145
扩散势垒　191
扩散速率　69,145,197
扩散通量　187,191
扩散系数(D)　188
扩散系数　71,187
扩散行为　166

L

喇叭口圆柱介孔　211
累积层　97,149
离子半径　116
离子传导　157
离子点缺陷　156
离子价态　116
离子键　14
离子类固体材料　14
离子氧　38
离子注入　20
李斯特菌　324

理论检测限　56
理想内部响应值　202
利用系数　242
粒径　70
连续流　189
两亲性嵌段共聚物　78
两性半导体　14
量子阱　103
临界温度　60
灵敏度　1,3,54,225
零维缺陷　16
硫化氢(H_2S)　313
硫化氢传感器　314
六方介观结构　8
六方晶系　16
六方纤锌矿结构　152
氯气(Cl_2)　328
裸露晶面　155

M

脉冲激发气体传感器　208
脉冲加热　208
脉冲驱动气体传感　225
脉冲驱动气体传感器　234
密度泛函理论（density-functional theory, DFT)　175
密度泛函理论(DFT)计算　249
密度泛函理论　12,117
幂律关系　202
幂律规　209
幂律规则　167,180,200,203
幂指数　182,202
面缺陷　16,17
面心立方(fcc)结构　135
敏感材料　1,50,244
敏感层　43,117
敏感元件　1
敏化机制　102
敏化作用　45
模块化　14

模式识别（PARC)技术　266
模式算法　263

N

纳米晶　10
纳米晶表面配体　77
纳米晶态　66
纳米晶体　75
纳米器件　32
纳米铸造　78,133
内部缺陷　152
内扩散　197
内扩散过程　195
内扩散效用因子 η　195
能带　16
能带分离　18
能带结构　14,17,91
能带理论　225
能带弯　102
能带弯曲　90,91,97
能级分布　17
能级位置　68
能垒高度　39
柠檬酸盐基团　75
浓度梯度　187

O

偶联效应　139
耦合的 DR 过程　194
耦合的扩散-反应（diffusion-reaction, DR)　167

P

判别函数分析(DFA)　268
配位环境　117
碰撞理论　188
平均活化能　187
平均自由程　145,188,210
平面缺陷　156
平行路径　129

Q

气-固界面 194
气固界面 48
气-固界面的相互作用 166
气-固相互作用 166
气敏材料 3
气敏催化 151
气敏催化反应 48
气敏机制 251
气敏性能 12
气凝胶 120
气溶胶辅助(AA)CVD 82
气体传感 1
气体传感机理 30
气体传感器 1
气体扩散 60,135,143,205
气体扩散机理 166
气体敏感机制 1
气体平移模型 190
气体渗透性 68
气体吸附和解吸 21
气体吸附-脱附 55
气体响应 42
气体响应值 204
气相色谱-质谱(GC-MS) 264
气-液界面 194,196
迁移率 18,180
嵌段共聚物 75
嵌段共聚物协同共组装 106
羟基 175
亲水表面基团 74
氢传感 33
氢键 75
氢键作用 180
氢气(H_2) 320
氢气传感器 227,320,321
取向纹理 154
缺陷 14,106,156
缺陷电离能 157
缺陷态 157
缺陷型氧化物 155

R

热平衡载体 19
热稳定性 59,79,107
热注射法 72
人工神经网络(ANN)分析 269
人工神经网络 274
人工智能 166
溶剂 72
溶剂挥发 78
溶剂挥发诱导共组装策略 107
溶剂挥发诱导取向共组装 97
溶剂挥发诱导自组装(evaporation-induced self-assembly,EISA)手 207
溶剂热法 73
溶胶-凝胶法 20,71
溶液挥发自组装 78
柔性传感器 59
软模板(两亲性嵌段共聚物) 207
软模板法 78

S

三甲胺(TEA) 148
三甲胺(TMA) 275,324
三甲胺(TMA)传感器 95
三甲胺 324
三甲胺气体传感器(Au/WO_3) 10
三维立方 133
三维立方 $Ia3d$ 型 143
三维立方结构 143
三硝基苯酚(PA) 330
三硝基甲苯(TNT) 330
三乙胺 91
三元复合氧化物 134
烧结陶瓷管 243
深度学习 275
渗透模型 196
生物标志物 324

施主(donor)杂质 18
施主能级 19
湿度传感器 179,245
湿度耐受性 8
十六烷基三甲基溴化铵 78
石墨化孔壁的 33
石墨化碳骨架 136
石墨烯 118,180,192,232
石英晶体微量天平(QCM)传感器 264
实时气体监测 4
食源性致病菌 145
势垒高度 96,100
势能面 155
室温传感 12
室温传感器 55
室温气体传感器 228
室温氧气传感器 305
受体 15
受体功能 242
受体缺陷 157
受体型掺 150
受限界面胶束聚集组装 146
受主(acceptor)杂质 19
受主能级 19
树状图 268
双介孔结构 143
双金属合金 49
双选择性机理 49
双重脉冲模式 208
双重选择性 49
水动力学参数 196
水解和缩合反应 72
水热法 73,98
水中毒机制 43
顺序前进法(SFS) 280
瞬态响应 6,237
隧道模型 181

T

碳纳米管 118,241

碳族材料 128
特征提取 266
梯度微阵列电子鼻 270
体电阻控制型 30
体积缺陷 16,17
体敏感型传感器 241
体相 Poiseuille 流 189
体相分子扩散 190,191
体相分子扩散系数 192
跳跃模型 192
图案化自组装单层(SAM) 247
图形分析 267
图形分析方法 268
团聚效应 142
脱附 168,209

W

外扩散 197
外扩散过程 194
网络分析 267
微/纳米结构 8
微机电系统(MEMS) 234
微机电系统(MEMS)传感器 243
微机电系统(MEMS)技术 244,246
微加工技术 241
微晶尺寸 66,152
微孔 143,190
微热板结构 245
微乳液 80
微生物挥发性有机化合物(MVOC) 9
微生物挥发性有机物(MVOC) 276
微相组成 128
微型气体传感器 234
微制造技术 252
位错 157
位错缺陷 17
温度依赖性 185
稳定性 60
稳态电阻 60
无创诊断 111

无定形　60
无监督技术　267
无线传感　4,111
物理吸附　21,167,168,171,179,209
物联网　166
误差逆传播神经网络（BPNN）　275

X

吸附/解吸能　68
吸附　142,168,209
吸附饱和　168
吸附焓　169
吸附和解吸速率　152
吸附机理　167
吸附剂　169
吸附-解吸动力学模型　21
吸附-解吸行为　55
吸附能　4,175,227
吸附速率　226
吸附位点　37,57,169
吸附氧　39,116
吸脱附过程　185
吸脱附机理　169
稀土金属　115,131,133
限域效应　136
线缺陷　16,17,156,157
线性DR模型　198
线性判别分析（LDA）　269
相对湿度（relative humidity，RH）　178
响应/恢复的动力学　184
响应/恢复时间　6
响应/恢复瞬态　184
响应/恢复速率　184
响应　167
响应和恢复动力学　185
响应-恢复时间　60
响应-恢复速度　225
响应-恢复速率　40,56
响应或恢复时间　30
响应机制　1

响应性　3
响应值（R_a/R_g）　11
响应值　54,182,184
硝基爆炸物　330
小型化　166,241,244,251
肖特基结　45,102
肖特基缺陷　156
肖特基势垒　21,67,230
效用因子　202,235
协同效应　6,120
斜角沉积（OAD）　272
嗅觉算法（all-feature olfactory algorithm，AFOA）　274
悬浮膜　245
悬浮栅场效应晶体管　242
悬浮栅极　237
选择性　1,3,56,263
循环稳定性　6

Y

亚稳态　30
亚稳态方铁锰矿相　154
亚稳相　53
亚阈值摆幅　252
氧分压　156
氧负离子　31,175
氧化还原/再生机理　170
氧化还原反应　15
氧化还原过程　185
氧化气体　2,129
氧化锡半导体传感器　244
氧化型半导体　14
氧化性气氛　15
氧化铟　40
氧空位（O_v）　43
氧空位　12,69,113,116,128,133,134,150,156
氧浓度电池　241
氧气（O_2）　304
氧气传感器　304

氧缺陷　133
氧势能　155
氧吸附　167,209
氧吸附机理　33
氧吸附模型　171,174
氧阴离子　40
液化焓　169
液化石油气(LPG)　322
液体效用因子(η_L)　197
一步协同组装　8
一级动力学　184,185,210
一维缺陷　157
一氧化氮传感器　7
一氧化碳(CO)　311
一氧化碳传感器　311
乙醇　287
乙醇传感器　292
乙醇气体传感器　287
乙炔(C_2H_2)　328
乙炔传感器　329
异相掺杂　107
异质结　15,90,166,233,249
异质结构　1
异质结界面　148
异质结有机场效应晶体管(OFET)传感器　238
异质界面　15,148
溢出效应　10,47,102,104
溢流效应　166
硬模　98
硬模板　133
硬模板脲醛树脂　207
硬球模型　188
有机染料分子　228
有机-无机界面　76
有效扩散率($D_{A,e}$)　195
有效载流子浓度　68
有序超晶格　77
有序介孔　143
有序介孔材料　78,210
有序介孔金属氧化物　78,143

有序介孔氧化钨材料　146
阈值电压　252
元素掺杂　15,90
原位掺杂　133
原位合成　98
原位气相色谱仪-质谱(GC-MS)分析　113
原位协同组装　117
原子层沉积(ALD)　272
原子层沉积　98
原子利用效　49
原子取代　116
约束性扩散　188,190,204

Z

杂原子掺杂　292
杂原子掺杂工程　8
杂质能级　18
载流子　2,91,128,168
载流子分离　90
载流子复合　18,20,228
载流子扩散　19
载流子密度　50,116
载流子浓度　141
载流子漂移　19
载流子迁移率　19,154
载流子势垒　66
在线监测　232
增敏效果　57
栅极　252
栅极电压　237,254
栅极金属　237
窄颈区域　142
折射率　229
正丁醇　328
正态分布　269
直接带隙半导体　255
直接间隙半导体　18
指纹信息　266
智能便携式传感器　117
智能化学传感器阵列系统　263

智能气体传感器　8
滞膜模型　194,196
中空金属氧化物　74
主成分分析(PCA)　268
助表面活性剂　80
紫外光 LED　225
紫外光发光二极管(LED)　227
自扩散　187
自由程(λ)　188

自由电子　20,47,91
自由分子　189
自由扩散　188
自由载流子　19
自组装　75
总挥发性有机化合物(TVOC)　6
最佳测试温度(T_{opt})　205
最佳工作温度　204

彩 图

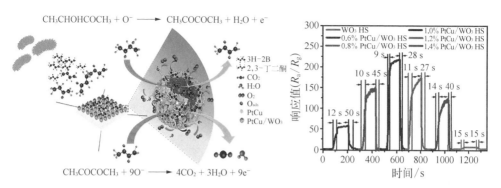

图1.6 双金属 PtCu 纳米晶敏化 WO_3 空心球高效检测 3-羟基 2-丁酮生物标志物的示意图(a)和不同含量 PtCu 纳米晶敏化的 WO_3 空心球对 3-羟基 2-丁酮的响应-恢复曲线(b)

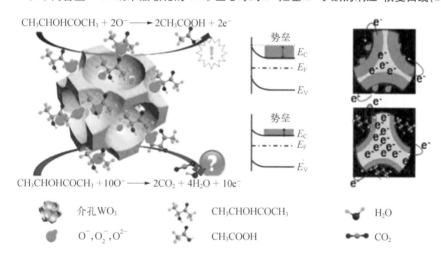

图2.3 检测暴露于空气和目标气体-空气混合物中的 3-羟基-2-丁酮的介孔 WO_3 基传感器的传感机理示意图

E_V: 价带边;E_C: 导带边;E_F: 费米能级

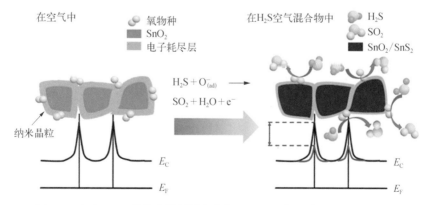

图2.4 介孔 SnO_2 基传感器暴露在空气和 H_2S 空气混合物中的传感机理

E_C: 传导带边缘;E_F: 费米能级

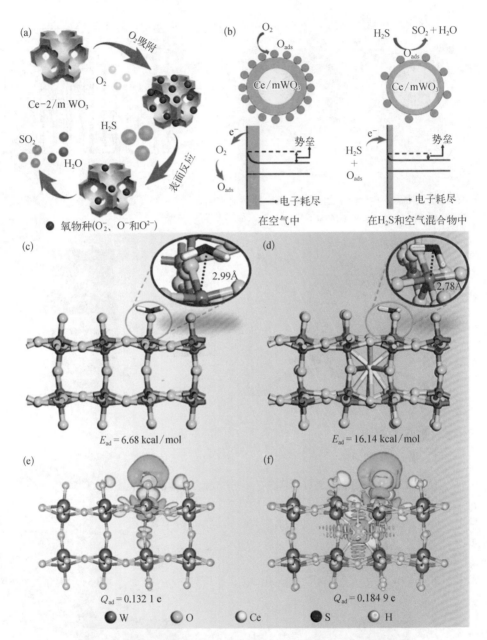

图 2.15 基于 Ce 掺杂有序介孔 WO_3 的 H_2S 传感器机理示意图和 DFT 计算

(a) 用于 H_2S 检测的 Ce-2/mWO_3 基气体传感器的传感机理;(b) 暴露于空气和 H_2S 空气混合物中的 Ce-2/mWO_3 敏感材料的能带结构和电子转移过程;掺杂 Ce 的 WO_3 传感器的 H_2S 吸附能和电荷转移行为的 DFT 计算:(c)(d) H_2S 分子在 WO_3、嵌入 Ce 的 WO_3 的最佳结构上的吸附构型和能量;(e)(f) WO_3、Ce^{4+} 嵌入 WO_3 的 H_2S 分子吸附的微分电荷密度以及电子密度,青色和黄色分别表示增加和减少

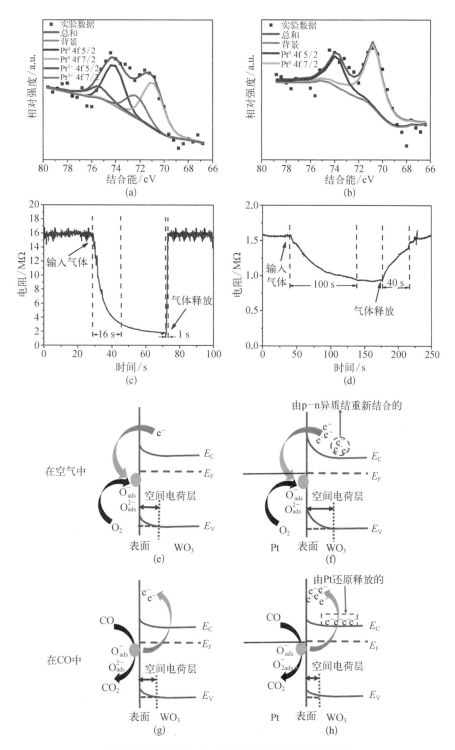

图 2.17 WO₃/Pt 的 CO 传感机理研究和示意图

(a)、(b) 分别为在 125℃ 及 57% 相对湿度下对 100 ppm CO 进行 5 次传感测试后,在空气中和在 CO 中冷却至室温后 WO₃/Pt 的原位 Pt 4f XPS 光谱;(c)、(d) WO₃/Pt、WO₃ 在 125℃ 及 57% 相对湿度下向 100 ppm CO 的动态电阻过渡特性;(e)~(h) 介孔 WO₃/Pt 和介孔 WO₃ 传感机理示意图

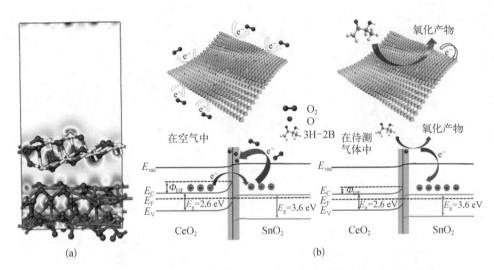

图 5.20 多孔 CeO_2/SnO_2 纳米的电子结构及其对 3H-2B 的传感机理示意图

(a) CeO_2/SnO_2 纳米片中电荷密度差异的侧视图,其中红色和蓝色区域分别表示电子积累和耗尽,Sn、Ce 和 O 原子分别用灰色、白色和红色标记;(b) 多孔 CeO_2/SnO_2 纳米片对 3H-2B 的表面传感反应示意图以及相应的传感机理图

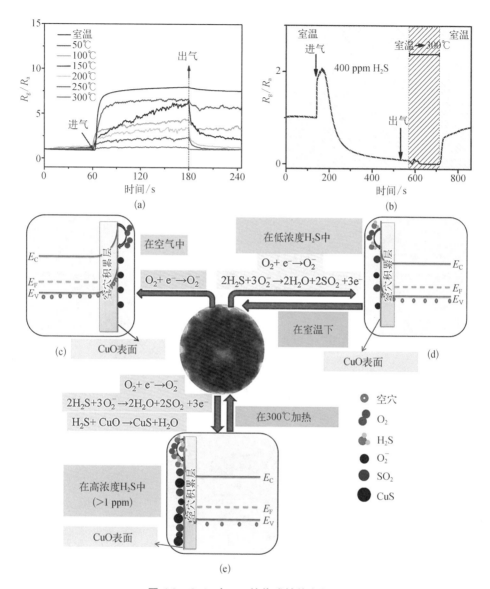

图 6.8　CuO 对 H_2S 的传感性能和机理

(a) CuO 传感器在 10 ppm H_2S 下响应值与工作温度的关系;(b) CuO 传感器在 400 ppm H_2S 下响应值随时间的变化;(c)~(e) CuO 传感器与不同浓度的 H_2S 产生的不同传感机理,其中(c)为在空气中,O_2 分子吸附到 CuO 表面形成氧负离子,产生 HAL,(d)表示注入低浓度 H_2S 时,H_2S 与吸附在 CuO 表面的 O_2^- 反应,减少了 HAL,响应值提升,(e)表示注入高浓度 H_2S 时,除了 H_2S 氧化反应,H_2S 还与 CuO 反应而在表面产生了一层 CuS,响应值下降

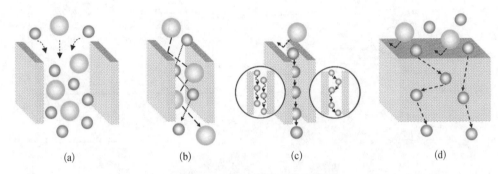

图 6.17 气体扩散机理的示意图

(a) 体相 Poiseuille 流；(b) Knudsen 扩散；(c) 尺寸限制的扩散（左小图：表面模型，右小图：气体平移模型）；(d) 固态扩散机理

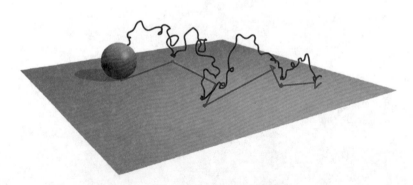

图 6.19 分子表面扩散的示意图

该图将分子固定在表面的时间与在表面上方发生体相扩散的时间结合起来，黑色曲线是真实的三维分子轨迹，而品红色折线是穿越表面的有效二维轨迹

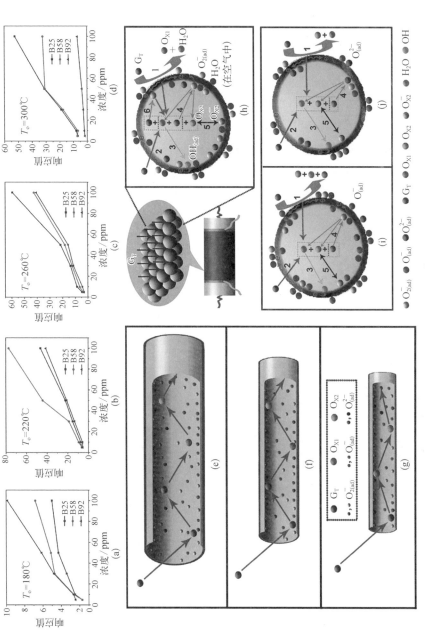

图 6.31 SnO₂ 空心微球丙酮传感性能及机理研究

(a)~(d) 丙酮在不同温度下的浓度-响应曲线,三种材料 B25、B58、B92 以其比表面积来命名;(e)~(g) 目标气体在不同孔径下扩散的示意图,其中(e)表示孔径较大、(f)表示孔径适中、(g)表示孔径较小;(h)~(j) 不同温度下 SnO₂ 空心微球下目标气体扩散的示意图,其中(h)表示<150℃、(i)表示 150~200℃、(j)表示>200℃

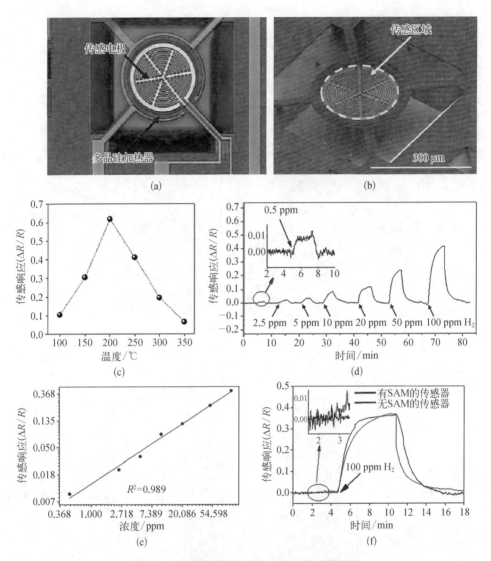

图 8.5 带有悬浮加热板的 MEMS 传感器芯片

(a) 光学图像;(b) 制造的微孔板传感器芯片的 SEM 图,其中感应区域被环形加热器包围,导线被一层 SiO_2 隔离,电极暴露在外以进行化学电阻连接;(c) 同一传感器对 200 ppm H_2 的温度相关传感响应 (200℃是最佳工作温度);(d) 传感器在 200℃下对 H_2 的响应,浓度范围为 0.5~100 ppm,插图显示传感器可溶解于 0.5 ppm H_2;(e) 传感响应与气体浓度之间的对数线性关系;(f) 有 SAM(红色)和无 SAM(黑色)的传感器对 100 ppm H_2 的响应,插图为传感器基线,显示了传感器的噪声基底

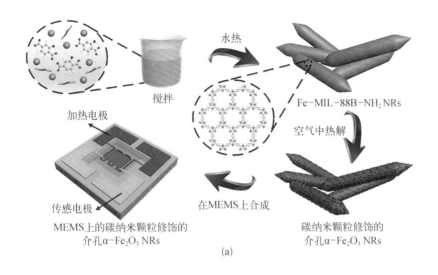

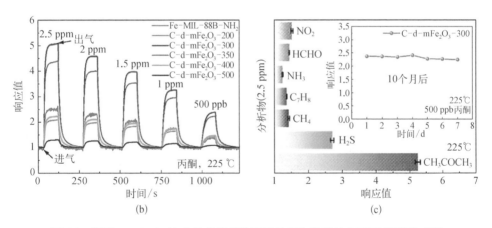

图 8.8　基于 α-Fe$_2$O$_3$/C 介孔纳米棒的高灵敏度和稳定的 MEMS 丙酮传感器

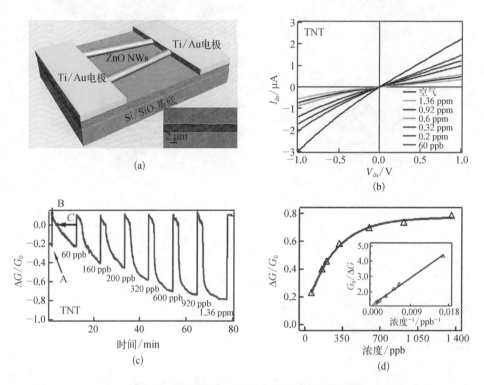

图 8.12　ZnO NW-FET 对 TNT 的传感性能

(a) ZnO NW-FET 结构示意图;(b) 在空气中和不同 TNT 浓度下的 I_{ds}-V_{ds} 曲线;(c) ZnO 纳米线化学传感器对 TNT 的传感响应;(d) 归一化电导变化相对于 TNT 浓度(C),使用 $S=1/(A+B/C)$ 进行拟合

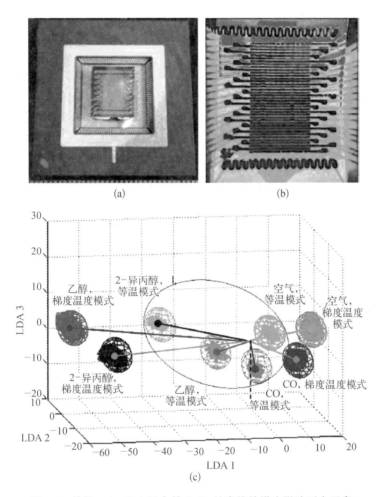

图 9.6 使用 KAMINA 平台的 SnO₂ 纳米线的梯度微阵列电子鼻

(a) KAMINA 芯片 SnO₂ 纳米线传感器;(b) 沿着电极阵列施加温度梯度,520(绿色区域)~600K(红色区域)芯片的 IR 图像;(c) 基于 SnO₂ 纳米线的微阵列在暴露于目标气体(2~10ppm)时的电导率模式的 LDA,其中分类区域对应于 0.999 9 置信水平的正态数据分布,微阵列分别在 580K(椭圆内的恒定 T 区域)均匀加热和 520~600K(梯度 T 区域)的温度梯度下操作

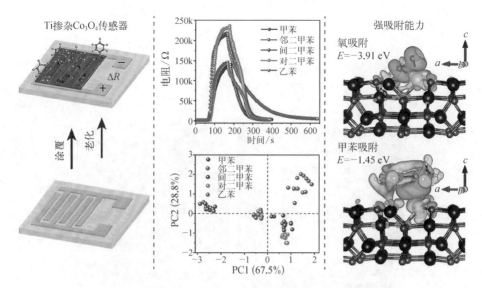

图 10.11 基于 Ti 掺杂 Co_3O_4 传感材料的 BTEX 传感器

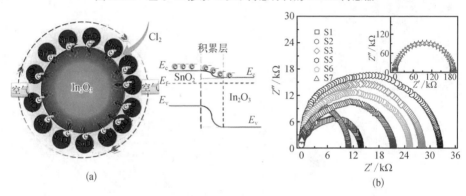

图 10.31 In_2O_3/SnO_2 异质结传感机理及交流阻抗研究

(a) In_2O_3/SnO_2 异质结微观结构机制的示意图;(b) 基于 S1、S2、S3、S4(插图)、S5、S6 和 S7 的传感器的交流阻抗谱

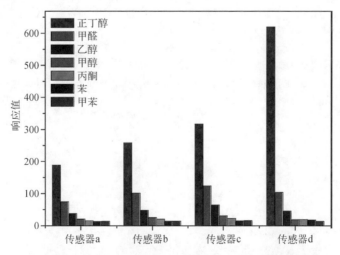

图 10.32 SnO_2 气体传感器在 150℃下对不同气体的响应